THE ALT-AZ INITIATIVE

Telescope, Mirror, & Instrument Developments

Description of Cover Pictures

The Alt-Az Initiative was founded in early June 2007 to develop low cost, 1 to 2 meter portable telescopes. Left to right: Russ Genet, Howard Banich, Richard Berry, Mel Bartels, Dan Gray, and the Lollipop, Dan's 14 inch robotic alt-az telescope with built-in field de-rotator.

Ten members of the Alt-Az Initiative stand beside the 8 meter Gemini telescope on Mauna Kea. Their "insider's" tour explored possible tech transfers from large to small alt-az telescopes. Gemini's Director, Doug Simons, led a tech transfer panel discussion at the conference that preceded the tour.

California Polytechnic State University students Josh Schmidt, Michelle Kirkup, Matt Swanson, and the Cal Poly 18 telescope. They designed the optical tube assembly while other students designed the fork. Other initiative members designed and built, the direct drive motors, their controls, the mirror supports, and the primary and secondary mirrors.

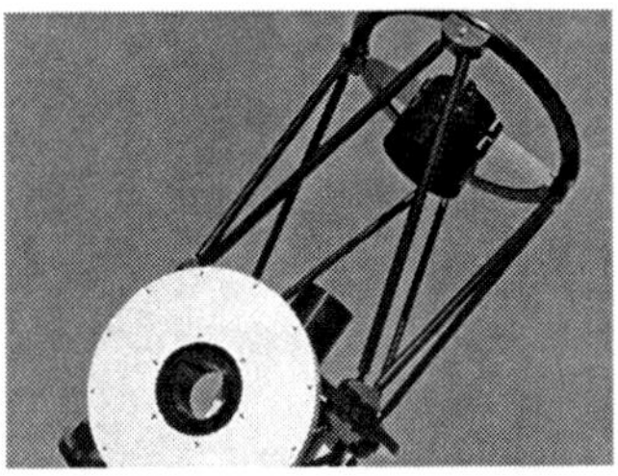

The PlaneWave Instruments 0.7 meter corrected Dall Kirkham CDK 700 alt-az telescope is—with its axial flux direct drive motors—the "descendent" of the Cal Poly 18 telescope. The CDK 700 completes the three-step tech transfer journey from expensive mountaintop direct drive motors to inexpensive Alt-Az Initiative telescope demonstration to affordable production telescope.

Lightweight, low cost, meter-class mirrors are the enabling technology for portable 1 to 2 meter telescopes. This "Starstone" foam glass composite mirror was made by Initiative member Andrew Aurigema. Other Initiative developments: meniscus, carbon fiber composite, glass and plastic replica, pneumatic Mylar, and spin-cast epoxy mirrors.

Several Initiative members are concentrating on "light buckets:" non-imaging, on-axis "flux collectors" for photometry and spectroscopy of relatively bright stars. Shown is Hanbury Brown's historic stellar intensity interferometer in Narrabri, Australia. A modern array of 2 meter light buckets could provide images of the surfaces of nearby stars.

Advance Praise

In the 1960's, amateur astronomers were blown away when John Dobson introduced low cost portable Newtonian telescopes on an alt-az (Dobsonian) mount. Now *The Alt-Az Initiative: Telescope, Mirror and Instrumental Developments* describes the development of, again, low cost, light weight research telescopes but in the realm of 1- 2-meter apertures. I'm confident the information in this text will lead to a similar groundswell of interest as occurred in the Dobsonian era.

Thomas G. Frey
Emeritus Professor of Chemistry, California Polytechnic State University

Congratulations on pushing the state-of-the-art of large, low-cost alt-azimuth telescopes for amateur and professional use. For our part, Tele Vue is committed to assisting your efforts through continued development of Paracorr coma-correctors and wide-angle eyepieces to maximize both the visual and imaging capabilities of your exciting instrument initiatives.

Al Nagler
Chairman and CEO, Tele Vue Optics

This book is a must read for people building or designing a small to medium sized telescope. The authors have successfully demonstrated how productive amateur R&D can be when you bring passionate people together to work on different aspects of a large problem. I look forward to the upcoming era when inexpensive meter (and larger) class telescopes are in common use by researchers, educators, and amateur astronomers.

Josh Walawender
Astrophysicist, University of Hawaii, Institute for Astronomy

Thankfully we have editors like Russ Genet and his two associates, Jo Johnson and Vera Wallen, who record the exciting developments in amateur telescope making as they occur. I am grateful to you three editors for your previous book, which described my own small contribution, and I am excited to read about the other continuing contributions.

Thomas J. Johnson
Founder of Celestron

With tremendous energy and enthusiasm, Russ Genet and company have assembled a diverse group of individuals who are dedicated to making the tools for astronomical research accessible to nearly everyone. The Alt Az Initiative may be one of the greatest revolutions in astronomical research!

Douglas Arion
Professor of Physics and Astronomy, Carthage College

The Alt-Az Initiative creates synergy from contributions by astronomers, engineers, students, and technophiles who are professionals or enthusiastic advocates with the desire to both create new telescope technology and to greatly reduce the cost of producing these systems. The thinking outside of the box approach they embrace is sure to produce innovative results.

Rich Williams
Founder and CEO, Sierra Stars Observatory Network

As a manufacturer of foamed glass from recycled glass I can honestly say that I hadn't expected to be involved in a project like this. The Alt-Az Initiative is a perfect example of a collaborative effort from a number of disciplines working together to solve a man made problem of disposing of glass on earth while simultaneously expanding our knowledge of the universe by using recycled glass to make mirrors to observe distant galaxies. It has been a pleasure working with the people in the Alt-Az Initiative on lightweight mirrors--they should be proud of their accomplishments.

George Morandin
CEO Growstone LLC

In *The Alt-Az Initiative*, theories and concepts that have only recently established themselves in the realm of amateur and professional telescope making are put to the test and pushed to intriguing limits in their application to the age-old alt-az mount. As they have with me, the people who have presented their ideas in this book will inspire the telescope experimentalist in you.

Danyal Medley
Principal Engineer, Celestron

The Alt-Az Initiative is a timely book that should improve and increase the amount of astronomical research at small- to mid-size universities, as well as by ambitious amateurs. This all feeds into the larger goal of expanding the scope of astronomy research across the nation. This work truly is bringing astronomy to the masses, for both research and education.

Paul Hardersen
Associate Professor, University of North Dakota Department of Space Studies

THE ALT-AZ INITIATIVE

Telescope, Mirror, & Instrument Developments

Edited by

Russell M. Genet, Jolyon M. Johnson, and Vera Wallen

Foreword by

Richard Berry

Collins Foundation Press

The Alt-Az Initiative: Telescope, Mirror, & Instrument Developments

Published by the Collins Foundation Press
4995 Santa Margarita Lake Road
Santa Margarita, CA 93453
www.CollinsFoundationPress.org

Editors - Russell M. Genet, Jolyon M. Johnson, and Vera Wallen

Includes bibliographical references and index

Cover and book design by Cheryl Genet
Production Manager - Cheryl Genet
Typeset in InDesign - Palatino Linotype 10.5

ISBN 978-0-9788441-4-1

First printing May 2010
Printed by Sheridan Books in the United States of America

Dedication

We dedicate this book to John Dobson. His clever design for lightweight, low cost, easily built, large aperture, alt-az telescopes launched a revolution.

Contents

Preface

The Alt-Az Initiative Spirit

This book is the second in a planned four volume Alt-Az Initiative series. The first book, already published, is *Small Telescopes and Astronomical Research.* It discusses small telescopes (less than 1 meter in aperture) and their use in research by professional and amateur astronomers as well as undergraduate and even high school students. This present volume complements the first book by concentrating on the ongoing development of lightweight, low cost alt-az telescopes in the 1 to 2 meter range that combine the portability of Dobsonian telescopes with computer control and other high tech features of giant mountaintop alt-az telescopes. The two forthcoming books have been tentatively entitled *Light Bucket Astronomy* and *Robotic Observatories.*

In this present volume, the key word is *development.* This is not a how-to handbook that describes proven solutions to difficult problems. Rather it is a snapshot in time (2007-2010) of parallel developments addressing a common problem: how to design and build useful, low cost, lightweight, portable, relatively easily constructed 1 to 2 meter telescopes. This is a daunting challenge. Without doubt some of the developmental solutions described herein will fall by the wayside—intriguing ideas that didn't pan out—while others, hopefully, will succeed.

The driving spirit of this book is development, of dreaming up possible solutions to challenging problems and then testing them out in the real world. Failure is an option—a necessary option. What fun would it be if success was assured, if the outcome was known in advance?

Welcome to the little known world of research and development (R&D) as a hobby. Amateur R&D, however, is only amateur in the sense that it is being done for the fun of it, not for pay. Unlike their paid R&D counterparts in industry, academia, and the military, amateur R&Ders need not concern themselves with grants, patents, and secrecy. They freely share their ideas, kibitz on one another's projects, and, akin to reading thriller novels, they follow the twists and turns of their fellow hobbyist's developments. Which of the competing possibilities will fall by the wayside? Which, against all odds, will vault to fame and renown?

These hobbyists are definitely not amateur in the sense that they don't know what they are doing. They do. They are engineers, astronomers, and technicians. Some of the participants are key players in the development of large telescopes and can't resist, on the side, helping to solving truly challenging small telescope problems.

Many of the members of the Alt-Az Initiative are amateur telescope makers (ATMs). They have designed and built small telescopes as a hobby for years. The lure of designing and building truly large telescopes is, for some, irresistible. It is a disease with a name—aperture fever. They dream of monster telescopes hauled in the backs of their sport utility vehicles to favorite dark-sky sites to personally spend all night checking out distant galaxies in living color (it takes lots of photons to activate the eye's cones, the less sensitive but color receptors toward the middle of the retina).

Other Alt-Az Initiative members are photon-starved scientists. Their research requires large telescopes and months or years of observational time—an impossible combination unless they have their own telescopes. Many research programs do not require high quality images; in fact they don't require any images at all. Only on-axis, low cost, light flux concentrators are needed. Research areas such as high speed photometry, near IR aperture photometry, and moderate resolution spectroscopy come to mind. These research astronomers have also been infected with aperture fever. They lust after big cheap telescopes, seeing no reason to pay top dollar for high image quality, general purpose telescopes. Their dreams of lightweight, dirt cheap telescopes don't stop at 2 meters. They rub their hands in anticipation when email or workshop discussions turn to arrays of 2 meter telescopes which could be used, for instance, as an intensity interferometer to image the surfaces of nearby stars. Averse to staying up all night, these scientists envision photons gathered by the bucketful by ever faithful robotic telescopes while they sleep.

Shortly after its inception, the Alt-Az Initiative issued a challenge to create a transportable 1.5 meter telescope within three years. As you will see in the following pages, we are at the cusp of meeting that challenge. It is being brought to fruition by an eclectic, world-wide group of creative minds who have argued and created, lectured and designed, and supported and challenged each other. No single individual held the key, but the group has found the answers and pushed their vision toward reality.

Inside you will read how everything from the ground board to the optics has been questioned anew. The most common discussions started with "what about?" and "have you tried?" Old concepts were abandoned and nothing was taken for granted. What should a drive system do, must optics be made of glass, what is portable?

Responses to such questions are found in the chapters but not all of them are final. This is a book about R&D—the process continues. Some of the chapters were rewritten more than once as successes or failures became obvious. This is a snapshot of the springboard of genius caught by the camera at full compression about to launch a new way of making telescopes. You can make that jump with us—add your ideas to the mix.

This book is divided into a Foreword and four parts. The Foreword, written by Richard Berry, lays the foundation for the entire book by providing not only a history of alt-az telescopes, but by examining the "enabling technologies" that the Alt-Az Initiative needs to address if it is to succeed. Richard, who edited both *Astronomy* magazine and the *Amateur Telescope Making Journal,* was an active participant in the Dobsonian revolution and, with David Kriege, wrote the definitive book on the topic, *The Dobsonian Telescope: A Practical Manual for Building Large Aperture Telescopes.*

Part I is devoted to the Alt-Az Initiative itself. An Astro 2010 Position Paper, written by Initiative members for the [US] National Academies of Science, describes how numerous low cost 1 to 2 meter robotic telescopes could use their generous time allocations and low operational costs to provide the in-depth, movie-like time-series follow-ups to the numerous "snapshot" discoveries being made by the mountaintop giants. Large and small telescopes complement each other. This section also captures the "historic" flavor of early Initiative workshops and visits to large mountain top alt-az telescopes.

Part II describes the Initiative's telescope developments, starting with a reprint of the Initiative's very first published paper, and then moving on to describe a number of telescope developments in depth. Howard Banich suggests that "Moving a big telescope doesn't have to be difficult." His principles have been applied to the design of a 1 meter telescope that rolls (with top truss removed) into the back of a Jeep Cherokee.

Part III of the book—11 chapters in all—is devoted to lightweight, low cost mirror developments—the key enabling technology for large portable telescopes. Initiative members are developing mirrors across a broad spectrum: thin meniscus mirrors made from slumped, low cost plate glass, some with active figure adjustments; foam/plate glass composite mirrors; carbon fiber mirrors; mass produced mirrors replicated from master molds without optical figuring; mirrors made from thin sheets of aluminized Mylar curved by a partial vacuum on one side; and epoxy spin-cast mirrors that naturally form a parabola.

The fourth and final part describes work by Initiative members on instruments, techniques, and software—the "science end" of things. Instruments, such as the Bath interferometer, are used in making mirrors. Other instruments are used to make photometric or spectroscopic measurements. Software is used not only to reduce the data but to control the instruments and even entire observatories. Not quite far enough along yet to be included in this book were the near infrared aperture photometers being developed by Greg Jones and August Johnson. Imaging NIR cameras are still quite expensive, so there is still a role for photodiode "one pixel" photometers that can be fed by non-imaging, on-axis, light bucket flux concentrators.

The editors are pleased to thank the chapter authors, who were a delight to work with, generous in sharing their scientific and engineering knowledge, and patient with multiple revisions. Several chapters in the book were based on previously published material. We are grateful for permission to utilize material from the National Academy of Sciences (chapter 1), the Society for Astronomical Science (8, 15, and 28), *Amateur Astronomy* magazine (4, 5, and 7), and *Astronomy Technology Today* (3). Published simultaneously with *Amateur Astronomy* were chapters 9, 10, 11, 12, 18, and 20. The editors extend their thanks to Cheryl Genet, Managing Editor of the Collins Foundation Press, and Dwight Collins, its Chief Executive Officer, for their hard work and unfailing encouragement. This book was printed digitally at Sheridan Press and we thank Kathy Brown and Dave Erickson for guiding us through this new "high tech" approach to the printed word.

Russell M. Genet, Jolyon M. Johnson, and Vera Wallen
May 2010

Foreword

The On-Going Evolution of Alt-Azimuth Telescopes

Alt-azimuth telescopes came first. Early instruments were mounted on simple stands that swiveled around and up to the stars. Astronomers hoisted long-focus refractors by ropes and pulleys. With the eyepiece anchored by a stand, the astronomer could follow a star in azimuth and elevation. Herschel's great reflectors and later the Leviathan of Lord Rosse were carried in great alt-azimuth stands. Without a doubt these cumbrous mounts worked, but they worked successfully only in the sense that a skilled user could coax them to operate well enough to find and follow stars for short periods of time. In a time of explosive discovery, with plenty of low-hanging fruit just waiting to be plucked, astronomers like Messier and Herschel did not need precise tracking. They needed only to point, glimpse, and deduce a rough position.

The equatorial telescope first appeared on the scene in 1625, supporting Père Scheiner's helioscope. The first clock-driven equatorial seems to have been constructed by the French astronomer Jean Dominique Cassini, the mount a curious cross between an armillary sphere and telescope stand. By the end of the 1700s, equatorials that to our modern eyes look terribly crude were in use both on the Continent and in England. The first equatorial we would recognize as truly modern was the 240 mm equatorial refractor of Dorpat, built by Fraunhofer for the great observer Wilhelm Struve in 1824. Because it could hold a star nicely on the crosshairs at 700x for long periods of time, the clock-driven mount enabled Struve to perform precise micrometer measurements of thousands of double stars. The great refractors of the 19th Century, culminating in the Lick and Yerkes refractors, are direct lineal descendents of this pioneering equatorial. What the Messiers and Herschels discovered, the Struves and Bonds would measure.

By the middle of the 20th Century, respectable astronomers would not even think of using an alt-azimuth mounting. Precise equatorials like the 200 inch reflector on Palomar Mountain, the 120 inch on Mount Hamilton, the 48 inch Schmidt camera, and scores of smaller equatorially mounted telescopes allowed ambitious astronomers to take high-resolution spectra at a coudé focus, measure starlight with exquisitely sensitive photomultiplier tubes, survey the entire sky in two colors, and even ride at the prime focus while exposing photographic plates to the light of receding clusters of distant galaxies for hours on end. The beauty of the equatorial is that by driving the telescope with a small motor in one axis only, the stars stood still. The focus of the telescope became a laboratory, a place where the sensitive emulsion, the slit of a spectrograph, or the diaphragm of a photometer could steep in a steady beam of starlight as long as necessary.

Amateur astronomers, eager to emulate their professional brethren, followed suit. The equatorial mounting was an *enabling technology* that solved two crucial requirements for serious visual observing. The first: with a clock drive you could follow an object for hours on end; the second: with setting circles, you could locate objects without a tedious search. And even if your telescope didn't have a clock drive, you could clamp the declination axis and then follow an object by gently pushing the telescope in right ascension only. The growth of amateur astrophotography reinforced the hold that equatorials held over amateur thinking. If you examine the ads in any copy of *Sky & Telescope* from the 1960s, you will understand how thoroughly equatorial-centric amateur astronomy had become.

But the times they were a-changin'—at least for the professional astronomers. Equatorial mounts, after 130 years of dominance, began to give way to alt-azimuth mountings. The driving factor was cost. Compared to an alt-azimuth telescope, the equivalent equatorials were bulky, heavy, and hard to house. Their size and weight drove up the cost of fabrication, and their huge slow-cooling domes suffered from seeing effects. In 1979, the Multiple Mirror Telescope went into operation with a computer-controlled alt-azimuth mounting, an innovative optical system with six primary mirrors and a square rotating enclosure with an enormous opening. The success of the MMT signaled the end of the equatorial era for big professional telescopes.

What made the MMT so important was that it demonstrated five key enabling technologies: an alt-azimuth mount controlled by computers, lightweight egg-crate mirrors, multiple small mirrors acting as a single large primary, active computer co-alignment of the small primaries, and a compact enclosure with large shutters that allowed rapid flushing and cooling to ambient temperature. The root-mean-square pointing error was 0.88 arc seconds. The 24 bit shaft-angle encoders mounted directly on the telescope axes plus tachometers mounted on the motor shafts controlled two DC torque motors with sufficient dynamic range not only to track at normal sidereal rates but also to slew at 90 degrees per minute. In 1979, three cutting-edge technologies—encoders, computers, motors—had been harnessed in a large professional telescope.

Some amateur astronomers took note. Although the amateur community consists in large part of hobbyist "stargazers," it also includes many engineers, technicians, and scientists who serve as the "movers and shakers" of the amateur astronomy and amateur telescope-making movements. In the early days of amateur telescope making, skilled machinists and mechanical engineers in the industrial northeast built precision equatorials with accurate clock drives. Amateurs picked up optical skills from professional and semi-professional optical workers through Albert G. Ingalls' (affectionately known as "Unk Ingalls" by his readers) columns in *Scientific American*. The first Schmidt cameras used at professional observatories in the United States were built by amateurs. Technology transfers—a grand way of describing clever guys talking among themselves—surely took place in both directions.

But alt-azimuth telescopes that could point and track posed a special problem for the amateur community. Large observatories and university consortia had access to significant funding, so the cost of adding a few minicomputers ($50,000 each) and encoders ($30,000 each) was an incidental expenditure—especially because the smaller footprint of an alt-azimuth mounting could knock a few millions off the cost of the dome! And although the surplus market was still crammed full of WWII and Korean War hi-tech discards, it would be another decade before surplus encoders and motors would appear on the shelves of C&H Sales for 0.1% of their original government cost.

So the spread of alt-azimuth telescopes in amateur astronomy followed a very different path. What the amateur world lacked—and wanted—was *APERTURE*. In 1980, a typical amateur telescope was a 6 inch reflector. Ten inches was a serious instrument for a personal observatory, and an astronomy club might aspire to a 12 inch Newtonian or 16 inch Cassegrain.

In the early 1970s, San Francisco sidewalk astronomer John Dobson came up with a radical redesign of the home-built telescope by focusing on the bearings. The classic amateur equatorial mount had shafts typically 1 to 2 inches in diameter running in babbit-metal or ball-race bearings. Because the bearings turned freely, it was necessary to control the motion, usually with a slip-clutched gear in right ascension and a friction clamp in declination. This technology was reasonably effective for 6 to 16 inch equatorial Newtonians, but it failed miserably when translated to alt-azimuth mountings. Dobson's innovation was to replace small, free-turning bearings with large-diameter bearings having lots of friction. Dobson's altitude bearings were typically 6 to 24 inches in diameter, and the azimuth bearings were made from inexpensive materials such as Formica countertop and Teflon plastic pads. The optics in Dobson's telescopes were rather ordinary *f*/7 Newtonians—although the mirrors, ground on war-surplus battleship porthole glass considerably thinner than the standard 1:6 ratio prescribed at that time for amateur telescope mirrors, caused many staid members of the ATM community to shake their heads in disbelief.

The signature of the Dobsonian telescope is its bearings. Because the bearings do not turn freely, the telescope is extremely stable. Left alone, it remains pointed at the same place in the sky. When the observer pushes on the tube with a few ounces to a few pounds pressure, the telescope moves in both axes simultaneously. If the bearings use materials with the right combination of kinetic and static friction—the classic combination being an Ebony Star Formica plate turning on virgin Teflon pads—the movement of the telescope has a characteristic "buttery" feel. Observers routinely report effortlessly tracking sky objects at 400× to 600×.

However, most telescope makers didn't even think about the bearings because they appeared to be trivially simple. What they focused on was the big, thin, light-weight mirrors. *APERTURE!* was the unique proposition, the attraction, the selling point. So the Dobsonian caught on in a big way. Crude Dobsonian bearings replaced finely machined ball bearings because anyone with a jigsaw could make them, and they cost practically nothing. Because the alt-azimuth configuration was balanced and the equatorial counterweight was not needed, the new telescopes were lighter and more portable. It was suddenly just as easy to build a 16 inch Dobsonian telescope as it had been to build a pipe-mounted 8 inch equatorial. A generation of observers brought up on Walter Scott Houston's *Deep Sky Wonders* column suddenly found themselves in hog heaven, observing with more aperture than they had ever dreamed of owning!

As the editor of the small, special-interest magazine *Telescope Making*, I had the opportunity to attend gatherings of amateur telescope makers to see the latest innovations and chat at length with the innovators. When I saw an outstanding telescope, I took lots of photographs and interviewed the builder. I solicited articles, and did a lot of article ghostwriting, too. Thus it was that *TM* became the primary source for information about building Dobsonian telescopes. For those who have never seen it, *TM* was founded in 1978 and published quarterly through 1992, for a total of 46 issues. Today's *Amateur Astronomy* magazine is in many ways the direct lineal descendent of the old *Telescope Making*—even *Amateur Astronomy's* cover design is similar!

Thirty years have passed since amateur astronomers learned, via John Dobson and the Dobsonian design, that alt-azimuth telescopes can be capable instruments. Yet among amateur astronomers, the alt-az mounting has so far failed to displace the equatorial as completely as it has among professional astronomers. Let's back up and re-examine the alt-azimuth from first principles. We'll take a look at the enabling technologies that amateurs still need—as reported in the book you now hold in your hands—to catapult the alt-azimuth telescope to the forefront of amateur astronomy.

Paraphrasing Raymond Wilson, author of *Reflecting Telescope Optics* (Springer: 2000), a telescope must be constructed in such a way that it can be pointed at the desired object field. Furthermore, because the object is moving as a result of the earth's rotation, the telescope must provide the means to compensate for this movement. Finally, it should provide a high quality image of distant objects, which may be point sources or extended objects. So the three tasks are to:

1) provide means to point at a field,
2) provide means to track a field, and
3) provide a high-quality image of the field.

The speed, accuracy, and quality of pointing, tracking, and imaging vary greatly depending on the telescope and application. Let's consider how amateur astronomers have worked to satisfy each of these requirements in their telescopes.

Pointing

Pointing a Dobsonian telescope is simplicity itself: the observer locates the object visually in the sky, pushes the telescope to the general vicinity, then looks into a finder and pushes the telescope to center the object in the eyepiece. For bright, easy-to-find objects, this works extremely well—and for 30 years, it has been the way that virtually every Dobsonian alt-azimuth telescope has been pointed. For that matter, many equatorially-mounted telescopes are pointed the same way—moved manually by the observer to the field location.

Historically, the equatorial has boasted an enormous advantage: setting circles. From their inception in 1625, equatorial mounts were equipped with circular scales marked in right ascension with hour and minute divisions, and in declination with degree divisions. (Few telescopes made today are equipped with adequate setting circles, that is, circles sufficiently large and accurately marked that they allow an observer to point the telescope within a few minutes of arc of the desired location.) On a polar-aligned equatorial mounting, setting circles act as analog computers. An astronomer with a small, well-equipped equatorial telescope in the year 1900 could probably point his telescope as accurately and reliably as today's amateur can with the digital coordinate readouts found on today's telescopes.

Setting circles are classic examples of an *enabling technology*. An astronomer can point a circle-enabled equatorial telescope faster and more reliably than he or she can point an equatorial without setting circles. They enable the astronomer to find objects that would otherwise take time and energy to hunt down by star-hopping; the astronomer gets more done in less time. In today's world of digital gadgets, setting circles seem so simple—and so analog—that it seems a bit odd to call them a technology—but making good setting circles demands good machine-shop skills, and as a technology, they enable the astronomer who uses the "circles" to point the telescope with high precision.

But setting circles were certainly not ubiquitous. Most amateurs squinted into the finder and pointed the telescope manually. When Dobsonians appeared on the scene, most amateurs simply relied on already well-honed star-hopping skills to find the Messier objects. Setting circles were hardly necessary because Telrad finders became another key enabling technology for Dobsonian users. The Telrad made a Dobsonian easy for novice observers to point, and many observers with "Dobs" in the 20 inch class claim that a Telrad is just as easy or easier to use than a conventional finder telescope when it comes to finding familiar deep-sky objects. In the course of about five years, the Telrad displaced the time-honored optical finder on many telescopes. Small innovations can sometimes make a huge difference!

But amateurs with large-aperture telescopes usually want to hunt big game—and a Telrad or even a finder can't easily replace setting circles when it comes to zeroing in on faint Abell clusters of galaxies. Today's computers make it trivial to compute the altitude and azimuth of a celestial object, but in 1980, these equations posed a bit of a problem. The equations are:

$$ha = st - ra$$
$$alt = \arcsin(\sin(lat)*\sin(dec)+\cos(lat)*\cos(dec)*\cos(ha))$$
$$az = \arctan(\sin(ha)/(\sin(lat)*\cos(ha)-\cos(lat)*\tan(dec)))$$

where: *ra* is the right ascension of the object, *dec* is the declination of the object, *st* is the local sidereal time, *ha* is the hour angle of the object, *lat* is the observer's latitude, *alt* is the altitude of the object above the horizon, and *az* is the azimuth eastward from the north point. To compute the sidereal time, you need the current date, the local time, the time zone, and the observer's longitude.

But then a new technology appeared: the pocket calculator. *Astronomy*'s associate editor, Robert Burnham, had gotten one of the first HP-35 calculators made. And for the December 1978 issue of *Astronomy*, he wrote an article describing how to calculate elevation and azimuth coordinates. The appearance of this article was fortuitous; at the time, we had no idea how many amateurs would soon be interested in this seemingly esoteric knowledge.

I first saw alt-azimuth setting circles at the 1980 Riverside Telescope Maker's Conference. Dennis Donnelly brought an 8 inch *f*/8 Dobsonian with plastic-coated tape-measure refills wrapped around the wooden disks that served as the telescope's bearings. The bearings were sized so that 1 degree was 1/8 of an inch on the tape. Dennis had calibrated the circles on stars the night before—and the next morning in bright sunlight showed me Capella and Venus. To find an object, he entered the right ascension, declination, and time into his pocket calculator. Evaluating the equations took 29 seconds—and his program converted the altitude and azimuth angles to inches! After tweaking the optics a bit, Dennis also found Mercury, Jupiter, Saturn, Mars, Betelgeuse, Rigel, and Aldebaran in the daytime sky. I was impressed. It is one thing to look at coordinate transformation equations, but it's another thing entirely to see stars in the daytime sky by means of calculator and tape-measure setting circles.

Donnelly had primed me, so when an article titled "Alt-Azimuth Setting Circles" arrived at *Astronomy*, we published it in the April 1982 issue. The author, Paul Burke, provided the readers with a clear description of using a Dobsonian plus a pocket calculator to convert between celestial and horizon coordinates. Almost simultaneously, an article by Tom Dey in *Telescope Making #15* described how he had constructed a mechanically precise 17.5 inch Dobsonian with setting circles computed using an HP9825 minicomputer coupled to an HP7225A pen plotter. His article included the TI-59 pocket calculator program he used to get the altitude and azimuth angles. In addition, Tom's article described three different methods for aligning alt-azimuth setting circles. For these pioneering amateurs, operating a full-size personal computer outdoors or in the observatory was probably impractical, if not impossible. The pocket calculator had indeed become an enabling technology for pointing alt-azimuth telescopes.

Bill Burton described an interesting wrinkle in the July 1988 *Astronomy*. Instead of using setting circles, Burton employed a surveyor's compass and a precision inclinometer to set and measure his telescope's pointing angles. The Suunto compass and inclinometer feature a magnified scale that eliminates reading errors and could be placed anywhere on the telescope. To compensate for small errors, Burton simply adjusted the compass and inclinometer to give the correct readings. By then, pocket computers had replaced pocket calculators. Burton's Sharp 1250A allowed him to store the right ascensions and declinations of 225 objects, and he could load and save tapes with libraries of spring, summer, fall, and winter objects. The software, written in BASIC, took care of precession, but otherwise, it was a simple coordinate transform program.

Ultimately, however, setting circles have limited applications. They require a human to look up the celestial coordinates and move the telescope to the computed alt-az coordinates. Through the segment of the amateur telescope making community with engineering backgrounds, amateur astronomers gradually became aware of two more enabling technologies: stepping motors and shaft encoders. And a third technology—microcomputers—had galloped into the world's consciousness.

Stepping motors have a four-pole stator and a multi-pole rotor. By changing the current flow through the stator windings, the rotor will move one step forward or backward. For telescope control, the stepping motors of choice took 200 full steps per revolution, and by clever manipulation of the currents flowing through the stator windings, could take as many as 16 microsteps per step. Attached through reduction gears or belts to the right ascension and declination altitude axes of an equatorial mount, or to the azimuth and azimuth axes of an alt-azimuth mount, a stepping motor could point a telescope to the desired coordinates simply by taking the required number of steps. A pair of surplus 200 steps-per-rev Slo-Syn stepping motors costing $10 each replaced the need for a human to move the telescope.

Shaft encoders performed the inverse task: they replaced the need for a human to read coordinates from a setting circle. Inside shaft encoders is a disk scribed with equally spaced lines mounted on a shaft. Light from two LEDs is detected by two phototransistors placed half a line apart. As the shaft turns, the disk turns, and the pattern of light reverses. A simple logic circuit detects when the shaft has rotated one "tick" and the direction of rotation, and by counting ticks up or down, tallies the rotational position of the shaft. For amateur telescopes, the favored shaft encoder was the US Robotics Softpot with 4096 ticks per revolution, purchased new at $50 each. Attached directly to the axes of an equatorial or alt-azimuth mounting, these devices read a telescope's position to 360/4096 = 0.088 degrees, more than adequate for visual observing; attached via gears or belts, they gave correspondingly higher angular resolutions.

The seminal amateur stepping-motor driven computer-controlled telescope was constructed by Louis Boyd in 1984. Writing in *Telescope Making #22*, Boyd states, "The field of amateur-built microcomputer-controlled telescopes is in its barest infancy, so it is too soon to say what sorts of mounts and drives will eventually evolve and in the end prove to be most popular and useful." Boyd's 10 inch telescope was mounted on a polar disk equatorial mount and used bicycle-chain drives on both axes. Almost simultaneously, the publication of a book, *Microcomputer Control of Telescopes* by Mark Trueblood and Russ Genet, set forth the detailed theory of computer-controlled telescope operation. Trueblood, for his master's thesis, had worked out the equations for telescope control—an instance of "technology transfer" from the professional to the amateur realm. Trueblood and Genet worked closely with Boyd and were co-authors of Boyd's *TM #22* article. Their book was updated in 1997 and retitled *Telescope Control*—and it remains an essential text today.

I saw my first amateur-built computer-controlled alt-azimuth telescope at the 1987 RTMC. The builder was David Gedahlia, an electrical engineering sophomore, who built the 10 inch *f*/4.5 Dobsonian drivers with parts out of his junk box. "Does it work?" I asked him. "Yes, it works," he said, "I looked at eighty or a hundred objects last night." He hadn't even been counting. The control computer was an Atari 800XL that had been languishing on the floor of his closet. The stepping motors were reduced through surplus gearboxes to drive discarded tape-recorder capstans against 18 inch diameter wooden disks. The whole project struck me as a major breakthrough in bang-for-the-buck telescope performance.

As best I can tell, shaft encoders made their entrance on the amateur scene via commercially made add-on digital setting circle kits for commercial telescopes. The Meade CAT system, for example, "attaches in minutes" to standard equatorially-mounted Meade and Celestron telescopes. It featured a database of 8,199 objects, and sold for $1495. A simpler system, which read only right ascension and declination, sold for $495. Roger W. Tuthill, Inc. offered a competing add-on encoder system for $399. Meanwhile, Celestron had already introduced the Compustar in 1986. This equatorial was the first-ever stepper-motor driven commercially produced go-to telescope. Although it was discontinued in 1995, the Compustar has a fan club with hints, tips, software downloads, and bug fixes. Compustars sold for premium prices—$6,000 for the CC8 and $20,000 for the CC14. Other manufacturers soon attempted to emulate the Compustar using inexpensive motors with plastic gears instead of stepping motors—resulting in reduced performance but bringing go-to equatorial telescopes into the price range of many observers.

For alt-azimuth telescopes, a crucial (but heretofore missing) new technology fell into place. Although transforming celestial coordinates to horizon coordinates works perfectly well, it is unduly restrictive. Imagine celestial coordinates and the coordinate system defined by a telescope's axes as two spheres. Given two points on each sphere representing the same two stars, there are only two possible ways to rotate the two spheres into coincidence—and it is normally possible to determine which of the two is correct. In the February 1989 *Sky & Telescope,* the article titled "A New Concept in Computer-Aided Telescopes" by Toshimi Taki showed how to use matrix algebra to convert between native coordinates of an arbitrarily oriented telescope and celestial coordinates by centering on two stars. The mathematically adept quickly realized that three stars yielded an even better solution, and for *N* stars, they could perform a least-squares fit. Versions of the method described by Taki were already in—or soon appeared in—commercially made digital setting circles and telescopes. Almost all go-to telescopes today use the same or very similar algorithms. Although two-star and three-star alignment had undoubtedly been known years earlier, its disclosure to the amateur community served as yet another technology enabling the widespread adoption of the alt-azimuth mounting.

Tracking

Without tracking, as newly-hatched amateur astronomers quickly learn, an object begins to drift out of the field as soon a telescope is pointed at it. Equatorials have the advantage of simplicity over alt-azimuth telescopes. By clamping the declination axis and driving the right ascension axis at a constant rate, a set object stays centered in the field. Tracking with an alt-azimuth requires moving the telescope in both axes simultaneously. John Dobson's big high-friction bearings became popular because visual observers had only to push on the telescope and the bearings would turn as needed.

The first amateurs to equip alt-azimuth telescopes with stepping motors probably assumed that tracking would be easy: put the pointing algorithm into a loop that would update the mount's pointing every five or ten seconds. It didn't take long to realize—as the telescope bounced and jounced across the sky—just how unsatisfactory this solution was. Although a stepping motor resolution of 1 arc minute on the sky was adequate for pointing, it turned out to be far too coarse for tracking. Instead, the step size had to be 1 arc second or smaller, and the pointing algorithm had to operate at least 15 times per second! A further complication arose: to slew to a new position, the stepping motor had to step so quickly that 1) it missed steps and 2) its torque fell too low to drive effectively. Finally, it became apparent that even when tracking was good, the field of view in the telescope rotated. This might be okay for visual observing, but a rotating field ruled out effective photography.

Given these difficulties, amateur telescope makers pursued an alternative solution: the equatorial platform. Originally described by Adrien Poncet in the January 1977 *Sky & Telescope*, "a rigid body that contains three points, one acting as a fixed point and the other two constrained to move in a fixed plane, can only rotate around an imaginary axis perpendicular to the plane and passing through the pivot." Three years later, in the February and March 1980 issues of *Sky & Telescope*, Roger Sinnott gave a round-up of Poncet spin-offs ranging from small platforms for tracking a 35mm camera to Jack Ells' jumbo platform that carried his entire observatory! Alan Gee's observation, in the February article, that the Poncet platform can be derived directly from a classic Porter split-ring equatorial, was an important conceptual breakthrough for amateurs baffled by Poncet's geometric description: the Poncet platform was just a cut-down version of the familiar equatorial! With the concept behind the Poncet now spelled out clearly, for the next 15 years amateur astronomers who wanted a tracking Dobsonian telescope turned first to an equatorial platform.

For tracking, an equatorial platform offers substantial advantages over the stepping-motor-driven alt-azimuth. Like an equatorial, tracking is accomplished by a single motor running at a constant rate. To guide the telescope in right ascension, the motor rate can be changed. In declination, a simple tangent arrangement allows small movements. There is no field rotation. Articles exploring the Poncet platform and its many variations appeared in nearly every issue of *Telescope Making* published between 1980 and 1984. It is an exaggeration to claim that the Poncet platform killed the development of effective alt-azimuth tracking in the 1980s, but there is little doubt that equatorial platforms removed much of the incentive amateurs might have had to develop the required technologies.

As the 1990s began, CCD cameras distracted technically-oriented amateurs from thinking about alt-azimuth tracking. In quick succession I wrote three books: *Introduction to Astronomical Image Processing* (1991), *Choosing and Using a CCD Camera* (1992), and *The CCD Camera Cookbook* (1994). The pixel count of the Cookbook 245 CCD camera—242x378, or 0.09 megapixels—seems ridiculous today, but it was a big deal back then. Something like 7,000 amateurs built the Cookbook 245 CCD camera, and like me, they found dealing with one tricky new technology more than enough! I opted to do my imaging with a 6 inch Newtonian reflector on a safe and solid Byers 812 equatorial mounting.

Fortunately, not everyone succumbed to the fascination of CCD imaging. In 1992, Oregon amateur astronomer Mel Bartels computerized his 20 inch *f*/4.8 Dobsonian. When I met Mel at the 1995 Oregon Star Party, he had already solved most of the problems in alt-azimuth tracking. Mel's solution involved hardware to drive the stepping motors and—more important—software that knew how to run stepping motors fast enough to slew the telescope, yet also knew how to microstep the stepping motors for fine tracking. Later that summer, Mel and I made tracked exposures several minutes long with my Cookbook 245 on Mel's 20 inch telescope.

By 1995, the web had become an important means of communication among computer-savvy amateurs, and Mel was sharing plans and ideas with a growing on-line community. Eventually, Mel's hobby became BBAstroDesigns, Inc., freely trading information and selling circuit boards, stepping motors, and most importantly, finely-crafted software that imbued the system with enough embedded intelligence to point and track a telescope accurately.

Just what obstacles are there in tracking an alt-azimuth telescope? The first issue is field rotation. The orientation of the image rotates relative to the telescope. As with altitude and azimuth, the angular orientation of the field—called the parallactic angle—can be computed:

pa = arctan(sin(*ha*)/(tan(*dec*)*cos(*lat*)-sin(*dec*)*cos(*ha*)))

where *pa* is the parallactic angle. When an object is on the meridian, the parallactic angle is zero.

Another complication is that the rate of motion in altitude, azimuth, and parallactic angle changes continuously. In the northern hemisphere, after an object on the celestial equator rises in the east, its altitude increases steadily but the rate of change of altitude decreases until it reaches zero at the meridian, and then the rate of change of altitude becomes negative. The altitude decreases as the altitude rate of change becomes increasingly negative until the object sets in the west. If an object approaches the zenith, the speed of rotation in azimuth rises toward infinity. The equations for these rates are:

altdot = sin(*az*)*cos(*lat*)
azdot = sin(*lat*)-tan(*alt*)*cos(*az*)*cos(*lat*)
padot = -(cos(*lat*)*cos(*az*)/cos(*alt*))

where:

altdot is the altitude drive rate with respect to hour angle,
azdot is the azimuth drive rate with respect to hour angle,
padot is the parallactic angle rate with respect to hour angle.

Once a telescope is pointed at a field, the computer controlling the telescope can compute the drive rates and then pass control of the stepping motor to a microcontroller to maintain the computed rate. This reduces the load on the control computer. For guiding, the computer can monitor a CCD-based guide camera and apply tracking corrections as required. Thus stepping motors, though not ideal for driving telescopes, can be tamed through well-designed software.

However, around the turn of the century, another enabling technology entered the amateur telescope maker's toolkit: the DC servo motor. Like stepping motors, DC servo motors require special control circuitry, but unlike stepper motors, servos have a wide speed range and a flat torque curve. The popular Pittman 8000 Series are sold plain or with an integral shaft encoder providing 500 ticks per revolution. With appropriate control software, a DC servo can move backward or forward by just one tick (1/500 of a revolution) for fine guiding, or speed at 6800 rpm for rapid slewing.

Servo motors entered the amateur world in the fall of 2004. Mel Bartels, recognizing the potential offered by servo motors, was adapting his stepping-control software for the servos—but needed servo-control electronics. Mel contacted Dan Gray, and Dan made Mel a servo-control circuit board using Pic servo chips. Although it worked well, Dan realized that he could design a better controller. Immersing himself in the hardware and software, Dan delivered his first new boards to customers by early the next summer. Dan has incorporated as Sidereal Technology, Inc., and offers both mechanical components and telescope control software for all kinds of telescopes.

The next generation in alt-azimuth pointed and tracked telescopes is now very close. In this book, Dave Rowe describes yet another new technology now available for use in amateur telescopes: the direct-drive motor. Working in tandem with Dan Gray, Dave demonstrated a prototype direct-drive motor and control software at the 2007 Portland I conference. The motor, roughly 12 inches in diameter, consisted of two plywood disks. On one disk were glued strong magnets; on the other were glued hand-wound copper coils. Wrapped around the periphery of the moving disk was a length of Renishaw encoder tape. Under software control, the motor could position itself within one arc second of a desired position, or move in either direction at a steady pace.

Integrated into the California Polytechnic 18 inch demonstration telescope are two improved prototype direct drive motors capable of moving the telescope at 45 degrees per second, but setting to a desired position within one arc second. As the name implies, the direct drive motor is not geared, rollered, or reduced in any way: the motor is both the rotational axis and the driver. When power is removed, the telescope moves smoothly and freely, without friction. When power is applied, the telescope moves to the desired position with a slightly uncanny determination. Push on the telescope and the motor pushes back to restore the telescope to the correct position. Coupled with inexpensive ultra-high-resolution encoders now under development, direct drive motors have the potential to revolutionize amateur telescopes.

It may be another year or two before the Cal Poly 18 inch demonstration project becomes fully operational and proves its worth by acquiring science data. At the time of this writing, Dan Gray's "Lollipop" telescope comes closer than any I know of realizing a fully functioning amateur-class alt-azimuth telescope. At the 2009 Pine Mountain Observatory summer research workshop, in five nights the Lollipop collected over 4,000 images of the variable star XX Cygni while Dan was in the process of debugging his software.

Providing High-Quality Images

In the last 50 years, ground-based telescopes have increasingly given astronomers—both amateur and professional—sharper and better quality images. Professionals finally recognized that convection from warm mirrors, dome seeing, and local turbulence hampered many major telescopes. Professional observatories are now sited on the western mountain ranges of continents, at high elevation, with wide slots or doors to flush warm air from the telescope housing. Amateurs similarly have adopted open-truss designs and open-style mirror cells that allow optics to cool rapidly and completely. Every major professional telescope now in development has some form of active optics, and amateurs are experimenting with various forms of "fast guiding" and "lucky imaging" to capture fleeting moments of excellent seeing.

For amateurs, these developments increase the payoff from bringing a perfect wavefront to focus, but the desire to reduce the thermal mass and weight of the telescope often works in the opposite direction. Heavy as they were, classic "full thickness" mirrors could be counted on to hold their figure under the pull of gravity. Many of the large, thin-mirror Dobsonians suffered from astigmatism and potato-chipping due to thin mirrors supported by poorly conceived and sloppily constructed mirror cells. Thanks to fast computers, public-domain finite-element analysis software like PLOP, and a couple decades' experience in building excellent mirror cells, today's thin-mirror, lightweight Dobsonians give excellent images.

Thin mirrors have been around for a long time. The speculum-metal mirrors in William Herschel's 20 foot and 40 foot telescopes may have been the first. Lord Rosse, at his Birr Castle observatory, experimented with ways to reduce the mass of his 3 foot and 6 foot diameter metal mirrors. In an equatorial, pointing away from the meridian causes the gravity vector to roll and tip relative to the mechanical axis of the telescope mirror. Designing a cell to provide constantly changing lateral support as well as longitudinal support for a mirror with a 20:1 to 40:1 diameter-to-thickness ratio was complicated; it was more practical to put a full-thickness mirror (6:1) in the tube and finesse the cell details. And because the factors that go into a successful telescope were not fully understood, designers—rather than experiment—took the safe route of copying older designs. For these reasons, the conservatively designed telescopes of the early- to mid-20th Century tended to have full-thickness mirrors in relatively simple cells--even at the expense of long cool-down times.

The movement to alt-azimuth telescopes changed all that. Experimentation with mirror cells has been part of the alt-azimuth movement since the beginning. The alt-azimuth mount tips the mirror in one axis only, simplifying the forces on the mirror into tractable lateral and longitudinal components. In early amateur Dobsonians, a braided nylon strap sling handled lateral mirror support, and a crude nine-point floatation cell sufficed for longitudinal mirror support. For amateur telescope makers, a three-part series of articles written by Bob Kestner and published in *Telescope Making #12, #13,* and *#16* served as the Bible for amateurs interested in grinding, polishing, and figuring large thin mirrors. As commercial suppliers have developed techniques for working and supporting thin mirrors, the Dobsonian revolution has flourished.

Today we stand again at a crossroads between old proven techniques and new ideas that might reduce mirror weight while maintaining optical wavefront quality. Many of the chapters in the "Lightweight Mirror Developments" section of this book are about potential breakthroughs. As you read them, maintain an attitude of mixed enthusiasm, skepticism, and delight—for this is the spirit of the Alt-Az Initiative!

By bringing together individuals with diverse talents and outside-the-box ideas, by familiarizing them with a wide range of possibly enabling technologies, and by providing regular opportunities for the interchange of ideas and critiques, the Initiative has the potential to accelerate the development of a whole class of amateur-owned research-grade telescopes in the 0.7 to 1.5 meter class. If you are reading this book, and you feel the challenge of becoming part of an informal network of talented minds, then get in touch, contribute your time and ideas, and become one of the movers and shakers in astronomical technology.

Richard Berry
Handbook of Astronomical Image Processing and AIP4Win Software
Lyons, Oregon

Part I

THE ALT-AZ INITIATIVE

Chapter 1

The Small Research Telescope Challenge

National Academies of Science Astro 2010 Position Paper

Russell Genet[1, 2], David Rowe[3], Daniel Gray[4], Susana Deustua[5], Andrew Aurigema[6], Howard Banich[7], Mel Bartels[8], Richard Berry[9], Craig Breckenridge[10], Peter Chen[11, 12], Gary Cole[13], David Davis[14], Joe Haberman[15], Rick Hedrick[15], Arne Henden[16], Jolyon Johnson[2], Greg Jones[17], Tong Liu[18], Robert Nelson[19], John Ridgely[1], Greg Rohde[20], Kiran Shah[21], Thomas Smith[22], Vera Wallen[23], and Xiao-Hua Yu[1]

[1]California Polytechnic State University; [2]Cuesta College; [3]Sierra Monolithics; [4]Sidereal Technology; [5]Space Telescope Science Institute; [6]OTF Design; [7]Rose City Astronomers; [8]BB Astrosystems; [9]AstroPIX; [10]Dynamic Structures Ltd.; [11]Lightweight Telescopes Inc.; [12]NASA Goddard Space Flight Center; [13]Western Nevada College; [14]Toledo Scope Werks; [15]PlaneWave Instruments; [16]American Association of Variable Star Observers; [17]Eclipse Technologies; [18]Hubble Optics; [19]Sylvester Robotic Observatory; [20]Technical Marine Services; [21]Chroma Systems Pvt. Ltd.; [22]Dark Ridge Observatory; [23]Educational Consultant

Summary

Large telescopes excel in the detailed study of faint objects at the edge of the observable universe and high-resolution examination of specific objects. However, small telescopes continue their valuable role in astronomical research through time series, networked, and other observations that only large numbers of small telescopes can provide—tasks which are cost prohibitive for large telescopes. Small telescopes also continue to play a vital role in recruiting and training the next generation of astronomers and instrumentalists, and serve as test beds for developments of novel instruments and experimental methods.

There is a synergy between large and small telescopes—an appropriate balance between the two. Yet the rising costs of large, cutting edge telescopes—absolutely necessary for the advance of astronomical science—has made it difficult for governmental and large institutions to also invest in small telescopes. Currently, a tight economy makes it difficult for smaller institutions to invest in small research telescopes themselves because the cost is a substantial outlay for them. Yet we need to maintain the balance between large and small telescopes to preserve their synergy.

A solution to this dilemma is to greatly reduce the cost of smaller telescopes. This would make them much more affordable for smaller institutions. We believe such cost reduction is possible through the transfer and innovative transformation of technologies from large to small telescopes and by producing the resulting low cost small research telescopes in production quantity.

Technology transfer, innovation, and manufacturing in quantity form a natural evolutionary process, which is commonly seen in the application of new technologies. However, this process is often slow. It can be purposely accelerated in astronomy by developing bridging mechanisms between the large and small telescope communities, and by encouraging technology transfer initiatives, demonstration telescopes, and the production of affordable research telescopes.

The Alt-Az Initiative is an example of such purposeful acceleration. The Initiative is a self-funded, cooperative effort now well into its third year. The Initiative has utilized bridging mechanisms between the large and small telescope communities to launch over a dozen technical initiatives. Results from some of these efforts have already been incorporated in a demonstration telescope and the design and construction of a commercially produced telescope.

The Initiative, from its inception, has involved undergraduate engineering students in its telescope and instrumental developmental efforts. The Initiative has also sponsored student research with small telescopes that resulted in published papers, and student researchers have presented their findings at Initiative workshops and conferences. Undergraduate students have also co-chaired two of the Initiative's conferences and co-edited the Initiative's two books.

Looking toward the future, we believe it would be appropriate to encourage this Initiative and other, similar efforts that accelerate the technology transfer process to produce lower cost, more capable small telescopes and instruments in a timely manner. A greater number of affordable small telescopes would not only support scientific research directly, but would increase undergraduate involvement in both research and development. Such encouragement could help maintain an appropriate balance between large and small telescopes and hence foster their continued synergy.

Small Research Telescopes

Their Enduring Value

In the age of giant telescopes, are small telescopes still useful? Weaver (2003) points out that

> Both quantitative and qualitative arguments demonstrate the continuing importance of small telescopes to the astronomical endeavor. The quantitative arguments show that it is significantly less expensive per citation to use the smallest telescope

> that will accomplish the research. Both the quantitative and qualitative arguments show that the research accomplished by small telescopes is of continuing and lasting significance to the discipline as witnessed by their non-diminishing contribution to astronomy over the last century and the persistence of their citation histories.

Ringwald et al. (2003) suggests that

> Small telescopes can hold their own with larger instruments since more time is available on them. This makes possible monitoring campaigns, aerial surveys, and time-resolved campaigns, particularly if the telescopes are networked or automated—all difficult to carry out with larger telescopes, for which even small amounts of telescope time are in great demand.

The 2007 report of the Committee for Renewing Small Telescopes for Astronomical Research (ReSTAR) concluded that

> The science to be done with small and mid-size telescopes remains compelling and competitive in the era of big telescopes. Small and mid-size telescopes continue to produce innovative science themselves, and to provide precursor and follow up observations that enhance the scientific productivity of larger telescopes. Small and mid-size telescopes also enable scientific investigations that are not possible on larger telescopes.

The ReSTAR report went on to state that

> Small and mid-size telescopes contribute additionally to the discipline through their training and educational functions and as test beds for innovative new instrumentation and techniques.

The Challenge

Increasing Participation through Greater Affordability

Weaver (2003) recalls that

> Astronomy has a history of an essential synergy between small and large telescopes. This synergy can be maintained only if there is a reasonable number of well-maintained, well-instrumented smaller-sized telescopes.

Budget realities suggest, however, that both national and large-institution funding of expensive to build and operate but vital cutting edge large telescopes will make it difficult to also fund smaller telescopes in sufficient numbers to maintain an optimal balance between large and small telescopes. Further, during hard economic times, it will also be increasingly difficult for smaller institutions to fund small research telescopes themselves. So what is to be done? How can we maintain the balance and the synergy between large and small telescopes?

Meeting the Challenge

Lowering Costs through Advanced Technology and Quantity Production

In his book, *Excursions in Astronomical Optics,* Lawrence Mertz (1996), after noting that there is a "crying need for more observational facilities to outfit the throngs of young astronomers," offered a solution:

> The remedy, it seems to me, would be a proliferation of modest-sized telescopes in the 1.5 to 2 meter class that are sufficiently standardized for economy. The size is large enough to accomplish significant work, yet is small enough to become affordable if the design were standardized. All too frequently small institutions insist upon designing their own telescopes almost from scratch to fulfill their individual tastes. It should be evident that that course is hopelessly uneconomical, just as it would be uneconomical for each automobile to be custom designed. In essence, what is needed for observational astronomy is numerous ... relatively inexpensive telescopes in the 1.5 to 2 meter class. Even at this modest size there does not remain much question between altitude-azimuth (alt-az) and equatorial mounting; computers now make the former more economical. Alt-az has the further advantage of simplifying the primary mirror support since tilting is restricted to the elevation axis.

We can expand on two of Mertz's vital points. First, advances in technology, especially in computers and electronics—but also materials and other areas—can reduce telescope costs. Mertz illustrated this with an obvious example—shifting from equatorial to alt-az mounts. Second, true economy can only be achieved through quantity production. One-off telescopes will always be expensive. Although Mertz did not mention it, a third point could also have been made: large one-off telescopes with their sizeable development budgets often pioneer the new technologies which subsequently migrate to small production telescopes. These "tech transfers" are completed once clever ways are devised to drastically reduce costs so that the large telescope technologically-inspired breakthroughs can be incorporated economically into small telescopes in quantity production.

Two Telescope Tech Transfer Examples

Consider, as the first tech transfer example, alt-az telescopes themselves. The 5 meter Hale telescope, completed over a half century ago, remains the largest equatorial telescope ever built. There are now over 20 larger telescopes. Every single one is alt-az. Smaller telescopes are also heading down this same evolutionary pathway.

Until recently, most truly small (less than 1 meter) telescopes remained equatorial because alt-azimuth telescopes not only required constantly changing drive rates in both altitude and azimuth, but, to avoid image smearing, also required a field de-rotator—a significant complication. However, these complications did not apply to visual-only, manually operated, non-tracking alt-az telescopes. In their search for larger apertures at low cost, amateur astronomers recognized the inherent advantages of an alt-az over an equatorial configuration. Today, low cost, easily transported, and now mass produced "Dobsonian" telescopes are numerous. One can purchase quantity manufactured Dobsonian telescopes from 4 to 50 inches in aperture that cost from $100 to $180,000.

Mel Bartels (1994, 1999, and 2000) developed a very low cost control system for alt-az telescopes that employed a personal computer, a small electronics board, and three stepper motors for, respectively, altitude, azimuth, and instrument rotation. Soon Mel's company, BB Astrosystems, was selling the alt-az control system electronics for a very low cost. The user supplied the stepper motors and a personal computer, and computerized their Dobsonian telescopes. More recently, Dan Gray's Sidereal Technology has offered the electronics and software for a complete alt-az DC servo control system, including instrument rotation and focus, for about $1000. He is about to release a direct drive control system. Dan graciously suggested he was inspired by *Microcomputer Control of Telescopes* (Trueblood and Genet 1985).

For the second example of tech transfer from large to small alt-az telescopes, consider telescope weight, stiffness, and materials. As computers became more powerful, finite element analysis (FEA) resulted in increasingly lightweight yet stiffer large telescopes—leading eventually to today's highly optimized truss structures. Lighter weights lowered costs, while stiffer structures allowed control systems to more effectively counter wind gusts. Stiff and economical steel remained the material of choice for large alt-az telescopes.

Dobsonian telescopes also evolved to become much lighter, stiffer, and more truss like. They shifted from ordinary plywood and cardboard Sonatubes to hardwood veneer plywood and aluminum trusses. A recent small alt-az telescope, the 18 inch Cal Poly telescope (discussed further below), was designed by undergraduate engineering students using FEA and features a carbon fiber composite structure. Carbon fiber composites have about three times the stiffness to weight ratio of steel. Although carbon fiber is too expensive a material to use in large telescopes, they can be an advantage in smaller telescopes.

A good example of a modern, lightweight alt-az telescope is Howard Banich's (2004) 28 inch computer-controlled, all aluminum alt-az telescope which weighs about 330 lbs. Howard transports his telescope in the back of a VW Eurovan and can assemble or disassemble it in less than ten minutes. A somewhat similar telescope, *SpicaEyes,* is available commercially in apertures up to 32 inches.

It might be noted in passing that not all tech transfers have been from large to small telescopes. Full telescope and observatory automation, remote Internet access, and global networking have all been pioneered by small telescopes. Larger telescopes subsequently adopted aspects of these technologies. These are all clear cases of reverse tech transfer.

Two Separate Telescope Development Communities

Although large to small alt-az telescope tech transfer is taking place, the fact is that the large and small developmental communities are, in the main, quite separate, with limited between-community communications. Large alt-az telescope developments are generally undertaken by a combination of paid professional engineers and astronomers. These telescopes are built in high-bay machine shops equipped with large cranes. The cost of such telescopes now often exceeds $50 million, and the telescopes can weigh over 1 million pounds. Each telescope development is, typically, unique. Information is exchanged through the meetings and journals of engineering organizations such as the Society of Photo-optical Instrumentation Engineers (SPIE), and through professional astronomical meetings such as those organized by the American Astronomical Society (AAS). A classic large telescope design book is *The Design and Construction of Large Optical Telescopes* (Beley 2003).

The development of small, Dobsonian alt-az telescopes, on the other hand, has mainly been undertaken by unpaid amateur telescope makers (ATMs) as an avocation. It might be noted, however, that many ATMs are, by vocation, scientists, engineers, or technicians. These telescopes are being built at home or small manufacturing shops, often with simple tools. Costs are typically a few thousand dollars, and telescopes weigh a few hundred pounds. Although many ATM-built telescopes are unique, several small telescope manufacturers produce identical telescopes in quantity. Information is not only exchanged on-line through a variety of email lists and web sites, but also through annual regional conferences such as the Riverside Telescope Makers Conference (RTMC). The quarterly journal, *Amateur Telescope Making,* which was the primary published record for Dobsonian developments for many years, has now been superseded by *Amateur Astronomy. Sky & Telescope* and *Astronomy Technology Today* also feature small telescope developmental articles. A classic small telescope book is *The Dobsonian Telescope* (Kriege and Berry 2003).

Instrumentation

The past decade has seen great advances in the availability of research grade instrumentation suitable for smaller telescopes. These advances have been driven by a combination of evolving technology and by the emergence of a community of advanced amateur astronomers with modest budgets but very high standards.

This combination has allowed instrument makers such as SBIG, Optec, Shelyak, and others to bring to market products with capabilities that were previously available only to well financed institutions. Equally important has been the emergence of a software infrastructure (ASCOM) that unifies instrument automation and telescope control.

To give just a few examples of progress over the last decade: high quantum efficiency, science grade, imaging cameras have gone from 1.5 mega pixels (MP) to 16 MP. Small telescope spectrographs, a category that hardly existed a decade ago, now include R~17K Littrow and fiber fed Echelle instruments. The cost of polarimetric instruments has declined by a factor of 10 thanks to achromatic polymer waveplates from Bolder Vision Optik. And whole new instruments are now available such as the Optec SSP-4 that provide JH band infrared photometry. Small telescope near IR spectrographs and imagers are within sight.

These individual instruments are generally lightweight, well documented, essentially maintenance free, and have multivendor software support. Moreover, they were designed to match telescopes in the .3m to 1.5 meter range because that is where their customers are.

Dedicated Telescopes, Automation, & Operation and Maintenance Costs

General purpose telescopes, which have to meet a wide range of requirements, tend to be more expensive than dedicated telescopes to purchase, and certainly are more expensive to operate and maintain. The initial cost of a general purpose telescope at a major observatory, while it can be considerable, is typically much smaller than the long run operation and maintenance (O&M) costs over the telescope's lifetime. The cost of paying O&M staff for decades adds up. Furthermore, the cost of the instruments, substantial to begin with and expensive to upgrade over the years, can be a large fraction of the life cycle cost. The long-term burden of O&M costs at the national observatories became so high that a number of smaller scopes had to be closed because there was insufficient money to operate them and also build and operate new, larger, cutting-edge research telescopes.

Taking instruments on and off a telescope on a regular basis is costly in both manpower and equipment wear and tear. Dedicated telescope/instrument combinations, on the other hand, can be designed together for a specific mission area with the instruments permanently emplaced on the telescope. A computer-controlled selector moves the optical beam between instruments in a cluster. Dedicated telescopes may be the way of the future for smaller research telescopes, because they are cheaper to operate and maintain, particularly when they are operated in a service/robotic mode.

If a dedicated telescope with a permanent, computer selectable cluster of instruments is run robotically, care has been taken to obtain high reliability components, and maintainability issues have been carefully addressed, then O&M costs can be quite low. The National Science Foundation has traditionally funded the telescopes at colleges and universities, but not, of course, for individuals. However, if small, dedicated, robotic research telescopes could be made inexpensively enough, they might become affordable enough for individuals and small informal groups to independently purchase, operate, and maintain.

The Alt-Az Initiative

An Experiment in Purposeful Evolutionary Acceleration

Meeting the challenge of increasing the number of small research telescopes by lowering their costs through technology transfers and quantity production is naturally taking place across a broad front, albeit at a rather slow pace. This natural technical evolutionary process (Basalla 1988) can be accelerated, however, through purposeful tech transfer "bridging mechanisms" between the large and small alt-az telescope development communities. These mechanisms include enhanced inter-community communications, as well as the encouragement of tech transfer initiatives, tech demo telescopes, and quantity production of unusually low cost, small research telescopes. We consider, below, a pilot project "experiment" in the purposeful evolutionary acceleration of small research telescopes.

Initiative Overview

The Alt-Az Initiative was established in June 2007. The Initiative has been self-funded, with individuals and small firms donating time and materials. The Initiative focused, from its inception, on the development of "advanced technology," lightweight, low cost, 0.5 to 2.0 meter alt-az research telescopes. To provide even more focus, a specific goal, a "challenge," was established early on, to develop a lightweight, transportable, 1.5 meter alt-az research telescope. From its inception, the Initiative also concerned itself with instruments, dedicated scientific research programs, and automation (as well as remote access and networking).

Special emphasis has been given by the Initiative to those instruments and research programs which would most benefit from low cost 1 to 2 meter telescopes/instrument combinations. These include time series optical and near IR photometry, time series fiber-fed low and medium resolution spectroscopy, and high speed photometry.

Special emphasis has also been given to the involvement of undergraduate students, who have participated since the Initiative's inception in research, development, conferences, and publications.

Considered separately below is each of the five key elements of the Initiative:

Enhanced communications
Tech transfer initiatives
Tech demo telescopes
Transfers to quantity production
Student participation in research and development

Enhanced Communications

Although the bulk of the Initiative's communication is via email and telephone, Initiative members have gathered together in person at over a dozen workshops to date. The two founding workshops were held in June 2007 in Portland (Sidereal Technology) and San Luis Obispo (California Polytechnic State University). A large workshop followed in October in Dallas, and there have been a number of workshops since then, including workshops in Hilo with Gemini and Subaru engineers. A very productive workshop was held in June 2009, in conjunction with the American Astronomical Society's summer meeting in Pasadena, California.

Figure 1: Initiative workshop in Dallas, October 2007.

The Alt-Az Initiative has also sponsored two major conferences, one held in San Luis Obispo in June 2008 and the other in Hawaii in January 2009. These four and five day conferences, respectively, were well attended, and each conference included several workshops and some three dozen talks, many of which appeared as chapters in the Initiative's two books, *Small Telescopes and Astronomical Research* and *The Alt-Az Initiative: Telescope, Mirror, and Instrument Developments.*

Figure 2: Initiative Small Telescope and Astronomical Research (STAR) Conference in San Luis Obispo CA, June 2008 (both pictures above).

Figure 3: Initiative Galileo's Legacy Conference at the Makaha Resort, Hawaii, January 2009.

Figure 4: Galileo's Legacy was followed by an insider's tour of the 8 meter Gemini Telescope on Mauna Kea on the Big Island of Hawaii, January 2009.

Details on the Initiative's workshops, conferences, and publications, as well as useful references, links, etc. are available on the Initiative's web site, www.AltAzInitiative.org.

Tech Transfer Initiatives

So far, a dozen technical initiatives have been completed or are underway. We will consider by way of illustration, example initiatives drawn from three areas:

Telescope drives and controls
Lightweight primary mirrors
Active primary mirror control

Telescope Drives and Controls

During the early stages of the Initiative, several members visited Magdalena Ridge Observatory and were given a very thorough tour and explanation of the modern 2.4 meter alt-az telescope by its Chief Engineer, Elwood Downey, and its Lead Engineer from EOS Technologies, Kevin Harris. The 2.4 meter MRO telescope employs direct drives. The telescope itself acts as the bearings and "frames" for the azimuth and altitude motors. A ring of permanent magnets in both altitude and azimuth are the motors' rotors, and similar rings of coils that act as the stators. Magnetic forces between the rotor and stator move the telescope based on the output of high resolution, on-axis encoders, with the motion determined by a servo algorithm. There are no gears, friction rollers, belts, or any other mechanical reduction devices to introduce periodic errors or compliance that could reduce the telescope's stiffness and natural frequency. As noted in Bely's (2003) classic book, *The Design and Construction of Large Optical Telescopes*, "A direct drive, which eliminates all mechanical systems, is the ultimate choice."

While the stators, rotors, and computer-controlled electronics are all commercially available, they can be more expensive than the total cost of some modest-aperture telescopes. Initiative member Dave Rowe devised a very low cost, axial flux, direct drive motor for telescopes. Normal large-machine direct drive motors are *radial* flux motors. The radial "one ring inside another" configuration is energy efficient, but involves an expensive arrangement of magnets, coils, and soft steel to contain the flux. Rowe realized that energy efficiency was not important for modest aperture alt-az telescopes, and devised an *axial* flux motor, with a simple coil ring placed on top of a permanent magnet ring. The axial flux direct drive motor only costs about $300 per motor in magnets, wire, and steel, and is extremely easy to build.

However, direct-drive, direct-position-reading servo systems are much more difficult to control than ordinary brushed DC motors, and require not only high resolution on-axis encoders as inputs, but high-speed computation of the control system filters and a feedback algorithm. Initiative member Dan Gray at Sidereal Technology designed a low cost control system that handles two "brushless," AC synchronous, direct drive telescope motors. Gray also wrote all of the real-time embedded firmware for the controller and the PC-based command-and-control software.

Figure 5: The axial-flux direct drive motor devised by Initiative member Dave Rowe. An 18 inch diameter motor was readily assembled from parts and materials that cost only $300. This azimuth assembly includes the motor, bearings, and high resolution tape encoder and read head.

Lightweight Primary Mirrors

Both large mountaintop alt-az telescopes and their diminutive lowland brethren share a common obsession with lightweight primary mirrors. Without lightweight mirrors, telescope weights and hence costs rapidly get out of hand. For small telescopes, especially in the 1 to 2 meter aperture range, transportability and assembly are vital issues if 18-wheelers and cranes are to be avoided. Initiative members have been addressing the challenges of lightweight yet low cost primary mirrors for small alt-az research telescopes from several perspectives.

Several Initiative members have been investigating the use of foam glass as lightweight spacer/structural material between the top and bottom glass plates of a primary mirror. Pittsburgh Corning makes Foamglas, a rigid insulating material used in liquid natural gas ships, many industrial applications, and even under heated concrete runways in cold climates such as Alaska. Andrew Auregema has been machining concave surfaces in the tops of Foamglas blocks, while David Davis has been slumping and fusing glass plates to Foamglas in his kilns. Attendees at the 2009 conference in Hawaii were intrigued when David pointed out that Foamglas sandwich mirrors float on water. He then proceeded to toss a Foamglas mirror blank on the floor without damage. Soda lime Foamglas comes in a number of densities, and there is also a low coefficient of temperature expansion borosilicate version of Foamglas. More recently, another open cell foam glass material, Earthstone, has been employed with good results so far.

Initiative members are investigating the use of low shrinkage adhesives to fasten pre-slumped top plates (and flat bottom plates) to Foamglas cores as an alternative to kiln fusion. A group in Italy under the direction of Giovanni Pareschi is also working on Foamglas mirrors, and the Director of Development at Pittsburgh Corning, Steve Badger, has been generously supportive of these Foamglas mirror developments, as has George Morandin, CEO of Eartstone.

Initiative member Peter Chen has been developing very lightweight carbon fiber replica mirrors for several years, and is now achieving success with this approach. Initiative member Kiran Shah is experimenting with polyurethane replica mirrors. Shah's mirrors have a solid polyurethane front surface supported by a dense polyurethane foam body that is formed in a mold.

David Davis has been experimenting with slumped meniscus mirrors. He builds his own kilns and does his own grinding, polishing, and figuring by machine. Work is also continuing on the development of sandwich mirrors with glass spacers by Tong Liu at Hubble Optics.

Active Optical Correction

Initiative members Mel Bartels and David Davis are both interested in seeing how far thin, slumped meniscus mirrors can be "pushed" with respect to both aperture and thinness. Soda lime or borosilicate float glass can only be procured in thicknesses up to 25 mm. As apertures of meniscus mirrors this thin increase, however, an aperture will eventually be reached where, similar to the situation with mountaintop alt-az telescopes, corrective forces under computer control will need to be applied to the primary mirror if it is to retain its proper figure as the telescope changes its altitude and hence the gravity vector acting on the mirror.

In a visit to the Gemini North eight meter telescope, Initiative members discussed the Gemini's "voice coil" active mirror support system with Chris Carter, then Gemini's control system engineer (Chris recently joined the Thirty Meter Telescope developmental team). Chris suggested that for smaller telescopes, a voice coil support system to "tweak" the mirror into proper shape as altitude changed could be made at low cost. Two undergraduate electrical engineering students at California Polytechnic State University developed low cost electronic controls for voice coil meniscus mirror adjustment.

In both cases, our only concern is to make very low frequency (less than once per second) adjustment of the primary mirror to correct for the lowest order Zernike terms, especially astigmatism, as a function of altitude. The requisite adjustments can, we believe, be calibrated "off line" on a number of bright stars distributed in altitude with a coefficient-determination algorithm that minimizes the spread of the stellar image. During operation, the current altitude of the telescope will be noted, and the required settings interpolated from a lookup table. This approach is used by the eight meter Subaru telescope with recalibration of their look-up table only required about once a year.

Finally, there is a third approach we are considering for active optical correction, and this is the use of a semi-passive bimorph mirror. These mirrors consist of two thin disks, one of active piezo material coated with a thin layer of conductive metal on each side, and the other a thin front surface mirror. The two disks are bonded together with a low shrinkage adhesive. The back metal coating on the piezo disk can be divided into sections. When a high voltage (100 to 350 volts) is applied to the sections, the piezo materials in each section contracts or expands some amount, depending on the applied voltage and its polarity. The combination of these expansions and contractions can warp the thin mirror disc in a modal manner. A computer controls these voltages through a digital-to-analog converters and high voltage operational amplifiers.

Instead of using actuators to apply forces directly to the primary mirror itself, a much smaller, semi-passive bimorph mirror near the instrumental payload is distorted via applied voltages in such a manner as to cancel out the gravity-induced distortions in a lightweight primary mirror. These corrections can be applied in a manner similar to what was discussed above, i.e. via a lookup table versus altitude. A similar approach would utilize small piezoelectric actuators to bend a thin glass secondary or tertiary mirror. Edmund Optics has such a device that only costs $2,000.

Instrument Clusters

One of the Alt-Az Initiative's members, Gary Cole, has been focused on the concept of developing lightweight automated science instrument clusters to maximize the number of available observing modes on a single telescope. This in turn multiplies both the research and educational value of the facility.

The core of this project has been the development of a programmable, lightweight, 4 way instrument selector that can tie together, both optically and mechanically, a complete suite of science instruments including the new 50 mm CCD cameras. This device has recently entered production at Optec, Inc.

As an example, Cole's prototype 20 pound cluster provides: target acquisition, imaging, Sloan band photometry, deep R~400 survey Spectroscopy, R~10000 spectroscopy, JH band infrared photometry, and broadband dual beam imaging polarimetry. The selector and instruments operate as IP network devices. The data is gathered in FITS formats. A single laptop operates the entire cluster along with the telescope under the control of an automated observation scheduler.

Assuming a typical 1000 hr/year observing program and five year lifetime, the amortized cost of this instrument cluster is less than \$5/hour. The range of both research and education opportunities provided by an instrument cluster such as this one on a 1 meter platform is very exciting.

Figure 6: Left, 3-D drawing of the lightweight programmable instrument selector offered by Optec, Inc. Right, bench testing of the lightweight instrument cluster prototype.

Optec SSP-4 NIR JH Band Photodiode Photometer Automation

Currently, near infrared imaging cameras remain quite expensive (\$50K and up) because these cameras cannot be fully implemented in silicone chips alone, requiring a hybrid combination of materials. While efforts are underway to reduce the cost of these cameras, a much lower cost and easier to handle option for JH band near IR photometry is available in Optec's SSP-4 photometer for only \$3,000. The SSP-4, developed by Optec, in cooperation with the American Association of Variable Star Observers, is ideally suited for observing long-period Mira and other variable stars.

Initiative member Dan Gray has automated the filter changing and flip-mirror functions of the Optec SSP-4 to allow fully automatic operation. Dan and Initiative member Russ Genet made a test run on this modified SSP-4 at the University of Oregon's Pine Mountain Observatory in July 2008. Dave Rowe analyzed the resultant data. Initiative members Greg Jones and August Johnson are developing low noise near IR aperture photometers for J, H, Ks band photometry.

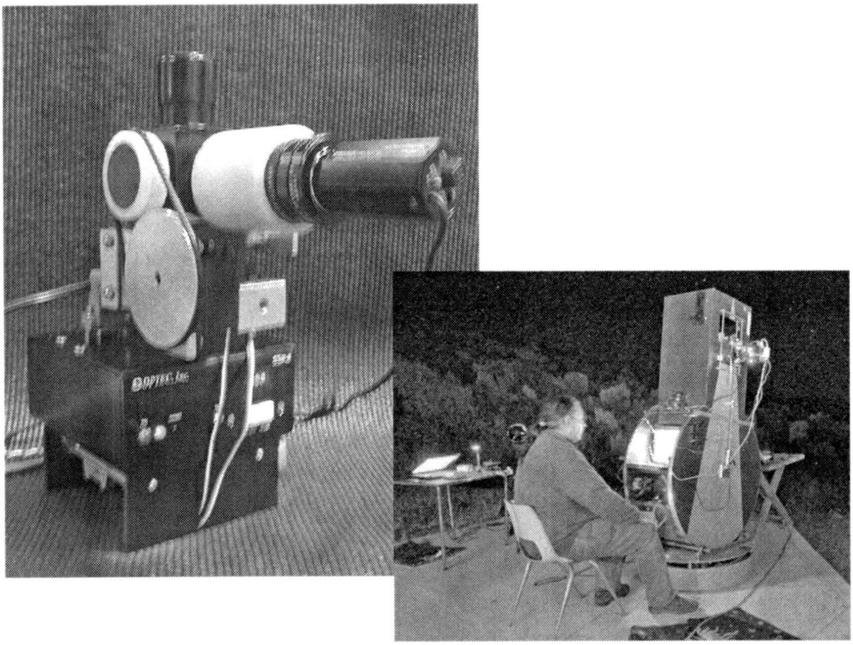

Figure 7: The Optec SSP-4 near IR photometer, left, modified for automated operation, was evaluated on Dan Gray's automated 14 inch alt-az telescope (right) in a test run at Pine Mountain Observatory.

Tech Demo Telescope

The objective of the Initiative's first tech demo telescopes was to combine together various elements from individual technical initiatives and to gain experience as a group in building and operating a complete system.

The modest aperture (18 inch) Cal Poly 18 (CP 18) alt-az telescope was designed and built not only as a technical demonstration, but also for eventual use by student researchers at California Polytechnic State University. The telescope's drive system has no gears, belts, or friction wheels; instead, as described earlier, direct drive motors and high resolution encoders are completely integrated into the bearing assemblies and telescope's structure. In altitude, for instance, a ring of permanent magnets is firmly mounted to the OTA while an opposing ring of coils is mounted on the inside of a fork arm. The electronic control system has been designed to operate these brushless motors in a high precision mode.

To achieve the highest possible closed loop servo bandwidth, the structure was designed—by Cal Poly students using finite element analysis as well as traditional analytic tools—to have an unusually high natural frequency. The direct drive system and stiff structure effectively counters wind gusts when such a telescope is operated out in the open or within a roll-off roof observatory.

Transfers to Quantity Production

PlaneWave Instruments CDK 700 Telescope

Design and construction has been completed at PlaneWave Instruments by Initiative members Rick Hedrick, Joe Haberman, and Allan Keller and their associates on a production 0.7 meter corrected Dall Kirkham telescope with Nasmyth foci. The telescope employs direct drive motors and control system electronics and software that are similar to and based on the Initiative's experience gained with the Cal Poly 18 demo telescope described above.

Renderings by Bob Peasely

Figure 8: The PlaneWave Instruments CDK 700 telescope. The Nasmyth configuration facilitates instrument changes without rebalancing, and also provides a convenient height for visual observing. Shown is an overall telescope drawing.

A Nasmyth optical configuration with two foci (optional second focus) was chosen to place instruments and eyepieces at a convenient height, to make the telescope insensitive to even large changes in instrumental weight (counter weighting is not required), and to minimize the telescope's moment of inertia in altitude. The corrected Dall Kirkham (CDK) optical system, designed by initiative member Dave Rowe, was chosen for its wide, well corrected flat field, generous back focus, and its relative insensitivity to secondary mirror lateral misalignment (the secondary mirror is spherical).

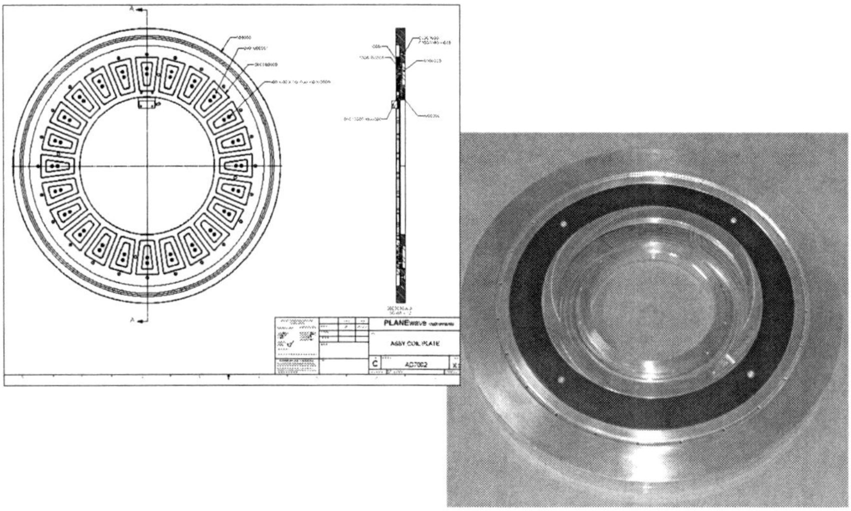

Figure 9: The CDK 700 azimuth drive assembly is more than a motor. It is a 3 phase 24 volt AC direct drive torque motor, with an integrated high resolution encoder, and an integrated 22.5" bearing. This assembly is both the base of the telescope and the drive. On the left is a drawing of the assembly that illustrates how compact it is, while a piece of the assembly being machined is shown on the right.

The 28 inch primary is relatively easy to manufacture, as is the 11 inch spherical secondary. The two-element corrector employs all spherical surfaces. The well corrected field supports the largest commercially available CCD sensors available today, and should continue to do so well into the future. The large, flat field is very useful for a variety of off-axis guider configurations.

Sidereal Technology Direct Drive Controller
The controller electronics and software for both the Cal Poly 18 and PlaneWave Instruments CDK 700 were designed by Dan Gray, and are manufactured by Sidereal Technology. For precision telescope motion, brushless direct drive motors must be operated in an AC synchronous, servo-control mode. The position of the telescope is sensed with ultra-high precision encoders on the two telescope axes, the position error is

calculated, and the requisite torque feedback is computed. The three phases of the motor are then energized based on this feedback torque by way of pulse-width modulated switches. This control system is ASCOM compliant, so it can be used with a number of overall telescope/observatory high-level supervisory software suites.

The 2.5 Meter Corrected Dall Kirkham Project
As an off-shoot of the Alt-Az Initiative, Caisey Harlingten and Craig Breckenridge are developing a higher-end 2.5 meter alt-az CDK telescope to be used at the university level. Borrowing many ideas presented elsewhere in this paper, the goal is to produce a singular design that can be produced in volume in order to lower manufacturing costs.

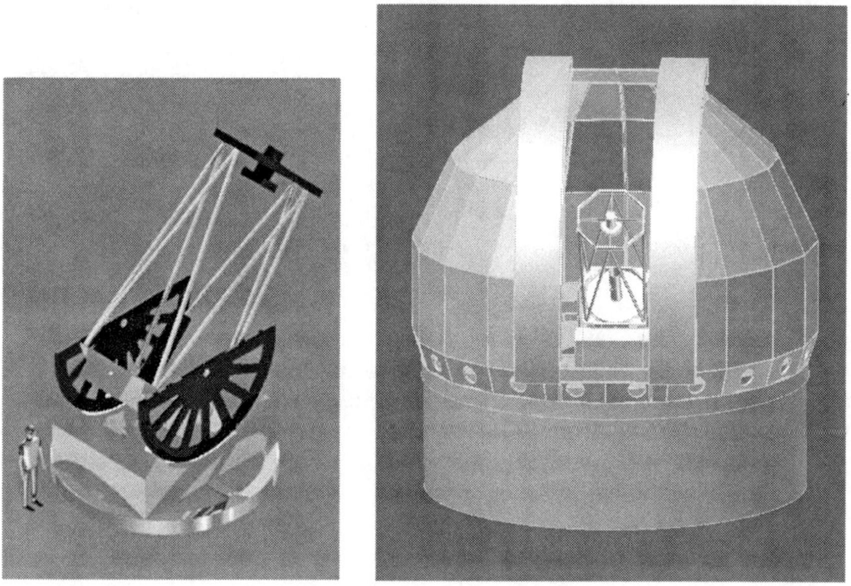

Figure 10: The 2.5 meter CDK telescope (left) and low cost enclosure (right). Remote operation is a key feature of the overall design.

The initial design includes the direct drive motors, dual position tertiary mirror, and Alt-Az concept with field rotators discussed elsewhere. The design is based on an optical configuration by Dave Rowe. Caisey and Craig plan to make telescopes more affordable for universities by utilizing a flat field of view and focal length that provides a good compromise between telescopes used for discovery and follow-up research. The design follows their philosophy of efficient thermal and mechanical design coupled with an equally efficient enclosure design. Both telescope and enclosure are designed to be operable remotely.

Undergraduate Student Research and Telescope Development

The Initiative is actively pursuing the vital connection between small telescopes and the training of a new generation of astronomers and instrumentalists. A number of institutions with limited budgets are developing research programs that utilize low cost telescope technology. To date, several dozen undergraduate students at Cuesta College have participated in astronomical research programs resulting in published papers. They utilize both local observatories and remote, robotic observatories to observe a wide range of celestial phenomena including visual double stars, eclipsing binaries, intrinsic variable stars, asteroid rotation rates, and exoplanet transits.

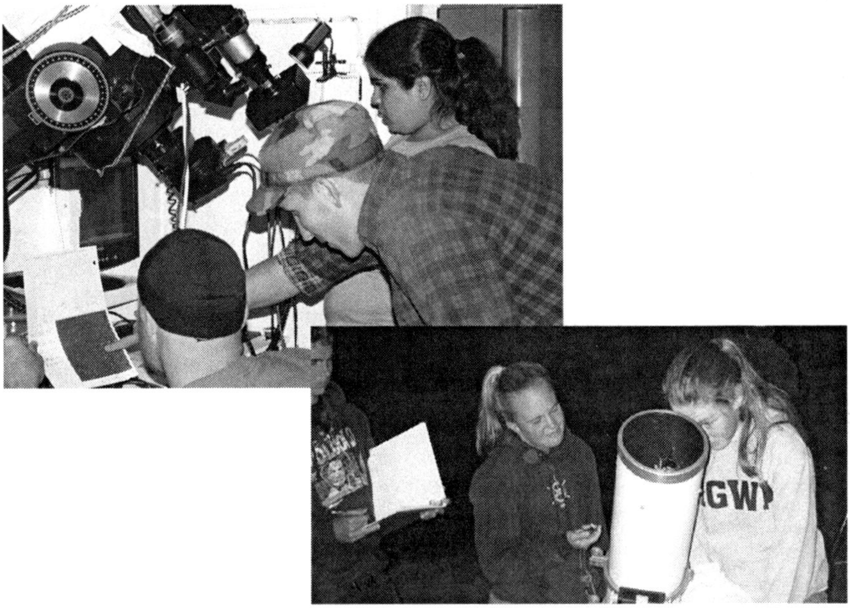

Figure 11: Cuesta College research seminar students observe an intrinsic variable star (left) and Arroyo Grande High School students (taking a Cuesta College research seminar) make astrometric measurements of visual double stars (right). Their results have been published in *Society for Astronomical Science Symposium Proceedings, Small Telescopes and Astronomical Research,* and the *Journal of Double Star Observations.*

Similarly, a dozen California Polytechnic State University, San Luis Obispo (Cal Poly) engineering (ME and EE) students have been involved in the design and fabrication of a research grade alt-az telescope, instrument rotator, and low cost, active primary mirror support electronics.

Figure 12: The Cal Poly engineering students present their design for an 18 inch Newtonian focus alt-az telescope at a critical design review (left). Two Cal Poly ME students fabricate the fork of the telescope. They designed the fork using finite element analysis (right). The foam core was later covered with multiple layers of carbon fiber cloth.

These research and engineering students have given presentations at multiple Initiative workshops and conferences, and have coauthored Initiative papers. One Cuesta College student, Jo Johnson, co-chaired two Initiative conferences and has co-edited both of the Initiative's books.

Acknowledgements

The Initiative members are pleased to acknowledge the kind support of a number of institutions, including California Polytechnic State University, Cuesta College, the Gemini Observatory, and the American Association of Variable Star Observers, as well as a number of companies including Sidereal Technology, PlaneWave Instruments, Sierra Monolithics, Hubble Optics, and Inca Corp. We also thank Cheryl Genet and Jo Johnson for assembling and preparing the graphics for this report.

Figure 13: Former Cuesta College student Jolyon Johnson co-chaired two conferences for the Alt-Az Initiative: the STAR Conference in California (left) and Galileo's Legacy in Hawaii. Cal Poly engineering student Michelle Kirkup receives a certificate of recognition from Professor John Ridgely at the STAR Conference for her participation in designing the Cal Poly 18 inch alt-az telescope (right).

References

Baily, Pierre, ed. 2003. *The Design and Construction of Large Optical Telescopes.* New York: Springer.

Banich, Howard. 2006. A portable 28 inch alt-az telescope. *Amateur Astronomy, 50,* 46-52.

Bartels, Mel. 1994. How to build a simple computerized alt-azimuth control system. *Observatory Techniques, 12,* 12-15.

Bartels, Mel. 1999. A computerized Dobsonian. *Amateur Telescope Making,* ed. Stephen Tonkin, 155-178. Springer.

Bartels, Mel. 2000. Keep up with the stars: a motorized telescope control. *Circuit Cellar, 119* (June), 18-25.

Basaila, George. 1988. *The Evolution of Technology.* New York: Cambridge University Press.

Genet, R., Rowe, D., Gray, D., Albertson, B., Chiu, W., Kirkup, M., Murphy, D., Ridgely, J., Schmitt, J., Swanson, M., Urban, R., Liu, T., Banich, H., Chelborad, C., and Keller, A. 2007. An 18-inch direct-drive telescope. In *Proceedings of the Society for Astronomical Science,* ed. Brian Warner.

Genet, Russell, Johnson, Jolyon, and Wallen,Vera. 2010. *Small Telescope and Astronomical Research.* Santa Margarita CA: Collins Foundation Press.

Kriger, David and Berry, Richard. 2003. *The Dobsonian Telescope: A Practical Manual for Building Large Aperture Telescopes.* Richmond VA: Willmann-Bell.

Mertz, Lawrence. 1996. *Excursions in Astronomical Optics.* New York: Springer.

ReSTAR Committee. 2007. Renewing small telescopes for astronomical research. Internet <www.noao.edu/system/restar/files/ReSTAR_final_14jan08.pdf> research. accessed 2 March 2010.

Ringwald, F. A., Culver, J. M., Lovell, R. L., Kays, S. A., and Torres, Y. V. A. 2003. The research productivity of small telescopes and space telescopes. *Bulletin of the American Astronomical Society, 35,* (4), 1063-1074.

Trueblood, Mark and Genet, Russell. 1985. *Microcomputer Control of Telescopes.* Richmond VA: Willmann-Bell.

Weaver, Bruce. 2003. Funding of the essential synergy between small and large telescopes. In *The Future of Small Telescopes in the New Millennium,* vol.1, ed. Terry Oswalt. Astrophysics and Space Science Library, vol. 287. Dordrecht, Germany: Kluwer Academic Publishers.

Chapter 2

Alt-Az Light Bucket Astronomy

Russell Genet, California Polytechnic State University
Bruce Holenstein, Gravic Inc.

Introduction to Light Bucket Astronomy

Light bucket alt-az telescopes are low optical quality light concentrators used in those areas of astronomical research where numerous low-cost photons are required but not an image *per se*. There is no more authoritative definition of astronomical terms than the *Cambridge Dictionary of Astronomy* (Mitton 2001):

> light bucket A colloquial expression for a *flux collector*. flux collector A telescope designed solely to collect radiation in order to measure its intensity or to carry out spectral analysis. No attempt is made to form an image, so a flux collector can have a more crudely figured reflective surface than a conventional telescope.

We have generously extended light bucket status to "imaging" alt-az telescopes, but only if the images are decidedly fuzzy, the fields of view are somewhat narrow, and the "images" are just used for simultaneous photometric measurements of the intensities of variable, comparison, and check objects.

Light bucket telescopes are, as we hereby further define them, low in cost, lightweight, and often pack easily for transport in small vehicles. Their instruments are also low cost, and the operations and maintenance of both the telescopes and their instruments can be performed by low cost non-specialized staff or no-cost volunteers.

Perhaps somewhat arbitrarily, we define an individual light bucket telescope as having an aperture between 1 and 3 meters. Light bucket telescopes can be divided into two major classes depending on their optical performance: (1) true on-axis light bucket flux collectors with many waves of aberration and (2) "fuzzy area" quasi light bucket telescopes with narrow fields of view and only a few waves of aberration. High optical quality telescopes are, of course, not light bucket telescopes.

Light bucket astronomy is advantageous in those situations where, compared to noise from the sky background, the noise from one or more other sources is dominant or, putting it the other way around, where the sky background is a small or nearly negligible source of noise. This situation can occur when (1) the object being observed is very bright, (2) the integration times are very short and hence photon arrival noise becomes important, (3) scintillation noise becomes a dominant noise source, (4) the bandwidth is very narrow or the light is spread out as in spectroscopy, resulting in significant photon arrival noise, or (5) noise from the detector is dominant (as it can be in the near infrared).

Some types of light bucket astronomy could benefit from an *array* of ultra low cost light bucket telescopes. Some interesting attributes of light bucket arrays include:

1. *Reliability:* The array elements operate separately and may provide immediate and independent confirmation of rare, transient events.

2. *Availability:* If a single scope in an array is removed from service due to a mechanical fault or mirror-aluminizing, the rest of the array will keep operating at virtually no reduction in performance. In other words, the array fails gracefully rather than all at once.

3. *Independence:* If the array is spread out over a substantial geographic area, clouds and other environmental factors will not disable (e.g. "cloud out") the entire array. Continuous coverage of the night sky might be possible if multiple continents are included in the geographic extent.

4. *Transportability:* If individual light bucket array telescopes are moveable, the array may be repositioned to avoid a storm or seek out an ideal location for an asteroid occultation event. GPS or laser ranging may be used to re-collimate the array.

5. *Expandability:* As funds and time allow, additional array elements may be added to increase the scientific reach of the telescope. One may also replace or upgrade (e.g. resurface) the mirrors in order to work on deeper sky objects.

Signal-to-Noise Ratio

The major factors that influence the overall signal-to-noise ratio (SNR) of program object measurements include the amount of both program object and background sky flux collected, scintillation, air mass, filter bandpass, and detector and readout noises. In order to understand some of the tradeoffs visually, the figure below displays figures of merit for the relative SNR for light bucket telescopes of 1.0 and 1.5 meter apertures relative to a 40 cm (16 inch) diffraction-limited Schmidt-

Cassegrain telescope as a function of program object brightness. A 40 cm SCT was chosen because we thought that its cost might be roughly equivalent to that of a 1 to 1.5 m light bucket telescope. Equation 20 of Holenstein *et. al.* (2010) was used for the calculations.

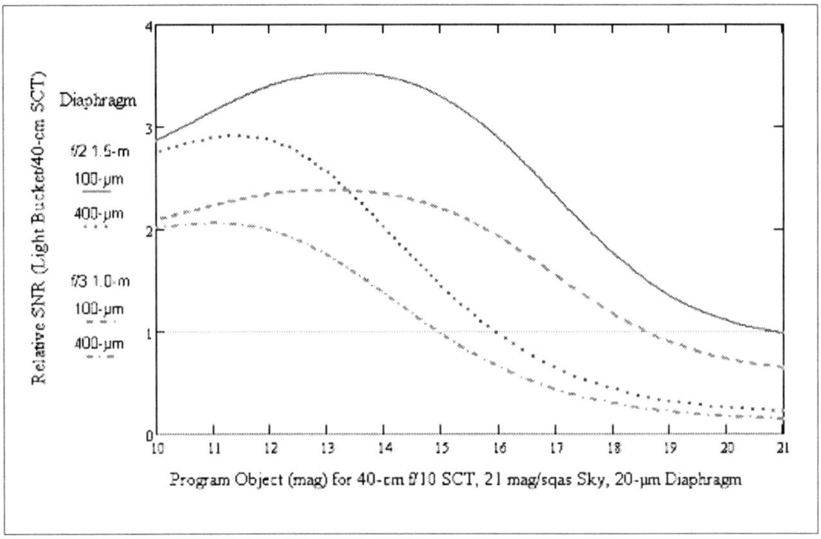

Figure 1: Relative signal-to-noise ratio of a light bucket telescope as compared to a smaller aperture but higher optical quality telescope.

Shown in the Figure 1 graph, the SNR of both 1.0 m *f*/3 and 1.5 m *f*/2 light bucket telescopes was divided by the SNR of a 40 cm *f*/10 diffraction-limited Schmidt-Cassegrain (SCT) telescope as a function of the luminosity of the program object. The background sky is 21 magnitudes/arcsec^2. The 40 cm telescope uses a 20 µm (1″) focal plane diaphragm. The illustrated cases are for 100 µm (7″) and 400 µm (28″) diaphragms for each of the 1.0 and 1.5 meter telescopes. A relative SNR of 1.0 means that the two telescopes have the same photometric performance. Scintillation is calculated according to Young (1967) at a 1000 meter elevation with an air-mass of 1.5. Amplifier and detector noise is modeled by an Optec SSP-5a with a Hamamatsu R6358 multi-alkali photomultiplier tube and a Johnson B filter.

For very bright program objects, scintillation takes over and limits the relative performance to a value that approaches the expected value based of the relative sizes of the telescope apertures. The relative SNR for mid-brightness objects peaks at a value which approaches the ratio of the example telescope apertures. For faint program objects, the traditional telescope in some cases outperforms the light bucket ones, depending on the size of the focal plane diaphragms. However, note that no allowance is made in the figure for program objects for which the diaphragm is unable to isolate the detector from field stars. Errors caused by program object tracking problems have been analyzed (Holenstein *et. al.* 2010).

The SNR crossover point of 1.0 depends on the relative size of the focal plane diaphragms and the apertures of the light bucket telescopes. For mid-brightness objects, the light bucket telescopes excel. For example, a relative SNR of 3 means that a 3 hour run with the SCT scope may be accomplished with the light bucket in one-third squared, or one-ninth, the time, *i.e.* 20 minutes. Another way of looking at the situation in this example is that the productivity of the light bucket is nine times higher than the SCT. Nine times the number of program objects may be observed at the same SNR level as the SCT is able to obtain per unit of observing session. But, it is not just a question of the economics of observing time. In many scenarios the more important factor is the SNR achievable in a set short integration time for fast changing objects. For example, if one wants to observe a flare star with 10 ms time resolution, a factor of three improvement in the SNR achieved may make all the difference in the world in the interpretation of the observation results.

The Hipparcos catalog contains 118,000 stars brighter than visual apparent magnitude 7.3. The Tyco-2 catalog contains 2.5 million stars to about V ~ 11th magnitude. For every magnitude fainter than that, one adds in about 300% more stars. As a result, there is rarely a shortage of interesting program objects at accessible apparent magnitudes compared to the background brightness.

The graph above is easy to adjust for observing sites where the background sky is brighter than 21 magnitudes/arcsec^2. For example, in a suburban city location, the sky brightness might be 17 magnitudes/arcsec^2. Since this setting is four magnitudes brighter, just subtract 4 from each value of the ordinate (*i.e.* the *X*-axis). So, 21 becomes 17 magnitudes/arcsec^2, 20 becomes 16 magnitudes/arcsec^2, and so on. Note that this adjustment is actually only approximate for the fainter objects due to the scaling relationship of the detector/amplifier noise.

Light Bucket Research Areas

At an Alt-Az Initiative workshop held in conjunction with the American Astronomical Society summer meeting in Pasadena in 2009, Arne Henden, the Executive Director of the American Association of Variable Star Observers, gave a talk entitled "1.5 Meter Light Bucket Astronomy." Arne outlined the key research areas where light buckets excel. These include high speed photometry, near IR photometry, time series spectroscopy, and fast cadence/high precision photometry.

One of the most interesting areas for high speed photometry is the interference (fringe) light curves produced when the Moon (or another solar system body) occults stars and data is obtained at milli-second intervals. Analysis of these fringe patterns can yield stellar diameters for nearby stars and discover new close binary stars with separations of just a few milli-arc seconds (Nather et al. 1970, Beavers et al. 1980, Ridgeway et al. 1980, Feckel et al. 1980, Schmidtke et al. 1984, Peterson et al. 1989, Richichi 1984, van Belle et al. 2002, Richichi et al. 2008, 2009). Obtaining a decent number of photons for each milli-second interval requires a large photon gathering area, although only on-axis light concentration is required if a high speed photodiode or photomultiplier is used as the detector. Brian Warner (1988) wrote the classic book on high speed photometry, while two more modern books are those edited by the trio of Don Phelan, Oliver Ryan, and Andrew Shearer (both 2008).

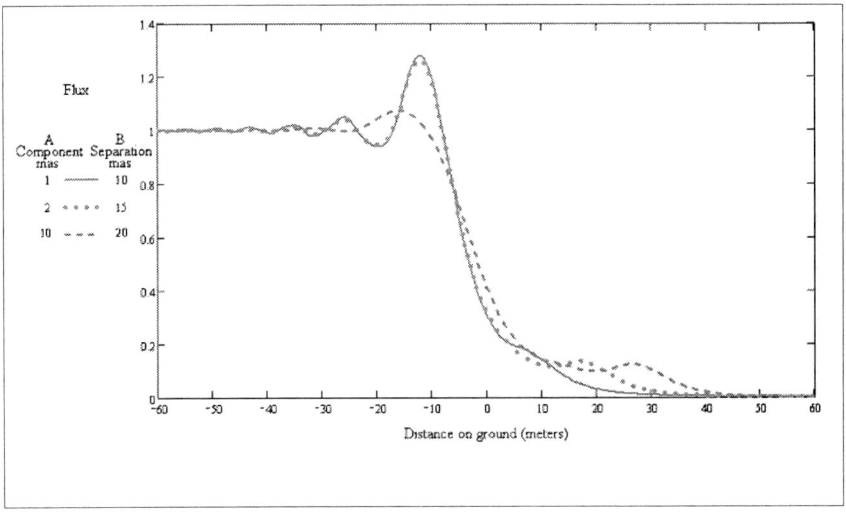

Figure 2: The diameters and duplicity of nearby stars can be determined through analysis of high speed photometry of lunar occultations.

Shown in Figure 2 are the theoretical diffraction pattern light curves (320-720 nm bandpass) for the lunar occultation of three stars: small with a 0.1 milli-arc second (mas) diameter, medium with 2 mas, and large with 10 mas (each with a small companion having just 10% of the luminosity of the primary). Limb darkening is set to zero and the location of the companions form the primary trails by 10, 15, and 20 mas respectively. The diffraction pattern sweeps over the ground in milliseconds, depending on the orientation of the moon's limb.

Near infrared (NIR) aperture photometry is another area where on-axis light bucket telescopes should shine (pardon the pun). NIR cameras are very expensive, and are likely to remain so for quite some time. Their complex electronics and liquid nitrogen cooling also makes them difficult to support. On the other hand, "one pixel" aperture NIR photometers use low cost silicon photodiodes and thermoelectric cooling. The Optec J/H band SSP-4 photodiode photometer costs only $3,000. As part of the Alt-Az Initiative, Greg Jones and August Johnson are developing very low noise J/H/K band photodiode photometers.

With light spread out in a spectrum, spectroscopy benefits from lots of cheap photons. At typical "small" telescope locations, seeing is typically several arc seconds. An on-axis spot size of 50 to 100 microns can fiber feed a medium resolution spectrograph to obtain time series spectra.

Some light buckets can produce slightly fuzzy images over a somewhat restricted field of view. This can be sufficient for fast cadence, high precision photometry of somewhat brighter stars. Here CCD camera integration times are a generous (compared to high speed photometry) 1 to 10 seconds, and the field of view is large enough to include nearby comparison and check stars.

A fast cadence is required when the object being studied changes its brightness fairly rapidly, and longer integrations would time-smear the details of interest. An example is an eclipsing cataclysmic variable, where the eclipsing star acts as a knife edge scanner cutting across the white dwarf or neutron star's accretion disk, thus revealing details of the disk's structure. Another example is an exoplanet transit where fast cadence, high precision photometric observations can yield high quality light curves from which very precise transit times can be deduced. Small variations in the transit times can reveal the presence of another planet in the system which advances or retards the transit timing, due to the light travel time effect, in a methodical manner.

The Future

Light bucket telescopes, as we have defined them, are in their infancy. The authors are both keen on making high speed (1 millisecond) observations of stars being occulted by the Moon. They plan, with their 1 meter light buckets, on observing the same occultations from opposite coasts of the United States. They are also interested in observing, at high speed, the occultation of stars by asteroids, trans-Neptunian objects, etc. Genet is planning on near IR photometric observations with an aperture photometer—the Optec SSP-4 and/or a photometer being developed by Greg Jones—as well as fast cadence, high precision "fuzzy" CCD photometry of eclipses of cataclysmic variables and exoplanet transits.

Larger aperture light bucket telescopes are on the drawing boards. How rapidly this field progresses will critically depend, of course, on the development of unusually lightweight, low cost, low optical quality mirrors. We expect that, depending on the specific mirror technology, an economically optimal individual mirror size may emerge—perhaps 1 meter for meniscus plate glass or foam glass mirrors. Four 1 meter mirrors could be combined to form a 2 meter equivalent aperture light bucket telescope that, being less than 8 feet wide, could be towed down the road in a lowboy trailer.

Seven 1 meter mirrors, left round or water jetted into hexes, would produce, respectively, 2.6 meter equivalent aperture versions of the Giant Magellan Telescope (GMT) or the Thirty Meter Telescope (TMT). Perhaps they could be dubbed the Mini Magellan Telescope (MMT) or Three Meter Mini (TMM). Such a light bucket telescope might be light enough to transport down the road, pulled by a pickup or SUV. Their core azimuth bases could remain intact as roll on/off assemblies, although on-site assembly of the rest of the telescopes would probably be required. If individual mirrors only weighed 50 lbs, then assembly by two persons without any mechanical assistance should be possible.

Arrays of Light Bucket Telescopes

We cannot resist mentioning the ultimate light bucket astronomy dream—an array of six or seven lightweight, low cost, sizeable (2 to 4 meter), on-axis light buckets that could, as an array, image the surface of nearby stars via intensity interferometry. Stellar intensity interferometry, based on a sub-nanosecond quantum-mechanical effect, was pioneered by Hanbury Brown many decades ago with a two-telescope system on a 188 meter circular track (Hanbury Brown 1956, 1974). Movement of the telescopes along this track kept the spacing between them constant and their baseline orthogonal to the incoming starlight. Changing the spacing between the telescopes provided the crucial interferometric information. Hanbury Brown used an analog correlator to compare the high speed signals from photomultiplier tubes at prime focus on the two telescopes.

Picture six or seven of the 2 to 4 meter light bucket telescopes described above, spaced out as an intensity interferometer array in somebody's generously proportioned back yard. Like the VLA radio telescope, the individual telescopes could be positioned to optimize specific observations. During periods of inclement weather, they could all be wheeled into a low cost storage barn.

Figure 3: R. Hanbury Brown's stellar intensity interferometer located in Narrabri, Australia. Galileo attempted to measure the diameter of Vega, but it took the Narrabri interferometer to finally make the measurement (about two milli-arc seconds). During periods of poor weather, the telescopes were moved along the track into a storage shed.

In the decades since Hanbury Brown's pioneering work, not only has the quantum efficiency and speed of detectors greatly improved, but very high-speed digital correlators are now possible. A revival of stellar intensity interferometry by use of Cherenkov telescope arrays has been contemplated (LeBohec and Holder 2006, LeBohec *et. al.* 2008) and a workshop on modern stellar intensity interferometry was held in Utah (LeBohec 2009).

Hanbury Brown derived the following formula for the SNR of his device:

$$SNR_{Hanbury\ Brown} = A\alpha\, n\, |\gamma|\, \left[\Delta f\, \frac{T}{2}\right]^{1/2}, \qquad (1)$$

where A is the telescope area; α is the photomultiplier quantum efficiency; n is the number of photons incident on the telescope per unit area, per unit time, and optical bandwidth; γ is the degree of coherence of the flux; Δf is the bandpass of the electronics; and T is the observing period. When multiple channels and path lengths are measured simultaneously, the following formula describes the overall SNR of the system:

$$SNR_{Overall} = \left[\binom{N_{Array}}{2} N_{Channels}\right]^{1/2} SNR_{Hanbury\ Brown}, \qquad (2)$$

where N_{Array} is the number of elements in the array, and $N_{Channels}$ is the number of simultaneous channels measured, and the noise is modeled as adding in quadrature.

The following is an estimate of the potential relative performance of a modern stellar intensity interferometer with a pair of 2 meter light bucket mirrors compared to Hanbury Brown's 6.5 meter aperture device. First, Hanbury Brown used an extremely narrow optical bandwidth for his observations, specifically a single interference filter, which was typically just 4 nm wide. There is no reason that the modern device could not use a low to medium resolution spectrograph and a high speed, higher quantum efficiency detector array containing one hundred or more high-speed (sub-nanosecond resolution multialkali PMTs or avalanche diode) detectors. Additionally, Hanbury Brown's analog electronics had a bandpass of about 60MHz. Modern electronics and DSP correlators can operate at least a factor of 100 times faster.

Using Equations 1 and 2, the overall SNR improvement due to these factors scales directly as the mirror collection area and relative quantum efficiency times the square root of the product of the relative bandwidth and number of channels, all other terms being held approximately constant. In this example case with two 2 meter telescopes only having 1/10th the area but with 100 channel detector/correlators operating at 100 times the speed with about three times the quantum efficiency, we have: (0.1) x (3) x (100 x 100)$^{1/2}$ equals a factor of thirty times better performance for the modern instrument. So, the much smaller (2 instead of 6.5 meters) modern intensity interferometer might operate three or more magnitudes fainter than Hanbury Brown's interferometer.

While Hanbury Brown was able to determine the diameters of 32 of the nearest stars, a modern array of a half-dozen or more light bucket telescopes with its many path length permutations could provide actual images of large nearby stars, complete with starspots and strange shapes as, for instance, red giants shed their mass. Using Equation 2, a seven-element array has up to 7x6/2 = 21 unique baselines (some may be degenerate depending on telescope placement, program object location, and other geometrical factors) which corresponds up to a 4.6 times improvement in the overall SNR with all other factors held constant. This corresponds to another factor of 1.5 magnitudes over the three plus magnitudes already calculated. Clearly, stellar intensity interferometry with light bucket telescopes deserves some close attention for further feasibility analysis. Also, note that the modern equipment might also be used to look for natural lasers, for example in hot systems such as Eta Carinae, through photon-correlation spectroscopy (Dravins and Germanà 2008).

Just to be complete, we decided to leave you with a diagram showing the photometric capabilities of an array of large light bucket scopes. The following figure shows the relative SNR for an array of seven 1.5 m *f*/2, 2.0 m *f*/1.5, and 3.0 m *f*/1 light bucket scopes with effective apertures of 4 m, 5.3 m, and 8 m, respectively, compared to a single diffraction-limited 8 m class telescope such as the Subaru or a single Gemini or VLA scope.

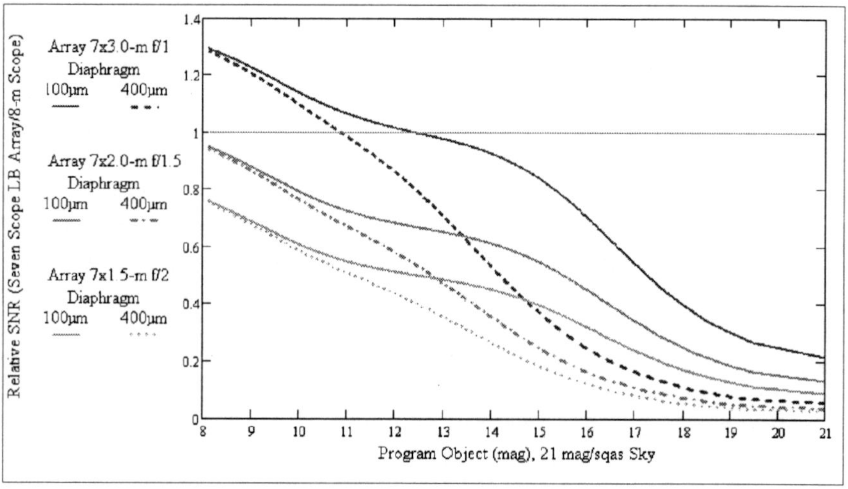

Figure 4: An array of multi-mirror light bucket telescopes can rival the performance of 8 meter telescopes in some types of observations.

In Figure 4, an SNR of an array of seven 1.5 m *f*/2, 2.0 m *f*/1.5, and 3.0 m *f*/1 light bucket telescopes was divided by the SNR of an 8 m *f*/1 diffraction-limited telescope as a function of the luminosity of the program object. The telescopes in the array are separated enough so that scintillation is minimized. The background sky is 21 magnitudes/arcsec^2. The 8 m telescope uses a 40 μm (1") focal plane diaphragm. The illustrated cases are for 100 μm (7") and 400 μm (28") diaphragms for each of the light bucket telescopes. A relative SNR of 1.0 means that the array and the 8 m telescope have the same photometric performance. Scintillation is calculated according to Young (1967) at a 3000 meter elevation with an air-mass of 1.5. Amplifier and detector noise is modeled by an Optec SSP-5a with a Hamamatsu R6358 multi-alkali photomultiplier tube and a Johnson B filter.

Each of the curves shows a degradation of the light bucket array performance compared to the 8 m telescope as the program object gets fainter. On the other hand, the array does much better for bright program objects. Starting at around 11th magnitude, the photometric performance of the 3 m light bucket array exceeds that of the 8 m telescope. The smaller 1.5 m array even has 60% or more of the performance of the 8 m telescope starting around 10th magnitude.

Conclusions

In this paper we defined light bucket telescopes as flux collectors which are low in cost, lightweight, and often pack easily for transport in small vehicles. Their instruments are also low cost, and the operations and maintenance of both the telescopes and their instruments can be performed by low cost non-specialized staff or no-cost volunteers. We further, somewhat arbitrarily, defined an individual light bucket telescope as having an aperture between 1 and 3 meters and comprising two major classes depending on their optical performance: (1) true on-axis light bucket flux collectors with many waves of aberration and (2) "fuzzy area" quasi-light bucket telescopes with narrow fields of view and only a few waves of aberration.

Some important conclusions about light bucket telescopes we pointed out are as follows:

- Light bucket astronomy is advantageous in those situations where, compared to noise from the sky background, the noise from one or more other sources is dominant or, putting it the other way around, where the sky background is a small or nearly negligible source of noise. When this condition is met, light bucket telescopes may produce measures with a superior SNR compared with a smaller traditional diffraction-limited telescope costing about the same amount. Alternately, the light bucket telescope may produce measures at a faster cadence at the same SNR level.
- Areas suitable for light bucket research include high speed photometry, near IR photometry, time series spectroscopy, and fast cadence/high precision photometry, lunar and solar system occultations, and polarimetry.
- Arrays of light bucket telescopes can extend greatly the ability to observe objects of interest at a desirable SNR. High-speed detectors and electronics, GPS and the use of lasers for collimation, may in some situations allow light bucket telescope arrays to challenge the scientific performance of giant professional telescopes at orders of magnitude less cost. Some projects possible with light bucket telescope arrays in the near future are not even attainable with the other instruments due to performance, costs, or competition for telescope time. An interesting example of this class of project is a revival of stellar intensity interferometry.

Acknowledgements

The authors wish to thank and acknowledge the following people who have inspired this work, including Arne Henden, Robert Koch, Stephan LeBohec, Dave Rowe, Charlie Warren, various members of the Alt-Az Initiative, and of course, the late R. Hanbury Brown. Additionally, we must express our appreciation for the patience of co-editors and production staff members Vera Wallen, Jo Johnson, and Cheryl Genet for putting up with our numerous changes and re-writes.

References

Beavers, W. I., Eitter, J. J., Dunham, D. W., & Stein, W. L. 1980. Lunar occultation angular diameter measurements. *Astronomical Journal, 85*, 1505-1508.

Dravins, D. and Germanà, C. 2008. *Photon correlation spectroscopy for observing natural lasers*. In *High Time Resolution Astrophysics*, eds. D. Phelan, O. Ryan, and A. Shearer. New York: Springer.

Fekel, F.C., Montemayor, T. J., Barnes III, T. G., & Moffett, T. J. 1980. Photoelectric observations of lunar occultations. *Astronomical Journal, 85*, 490-494.

Hanbury Brown, R. 1956. A test of a new type of stellar interferometer on Sirius. *Nature 178*, 1046-1048.

Hanbury Brown, R. 1974. *The Intensity Interferometer.* London: Taylor & Francis Ltd.

Holenstein, B. D., Mitchell, R. J., and Koch, R. H. 2010. Figures of merit for light bucket mirrors. In *The Alt-Az Initiative: Telescope, Mirror, and Instrument Developments*, eds. R. Genet, J. Johnson, and V. Wallen. Santa Margarita, CA: Collins Foundation Press.

LeBohec, S. 2009. *The Workshop on Stellar Intensity Interferometry: January 29 and 30 2009*, University of Utah, http://www.physics.utah.edu/~lebohec/SIIWGWS/ (accessed March 2010)

LeBohec, S., Daniel, M., de Wit, W. J., Hinton, J. A., Jose, E., Holder, J. A., Smith, J., and White, R. J. 2008. Stellar intensity interferometry with air Cherenkov telescope arrays. In *High Time Resolution Astrophysics*, eds. D. Phelan, O. Ryan, and A. Shearer. New York: Springer.

LeBohec, S. and Holder, J. A. 2006. Optical intensity interferometry with atmospheric Cherenkov telescope arrays. *Astrophysical Journal, 649, 399–405.*

Mitton, J. 2001. *Cambridge Dictionary of Astronomy*. Cambridge, UK: Cambridge University Press.

Nather, R. E. & Evans, D. S. 1970. Photoelectric measurement of lunar occultations. I. The process. *Astronomical Journal, 75*, 575-582.

Peterson, D. M., Baron, R., Dunham, E. W., Mink, D., Aldering, G., Klavetter, J., and Morgan, R. 1989. Lunar occultations of Praesepe II. *Astrophysical Journal, 98,* 2156-2158.

Phelan, D., Ryan, O., and Shearer, A. 2008. *High Time Resolution Astrophysics.* New York: Springer.

Phelan, D, Ryan, O., and Shearer, A. 2008. The universe at sub-second timescales. *AIP 2008 Conference Proceedings*, 984.

Richichi, A. 1984. Model-independent retrieval of brightness profiles from lunar occultation lightcurves in the near infrared domain. *Astronomy & Astrophysics, 226,* 366-372.

Richichi, A., Barbieri, C., Fors, O., Mason, E., and Naletto, G. 2009. The beauty of speed. *The Messenger, 135,* pp. 32-36.

Richichi, A., Fors, O., and Mason, E. 2008. Milliarcsecond angular resolution of reddened stellar sources in the vicinity of the Galactic center. *Astronomy & Astrophysics, 489,* 1441-1453.

Ridgway, S. T., Jacoby, G. H., Joyce, R. R., and Wells, D. C. 1980. Angular diameters by the occultation technique. III. *Astronomical Journal, 85,* 1496-1504.

Schmidtke, P. C. & Africano, J. L. 1984. KPNO lunar occultation summary I. *Astronomical Journal, 89,* 1371-1378.

van Belle, G. T., Thompson, R. R., and Creech-Eakman, M. J. 2002. Angular size measurements of Mira variable stars at 2.2 microns. II. *Astronomical Journal, 124,* 1706-1715.

Warner, B. 1988. *High Speed Astronomical Photometry.* New York: Cambridge University Press.

Young, A. T. 1967. Photometric error analysis: confirmation of Reiger's theory of scintillation. *Astrophysical Journal,* 72, 747-753.

Chapter 3

Small Telescopes and Astronomical Research

Russell M. Genet

California Polytechnic State University

Introduction

The trio of transportable Schmidt-Cassegrain equatorial telescopes, go-to telescope control systems, and affordable CCD cameras has revolutionized small telescope science, not to mention astrophotography. These portable systems—thanks to the high quantum efficiency of their CCD cameras—are, from a research point of view, the equivalent of yesteryear's massive mountaintop telescopes equipped with low efficiency film cameras.

Comprehensive robotic surveys are uncovering thousands of new pulsating stars, eclipsing binaries, asteroids, and potential transiting exoplanets every year. What are the periods of these objects? Which of the potential exoplanets are the real things, regularly transiting their parent stars? To answer such questions, each of these newly-discovered objects requires the dedicated hours, nights, and even weeks of follow-up observations that only modest-aperture telescopes can provide. Automation of, and remote access to, smaller telescopes is facilitating observations both locally and remotely, while global robotic networks are being formed to observe objects round-the-clock, passing them, as the Earth turns, from one longitudinal location to the next.

Low cost alt-az control systems, instrument rotators, and high-quality thin-disk mirrors of 20 inch aperture and beyond are just now beginning to be incorporated into portable alt-az telescopes with apertures approaching one meter. Similar to their huge mountaintop alt-az cousins, these new portable alt-az systems can precisely track and derotate objects, thus allowing long CCD exposures. Compared to equatorial telescopes of the same aperture, these new alt-az telescopes are not only lower in cost and more compact, thanks to their vertical/horizontal structural simplifications, but they are truly portable. Although they are never polar aligned, their precision instrument rotators allow long CCD exposures of many of the much fainter objects uncovered by the new surveys. With recently available off-the-shelf spectrographs, their superior light-gathering power will enhance both time-series and classification spectroscopy.

High school and undergraduate students with an interest in engineering and science are, in increasing numbers, joining the ranks of amateur and professional astronomers in developing and utilizing small telescopes for astronomical research. Whether developing new telescopes, instruments, or software, or conducting astronomical research, students gain invaluable hands-on experience in engineering and science while, as co-authors of published papers, their academic careers are given a boost. Assuring that students complete a research project within the confines of a single semester—including writing, reviewing, and submitting a paper for publication—is a challenging undertaking for educators. The Small Telescope Astronomical Research (STAR) Workshop not only provided a forum for instructors to exchange their insights and experiences, but to do so with a range of students.

Figure 1: A number of the workshop's participants during a lunchtime break. California Polytechnic State University in San Luis Obispo, CA, kindly provided the meeting room and refreshments. Orion Observatory and Sidereal Technology split the cost of the star-b-que.

The STAR Workshop was sponsored by the Research Scholar in Residence Program at California Polytechnic State University (Cal Poly), and was held June 22-24 on Cal Poly's campus in San Luis Obispo, California. Late June was chosen as a time when schools were out but summer vacations had not yet started in earnest. The workshop provided a forum where high school and undergraduate students, amateur and professional astronomers, educators, and the commercial designers and manufacturers of small telescopes could share their ideas in the rapidly expanding arenas of small telescope engineering development, scientific research, and undergraduate education.

Astronomy Technology Today

Tom Johnson, the founder of Celestron, opened the workshop's astronomy technology session with a historical sketch of how his development of a technique for manufacturing low-cost Schmidt corrector plates revolutionized the commercial production of small telescopes. Tom's son, Greg Johnson, described his involvement with the development of Celestron's computerized go-to capabilities.

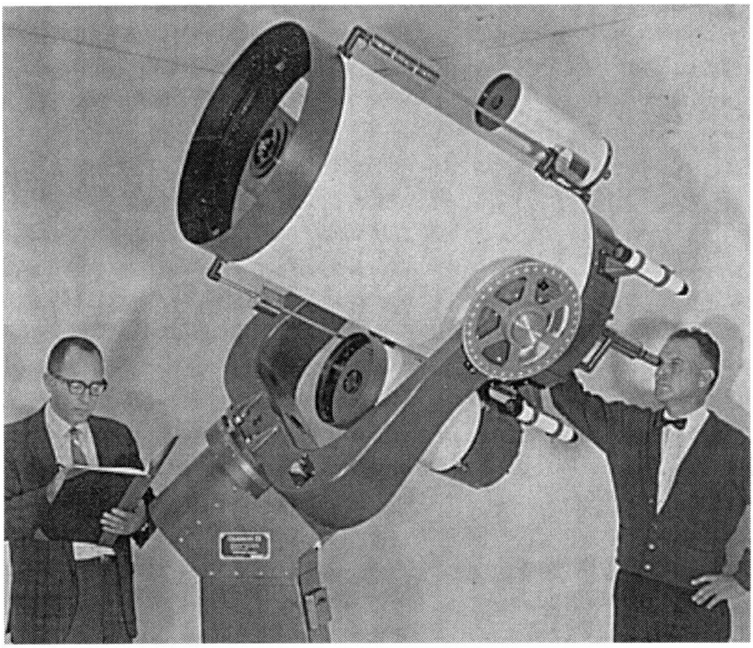

Figure 2: One of Tom Johnson's slides of the founding and early days of Celestron. A somewhat younger Tom is on the right, peering through a large-aperture Celestron Schmidt-Cassegrain telescope.

A significant portion of small-telescope astronomical research consists of continuous time-series CCD photometry of pulsating or eclipsing stars, tumbling asteroids, and suspected transiting exoplanets, as well as sputtering matter spiraling onto white dwarfs, distant microlensing events, and fading supernovae. Hundreds or thousands of exposures are taken of the same object, often all night long, night after night—a task which is an obvious candidate for automation. In the early 1980's, Lou Boyd and I developed robotic telescopes controlled by low-cost PCs, achieving our first fully automatic operation in 1983. Within a few years, our Fairborn Observatory featured a number of Internet-accessible telescopes at a normally unmanned observatory located on Mt. Hopkins in southern Arizona. Our split-ring, friction-drive, 32 inch telescopes were designed from scratch for dedicated robotic photometry. They had no eyepieces.

Automated split-ring equatorial telescopes dedicated to photometry have advanced considerably since these early Fairborn Observatory telescopes, thanks very much to the advent of CCD cameras (the Fairborn systems used aperture photometers equipped with photomultiplier tubes). At the workshop, Jerry Foote described the 16 inch telescope at the Vermilion Cliffs Observatory which he recently designed and built for his wife, Cindy Foote. This equatorial telescope features friction drives in both axes. A primary mirror made of Astrosital and carbon fiber trusses allow exposures to be made all night long without having to stop for focus adjustments. A Paracorr coma corrector, filter wheel, and SBIG ST-7 camera are located at prime focus.

Tim Brown described the 20 and 40 inch split-ring equatorial telescopes being developed at the Las Cumbres Observatory (LCO) for their global network. Their 20 inch telescope utilizes a modified Meade 20 inch Schmidt-Cassegrain optical assembly, while their 40 inch system is configured as a reflecting Ritchey-Chretien (R-C) system. Rather than using two stages of friction drive reduction, the LCO telescopes use high-torque "cartridge" servomotors and a single stage of friction drive. LCO is building 12 of their 20 inch systems and 6 of their 40 inch systems. These telescopes will be used in LCO's global network to provide follow-up observations of objects discovered in robotic surveys. Initial concentration will be on near-earth asteroids.

Automation requires carefully devised procedures, software, and equipment. Jerry Foote, in a one-hour overview, spelled out the basics of automated photometry. Jerry Horne overviewed the software available for automation, such as Doug George's Maxim DL and Bob Denny's DC3 scheduler and observatory controller.

As Schmidt-Cassegrain telescopes become larger, their portability falls off while their prices rise rather sharply. After 20 or so inches of aperture, thin-mirror reflectors with R-C figures or paraboloids with corrective optics near the focal plane begin to dominate "small" science telescopes. Parabolic mirrors, which are significantly less expensive than R-C mirrors, can be used with correctors such as TeleView's 2 inch Paracorr or the 3 and 4 inch Keller Wynn correctors to provide wide, undistorted fields. David Rowe described a corrected Dall Kirham (CDK) optical system he developed that utilizes a prolate ellipsoid primary, a spherical secondary, and two spherical corrector lenses near the focal plane. Being spherical, the secondary mirror is relatively insensitive to misalignment. This combination provides crisp, undistorted images across a 2 inch diameter field, fully supporting the latest large-format CCD cameras. Utilizing David's optical design, PlaneWave Instruments is manufacturing 20 inch CDK optical tube assemblies that employ carbon fiber trusses. David and several of his friends ground and figured a set of 42 inch CDK optics that employed a lightweight, cellular mirror blank. They plan on installing these optics in a folded, portable, alt-az CDK telescope.

Alt-Az Telescope Control

The emerging class of portable alt-az telescopes requires constantly changing drive rates and instrument rotation and, consequently, a sophisticated control system running complex software. Therefore, special attention was given in this workshop to alt-az control.

A short historical diversion is in order here. My involvement with microcomputer control systems began with the robotic telescopes at the Fairborn Observatory in the early 1980's. I coauthored a book with Mark Trueblood, *Microcomputer Control of Telescopes*, published by Willmann-Bell in 1985, that gave a complete set of equations for alt-az telescope control. (A revised version, *The Telescope Control Handbook*, was published in 1997 and is still in print). In 1994, Mel Bartels' article, "How to Build a Simple Computerized Altazimuth Drive System," appeared in *Observatory Techniques*, inspiring a number of other amateurs, such as Rusty Fletcher and Chuck Shaw, to build portable, computer-controlled alt-az telescopes. Mel supplied software and populated boards for stepper control systems from his BB AstroDesigns.com and was also working on a servo system. Dan Gray started a new company, Sidereal Technology, and designed, fabricated, and programmed a new D.C. servo telescope controller. Mel Bartels now sells these, along with his telescope control software.

The complete Sidereal Technology telescope control system was engineered and tested by amateur astronomers, for amateur astronomers. The full control system includes one or two Dual Servo Telescope Controllers, two or four servo motors (Alt/Dec, Az/RtAsc, Focuser, Rotator), and an FCC approved wireless handpad transmitter, with receiver complete with a built in auto-guider port. This system allows one to navigate the skies with amazing precision and ease with any ASCOM compliant planetarium software. There are over 100 installations of the Sidereal Technology Dual Servo Telescope Controller. It is a fully developed, easy to use system.

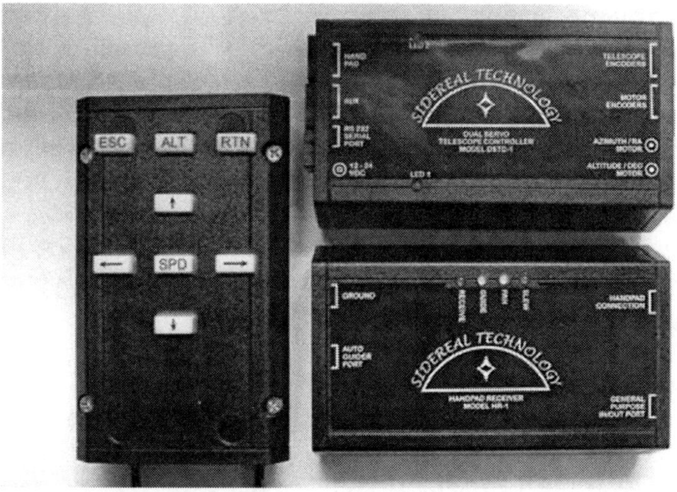

Figure 3: Sideral Technology Control System for 1 meter equatorial and alt-az telescopes.

Figure 4: Russ Genet, Howard Banich, Richard Berry, Mel Bartels, and Dan Gray pose by Dan's 14 inch alt-az telescope at the one-day pre-STAR workshop Portland alt-az meeting. The instrument rotator is on the other side of the telescope. Now known as Portland I, this workshop was the first, "founding" meeting of the Alt-Az Initiative.

The Sidereal Technology Dual Servo Controller is economical, very small, and packed with many features. It connects to two DC servo motors with integral encoders and to two high resolution telescope encoders.

It will track and/or guide your telescope without any computer connected, or provide tracking, guiding, GoTo, etc. when connected to either a computer or an Argo Navis™. The system is compatible with equatorial, German Equatorial, or alt/az mounts. It is versatile and easy to install on scopes of all sizes.

At the workshop, Dan demonstrated SiTech's low cost four-channel servo system. Dan also described his 14 inch alt-az telescope and its instrument rotator, and the recently completed computer-controlled 28 inch alt-az telescopes he and Howard Banich built using mirrors from Mike Lockwood and Kennedy Optics, respectively. I saw both of these telescopes a couple of weeks before the STAR Workshop at an impromptu one-day meeting on portable alt-az telescopes at Dan's shop in Portland. This meeting was also attended by Mel Bartels, Howard Banich, and Richard Berry.

Future Telescope Developments

The development of an emerging class of portable, computer controlled, general purpose, alt-az telescopes was discussed at the workshop and in depth immediately thereafter. Plans are being formulated for an 18 inch technology demonstration system that will feature SiTech's control system, servos and encoders in both axes, focus control, and an instrument rotator at the Newtonian focus. Truss tubes will be carbon fiber, and every effort will be taken to make the system unusually stiff, lightweight, and resistant to wind gusts.

Figure 5: Howard brought his 28 inch telescope to the Portland I workshop in the back of his SUV. The portability of this large-aperture telescope is amazing. The f/3.6 primary mirror was made by Kennedy Optics.

We envision this type of telescope being used for both teaching (visual, astrophotography, photometry, and spectroscopy) and research, including semi-automated all night time-series photometry. Emphasis will be placed on ease of setup and use, reliability, and transportability. Larger-aperture versions could employ a folded optical scheme such as the 42 inch CDK system described by David Rowe at the Workshop.

Figure 6: Howard Banich and Russ Genet stand beside Howard's portable 28 inch alt-az telescope, which features a Sidereal Technology control system.

Small Telescope Science

Small telescope science enjoys a rich, four century-long tradition. Research projects suitable for small telescopes abound. They include: the separation and position angles of visual double stars; asteroid and comet positions; searches for asteroids, comets, nova, and supernova; photometric variations over time of asteroids, intrinsically variable stars, cataclysmic variables, eclipsing binaries, exoplanet transits, and microlensing events; high speed photometry of asteroid and lunar occultations; and, with the larger of the "small" telescopes, spectroscopy time series variability and classification. These many possibilities and more are covered in Robert Buchheim's 2007 book, *The Sky is Your Laboratory: Advanced Astronomy Projects for Amateurs*. At the workshop, Robert kindly provided an overview of small-telescope science, followed by presentations from amateur, student, and professional astronomers in many research areas.

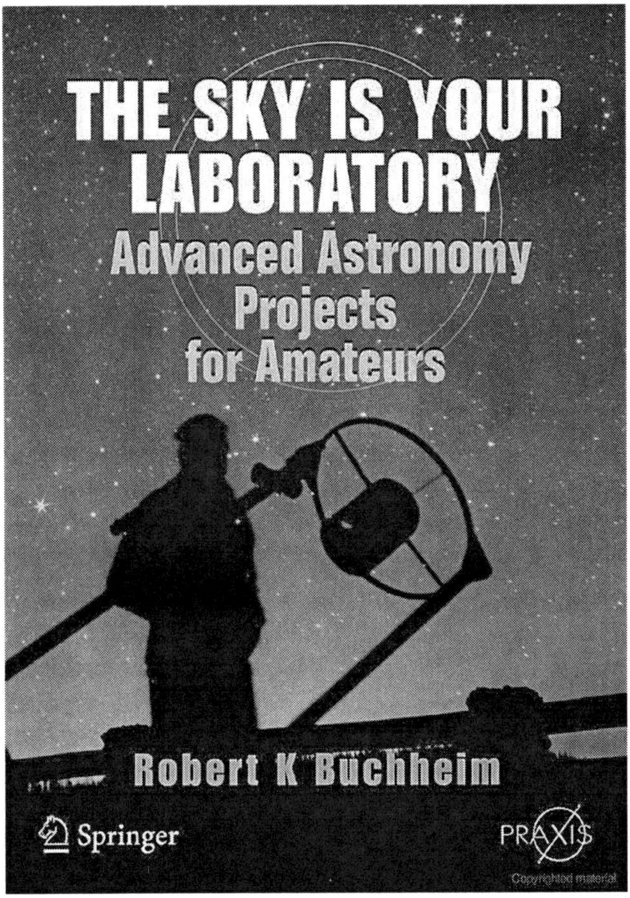

Figure 7: Robert Buchheim's book, *The Sky is Your Laboratory*, describes 18 research areas, their value, the observational and data reduction procedures, and venues for publication of the results. It facilitates the transformation from backyard stargazer to amateur scientist

Darrell Grisham and Jo Johnson, both students of mine at Cuesta College, presented, in their first-ever scientific talks, the results of their visual observations of the separation and position angles of double stars, published as papers in the *Journal of Double Star Observations*. Both used Meade laser-etched astrometric eyepieces. Darrell, with a truly *small* telescope—a 3 inch, manually operated vintage Tasco refractor mounted on a 6 inch diameter steel pier embedded in a ton of concrete—has achieved amazing precision. Jo built his own position angle indicator, making his observations with a 12 inch Meade LX-200 telescope at my Orion Observatory.

Figure 8: Jo Johnson, Russ Genet, and Darrell Grisham work through calculations for the separation and position angles of visual double stars for a presentation given by Jo and Darrell at the workshop.

Walter Morgan and Kent Okasaki, both long-time members of the International Occultation and Timing Association (IOTA), described their experiences with both lunar and asteroid occultations. Walter brought his 8 inch Meade LX-200 telescope, GPS receiver, Kiwi time inserter, and cam-recorder to the workshop's evening "star-b-que" at the Orion Observatory. He recorded several lunar occultations which he dutifully analyzed and presented at the workshop the following day.

Cindy Foote, an amateur astronomer for just two years—with her own 16 inch, research-grade, semi-automatic telescope situated at a very dark site—gave her first-ever science talk at the workshop. Cindy struck pay dirt as the co-discoverer of two transiting exoplanets with the Vermillion Cliffs Observatory's 24 inch telescope, which she was using while Jerry Foote built her 16 inch telescope. As a member of Peter McCullough's research group, she and several other amateur astronomers checked out potential transits from candidates generated by an array of 4 inch robotic telescopes located at Haleakala on Maui. For both of these discoveries, Cindy was the first member of the team to detect the transits. Their results will be published shortly in the *Astrophysicsal Journal*.

John Keller, who teaches astronomy at Cal Poly, gave an overview of asteroids. John, a recent graduate in Planetary Sciences from the University of Arizona, is looking forward to working with Cal Poly and Cuesta College students to make astrometric determinations of asteroid positions and also to obtain asteroid light curves with Cal Poly's 12 inch Meade LX-200 telescope and new SBIG CCD camera. He has already observed one asteroid (2 Pallas) with three Cuesta college students.

Michelle Ouellette, who also teaches astronomy at Cal Poly, provided an overview of pulsating stars. Similar to Cepheid variables, determination of the pulsation period of RR Lyrae stars provides an estimate of their luminosity and hence distance.

Figure 9: Cindy Foote's 16 inch telescope at Vermillion Cliff's Observatory. Her CCD camera is mounted at prime focus. The telescope's truss is made of carbon fiber tubes.

Cuesta College research seminar student Brittany McCrigler reported her observations of a pulsating star that she observed at Orion Observatory with other students this past spring. This star had been identified, in the MOTESS/GNAT survey of a strip of sky near the celestial equator, as a likely short-term variable. Brittany used MPO Canopus along with the Peranso analysis program to determine the star's pulsation period.

Figure 10: Greg Johnson looks on as Walt Morgan sets up to observe lunar occultations. Walt's system includes a GPS receiver, a "Kiwi box" time stamp inserter, and a camcorder as well as an 8 inch Meade LX-200 telescope.

Finally, Sara Martin, a professional solar astronomer at Helio Research, described the Martin Solar Polar Telescope designed and built by her husband, Douglas Martin. After taking early retirement from Caltech and the Big Bear Solar Observatory, Sara and Douglas, together with colleagues, established the nonprofit Helio Research Corporation for research and education in the astronomical sciences. The Helio Research laboratory for data analyses and the Martin Solar Telescope are located at their residence in La Crescenta, California which has proven to be a good site for solar observing.

Figure 11: Cuesta College students Brittany McCrigler and Jolyon Johnson at the workshop dinner. They gave their first ever science talks at the conference, and both had papers nearing completion (since published).

With funding from research grants, Sara and Helio Research employees (including students) use special filters for recording the Doppler shifts of activated filaments, erupting prominences, solar flares and other dynamic features on the Sun. In practice, the observations are taken by tuning each of two filters to different wavelengths around the Hydrogen alpha line in the solar spectrum. The first is a 1 Angstrom prefilter that tunes mechanically under computer control of a motor that tilts it in steps of a fraction of a degree. The second narrow-band filter is a special Fabry-Perot etalon that changes wavelength when high voltage is applied across the surfaces of a very flat, thin crystal of lithium niobate.

A good observing day consists of automated images typically recorded at 5 second intervals for periods of 5 or more hours. Measurements are made from "movies" of changing solar events, and research results are presented at professional meetings and published in conference proceedings and journals. Sara discussed several research projects on solar filament evolution and eruption which involved students in nearly all phases of the research.

Undergraduate Engineering and Science Education

Small telescope engineering development is particularly well suited to high school and undergraduate students. By designing and developing telescopes and robotic observatories, engineering students can hone their skills as they provide their schools with cutting-edge scientific research facilities.

Small telescope science is also well suited to high school and undergraduate students and can be properly supported by telescopes and instruments that are affordable and useful in many areas research. In one or two semester research courses and summer camps, students have proven their ability to conduct high-quality, published research. Through hands-on research, these students develop an appreciation for the true exploratory nature of science.

I have been associated with student research since 1980, when Douglas Hall (Vanderbilt University) and I (Fairborn Observatory) founded the IAPPP, an international society devoted to astronomical photometry. A number of years ago, while teaching astronomy at Central Arizona College, I initiated my first one-semester undergraduate astronomical research seminar. Using a robotic telescope at the Fairborn Observatory, my students observed a number of Cepheid variables, determined their periods, and reported their results at a meeting of the American Astronomical Society. My firm requirement for this and my subsequent research seminars has been that students—either singly or in small teams—plan, conduct, and complete a research project by the end of the semester. Their projects must include the review and submission of a paper to an astronomical journal or, alternatively, they can submit an abstract for a poster or talk they give in person at an astronomical conference (with a published abstract or proceedings paper). Besides holding research seminars, I have enjoyed sponsoring senior projects at Cal Poly where, over a more generous two semesters, students (singly) plan, conduct, and report on their projects in published papers or as talks or posters at formal conferences.

The STAR Workshop featured an "educational roundtable" discussion of the previous fall's research seminar at Cuesta College. Students had obtained time-series CCD photometry of nine MOTESS/GNAT survey stars, found two to be continuously variable, and had determined their periods. They had written a paper, as well as published abstracts/poster presentations at the American Astronomical Society and the Society for Astronomical Science. Student travel was funded by a local retired physicist, George Alers. The students had received awards, and had been featured in local newspaper articles and television news programs.

Figure 12: Star-b-que at the Orion Observatory. In the background are permanently-mounted 12 and 10 inch Meade LX-200 telescopes. In the foreground is Tom Frey's 18 inch Obsession he brought for the evening's star party.

After the Fall 2006 research seminar, a number of actions were taken to improve the upcoming fall seminar. These were discussed during the roundtable discussions. Student participation is being increased to include a team of advanced-placement physics students from Arroyo Grande High School advised by their physics instructor, John Baxter. A number of amateur astronomers from the Central Coast Astronomical Society were enlisted as both students and mentors. A spring astronomy research star-b-que and star party was held to bring the seminar's participants together well before the first class. Not willing to wait until fall, a number of students began their research, informally, in the spring, nearly completing three projects by the June workshop.

The greatest difficulty with the fall research seminar was the need to operate within the confines of a single semester. The students had to obtain many thousands of time-series photometric observations over dozens of nights, and learn how to and actually reduce this mountain of data. In addition, the students had to analyze and write up their results and send them off for outside review. Although they were successful, and submitted their final paper on the last day of class, it was a strain on both these students, their hard-working mentor, Tom Smith, and on me—their instructor and observatory director.

The roundtable discussion suggested two solutions to reduce the workload for students, mentors, and instructor. First, a larger variety of research projects should be encouraged, including ones that would be less time consuming. Students should only tackle advanced projects after they had gained experience with a more basic project and were assured of published results.

Second, a number of observatories, both local and remote, should be asked to provide reduced observations. Students would request the observations in an appropriate format, and would analyze and publish the results, including the observer as a co-author in their paper. While not making the observations themselves, students would be expected to understand how the observations were made and reduced.

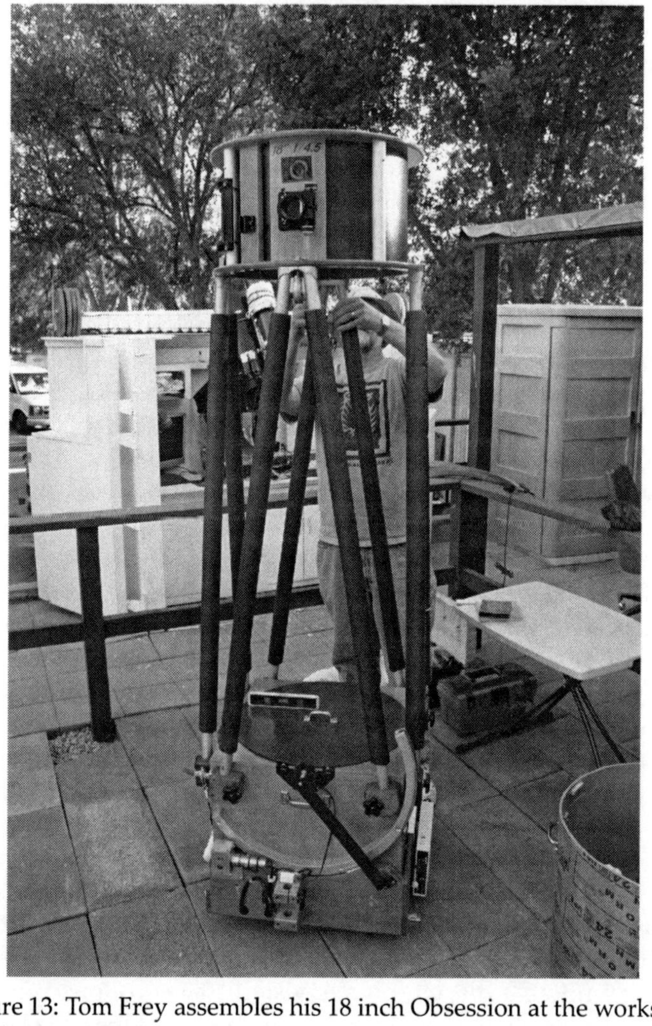

Figure 13: Tom Frey assembles his 18 inch Obsession at the workshop star-b-que. Tom is measuring the separation and position angles of visual binary stars.

Finally, telescope development and construction projects suitable for one-semester seminars and full-year senior projects were discussed after the workshop.

Small Telescope Development and Research Today

Small telescope development is entering an exciting new era. With the emerging development of portable computer-controlled alt-az telescopes with instrument rotators and thin but high-quality mirrors, "small" is being extended to 32 and perhaps, eventually, 40 inches or even larger. Spectroscopy here we come! Small has come a long ways since Galileo's 1.5 inch telescope!

Small telescope research has never been better positioned or more in demand. The conjunction of highly capable telescopes, and the overwhelming outpouring of new objects to research from the automated, big-telescope surveys, bodes well for the future of small telescope research. Not only can backyard amateur astronomers help meet this urgent need for scientific observations, but so can undergraduate and even high school students, aided not only by local amateur astronomy mentors, but by observations provided from remote amateur observatories.

Author's Note

This was the Alt-Az Initiative's first published paper. It appeared in *Astronomy Technology Today* in the summer of 2007 under the title of "STAR Workshop 2007: Small Telescope Astronomical Research, Technology, Science, and Edication." It has been included here for historical interest.

Chapter 4

The Portland III Alt-Az Initiative Workshop

Howard Banich[1], Russell Genet[2], and Richard Berry[3]

[1]Rose City Astronomers;
[2]California Polytechnic University, San Luis Obispo; [3]AIP4Win and HAIP

Introduction

The Alt-Az Initiative began in 2007 to facilitate the development of low cost, lightweight alt-az telescopes in the 0.5 to 2.0 meter aperture range. Large telescopes at mountaintop observatories switched from equatorial to alt-az telescopes decades ago, and have partially inspired the continuing evolution of Dobsonian telescopes. Our goal is to selectively translate design concepts from the world's largest and most sophisticated telescopes to affordable technologies that will transform amateur alt-az telescopes.

A new breed of modest aperture, hi-tech, computer-controlled alt-az telescopes is indeed emerging, and the Alt-Az Initiative is facilitating this emergence via workshops, conferences, a website (www.AltAzInitiative.org), and a Yahoo discussion group.

This article is an update of our progress and outlines our continuing efforts to make this new breed of large and affordable telescopes a reality. Our latest workshop was the third annual event in the Portland, Oregon headquarters of Dan Gray's Technical Marine Service, where he manufactures his Sidereal Technology products.

Figure 1: Portland III Alt-Az Initiative Workshop attendees stand in front of Dan's TMS facility. Don't let our cool demeanor fool you, it was hot!

The Portland III Alt-Az Workshop was held on August 1 - 2, 2009, preceded by the now annual BBQ on the evening of July 31. These dates coincided perfectly with a heat wave that brought temperatures well above 100 degrees Fahrenheit.

The pesentations and discussions of the Portland III workshop detailed our current progress but the selected summaries that follow are necessarily brief – for more information join the Alt-Az Initiative email discussion that can be found at www.altazinitiative.org.

Automation History

Peter Abrahams gave an excellent and occasionally humorous talk on the history of robotic telescopes. A fair number of the people involved with automating amateur scopes are a part of the Alt-Az Initiative and many were present at the Portland III workshop.

Figure 2: Lou Boyd and Russ Genet stand by the "Fairborn 10" robotic telescope on Mt. Hopkins in 1986. The Multiple Mirror Telescope (MMT) can be seen at the summit in the background. The entire observatory was automated and eventually grew to 14 telescopes including a two meter automated spectroscopic telescope.

"What's the telescope for?"

Richard Berry asked this question at the start of the Alt-Az Initiative and continues to ask it at every stage along the way. This is the all important question to ask and requires a detailed answer when designing and building a new scope – any scope, not just a relatively large one. It's equally important to answer as well whether the scope will be used for astrophotography, collecting data, for visual observing, or all three.

Figure 3: Richard Berry posing the big question for the conference.

For instance, a scope to be used for fast cadence, high precision photometry will have different optical requirements than one intended for astrophotography. Researching those requirements and how they translate into telescope design will play an important role in how successful the finished scope will be.

If data collection is the goal, knowing who will interpret the data is important because the data itself is useless unless someone can interpret, then work with it, and use it for published work. Working with a professional astronomer from the beginning will help keep the science mission of the scope as the central design theme of the scope and help insure its ultimate success.

Designing a scope that can be used for several missions is the more difficult and expensive path, but probably the most tempting one. Most of us want a scope that's equally effective collecting data, taking photos, or producing crisp images in an eyepiece. Many amateur and small professional scopes are made this way. Imagine converting a visual scope like a traditional Dobsonian so it can take good astrophotos or do high precision photometry. Although the basic design wouldn't change, the mechanical tolerances, stiffness, and precision of the drive would need to be significantly upgraded.

Light Bucket Science

While most amateur telescope makers (ATMs) strive for optical perfection, there are a few who purposely go the opposite direction. These contrarians prefer light buckets with an optical quality so poor one hesitates to refer to "images" at all. Unlike astrophotographers and visual observers who like crisp, sharply focused stars, stellar photometrists prefer somewhat fuzzy stars and may purposely defocus their telescopes to improve photometric precision and dynamic range—especially when observing brighter stars or fast paced events. They crave lots of cheap photons, preferring to spend money on larger apertures as opposed to higher optical quality.

During an Alt-Az Initiative workshop at the American Astronomical Society's summer conference in Pasadena in June 2009, Arne Henden, AAVSO Director, gave a talk, "1.5 Meter Light Bucket Science," that championed the cause of low cost, lightweight light buckets. Besides precision stellar photometry, light buckets are well suited to near IR aperture photometry. At the Portland III workshop August 2009, Greg Jones described his work on a near infrared aperture photodiode photometer housed in a cryostat. All the telescope's optical system needs to do is focus, on axis, most of the light from a star on a photodiode 1 mm or 0.5 mm in diameter. Off axis optical quality is of no concern. Near infrared astronomical cameras are very expensive and difficult to support, while aperture photometers such as the Optec SSP-4 and the Jones' photometer are affordable and relatively straightforward.

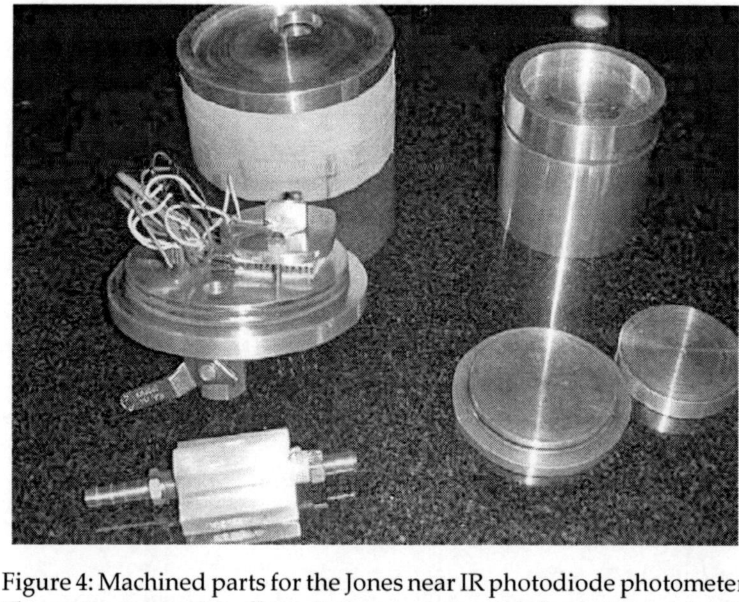

Figure 4: Machined parts for the Jones near IR photodiode photometer. The vacuum cryostat allows low temperatures to be maintained with a multistage thermoelectric cooler, thus reducing detector noise, the major noise source for photometry in the J and H bands.

Figure 5: Thin ¾" 75 lb. mirror from DOTI

Lightweight Meniscus Mirrors

What's a telescope without its primary mirror? A really cool tracking device that's no fun to look through and useless for gathering data. A traditional 1.5 meter mirror would weigh a ton and cost a fortune, so our group is exploring several non-traditional approaches.

Thin meniscus mirrors are lightweight, and if they are made from a slumped disk of ordinary soda lime float glass they can be quite inexpensive. Such mirrors are used extensively in less demanding optical applications such as flight simulators. Because they are so thin, they're also very flexible, and the resulting optical figure can be of low quality unless steps are taken to apply corrective forces to the back of the mirror in a manner somewhat similar to large alt-az telescopes (but on a much smaller, simpler scale).

Figure 6: Russ Genet leads a discussion on telescope design.

A similar approach was described by Russ Genet. He led a discussion on design options for a 1 meter light bucket photometric telescope he plans to build as a technology demonstration project using a thin ¾ inch, 75 lb. mirror from DOTI Optical, normally used for flight training simulators—see Figure 5 above. This was a continuation of an on-line discussion he began a couple months prior to the workshop, so a good deal of progress had already been made deciding on the final design.

The objective of the project is to show that high quality photometry can be done with a very lightweight, low cost alt-az 1 meter telescope—hopefully in an approach that could be scaled up to 1.5 or even 2.0 meters. A two element spherical aberration corrector will be used to improve the mirror's performance.

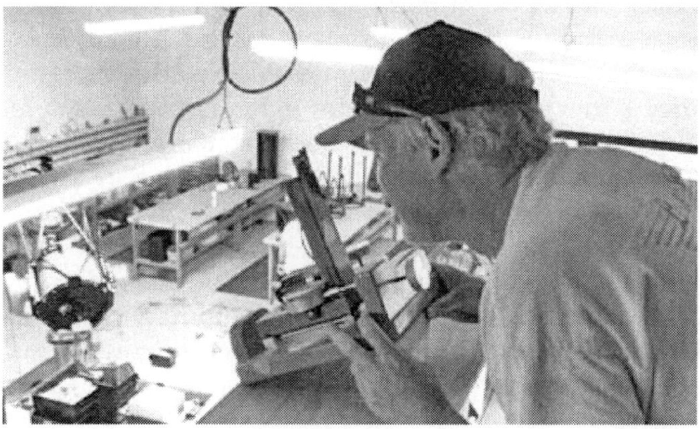

Figure 7: Joe Haberman testing the 1 meter meniscus mirror from DOTI.

Joe Haberman tested the 1 meter meniscus mirror from DOTI Optical in the shop at PlaneWave Instruments in Torrance, California, right after the Portland III Workshop.

Foamglas Tessellation

One of the more innovative approaches to large, lightweight primary mirrors is Foamglas tessellation. This approach uses small hexagonal mirrors that when combined make up one large mirror.

Dave Davis and Andrew Aurigema cut hexagon shaped pieces of soda lime glass which they glued to a piece of glass foam with the desired f-ratio curve ground into it. Foamglas, a product made by Pittsburgh Corning, is so lightweight it floats but it is exceptionally stiff and has nearly the same temperature expansion characteristics as soda-lime glass.

Figure 8: Lightweight mirror using Foamglas tessellation. Hexagonal pieces of soda lime glass are mounted on a lightweight substrate product by Corning.

This idea is essentially a low tech take-off of the Keck and Thirty Meter Telescope (TMT) primary mirror strategy that uses many small hexagonal mirrors that fit together to make one large mirror. In our case the small glass hexagons are glued to a Foamglas substrate and then ground, polished, and figured just like a monolithic mirror. This is in contrast to the high tech approach of using actuators, edge sensors, and computers to keep the individually made tessellations working as one optical surface similar to the Keck and TMT telescopes.

The first attempt was an 18″ mirror with 3″ tessellations of 1/8″ thick glass adhered to a chunk of foam glass with Gorilla Glue. David Davis was able to achieve a rough figure on the combined tessellations but a few of them were out of alignment. Even so, a star test was able to form two arc minute sized star images.

Figure 9: Dave Davis holding his first attempt at Foamglas tessellation. Joe Garlitz looks on with understandable skepticism.

The main problem with this first attempt was that the glass hexes ended up being only 1mm thick as a result of the optical work. Also, the hexes didn't fit flush to the glass foam because they had flat backs and the Foamglas was curved, causing an unequal distribution of adhesive.

The next attempt used thicker glass, and larger hex shapes that fit into individual pockets CNC'd into the glass foam that still follow the overall curve of the desired f-ratio of the completed primary. Andrew Aurigema used Milbond optical adhesive this time, which is more stable than Gorilla Glue. Note Figure 11 showing this Foamglas-tessellated blank floating in a fish pond. Optical work by Dave is progressing.

Figure 10: Thicker glass, and larger hex shapes that fit into individual pockets CNC'd into the glass foam.

Direct Drive

Gears, belts, friction drive rods, servo motors, stepper motors – who needs them? Anyone with a drive on their scope, that's who, but very likely that won't have to be the case for long. Direct drive does away with all the mechanical parts of a telescope drive that we've become accustomed to by essentially integrating them as part of the telescope itself. It would be just as accurate to think of the telescope becoming part of the direct drive motors (DDM). Many large professional scopes use direct drive motors on both the azimuth and altitude axis.

In 2008 Dave Rowe came up with the idea of making a direct drive motor which was purposely less energy efficient so it would be easier and less expensive to fabricate. The first rough prototypes worked so well that the 18″Cal Poly student project scope incorporated upgraded versions made of machined aluminum.

Figure 11: The 18″ DDM and scope.

Essentially a DDM consists of a series of coils and permanent magnets set up on opposing rings. Add current – and a DDM controller – and you have a telescope motor. As you can see from the photos opposite, a high precision DDM can be made with common materials. This one has a material cost of around $300.

PlaneWave Instruments is now producing its CDK700, a 28 inch alt-az corrected Dall-Kirkham Nasmyth focus telescope with direct drive motors and plans to produce the CDK1000, a 1 meter scope. Direct drive motors will soon become familiar as they are integrated into modest aperture, top end telescopes.

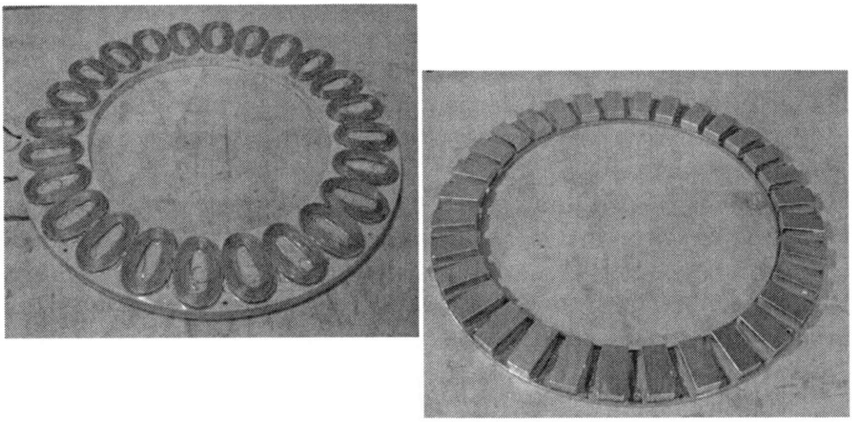

Figure 12: Direct Drive Motor prototype, which consists of cells and permanent magnets set up on opposing rings.

Figure 13: The metal disks on top of the prototype DDM in the above photo are a super high resolution encoder with a prototype of the SiTech DDM controller placed on top for display purposes. Don't let the plywood disks the prototype is made from fool you, this thing works.

There's more to direct drive as a high precision telescope drive than just the motors. Also required to complete the system are super high resolution encoders and a controller to handle the encoder and control the speed of the DDM. Dan Gray of Sidereal Technology is completing work on a DDM controller. Once it's available, the amateur telescope DDM revolution will begin.

Wheels

Large alt-az scopes need smoothly rolling wheels, especially in azimuth, to insure the scope moves with as little friction as possible so the motor can move the scope efficiently. Although the size and material of the wheels are important considerations, their profile is also an important design element.

Consider a wheel rolling on the flat circular track of an observatory dome or the azimuth axis of a large telescope. A wheel with a flat profile – meaning the contact area of the wheel – wants to roll in a straight line and so sets up a problem because the wheel will start to bind and skip at times. If the forces are high enough, the flat wheel may eventually tear itself or its circular track apart. This has happened at the professional level on the dome wheels of a major observatory (that will remain nameless). Even though amateur scopes are not likely to be heavy enough to tear themselves apart in this fashion, a mismatch between the wheel geometry and the radius of the track they turn on can cause binding.

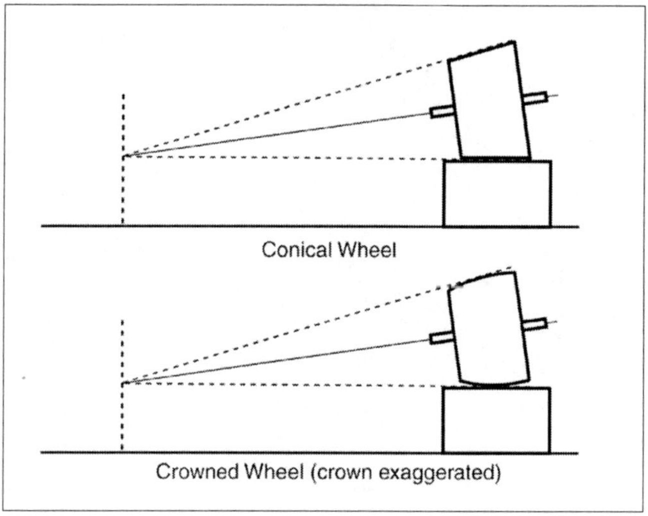

Figure 14: Illustration of conical wheels. Top shows an exaggerated design of a conical wheel to better illustrate the principle and the lower shows an example of a "crowned" conical wheel.

This means a wheel that turns on a circular track should have a conical profile. To imagine what a conical wheel looks like, imagine a paper cup turned on its side. When given a little push to roll on a table, the cup naturally traces an arc—that's what a conical wheel does. Matching the radius of a conical wheel to the radius of the circular azimuth or dome track will guarantee the minimum of rolling resistance.

A crowned wheel is also recommended so that the edge of the wheel (and thus its high contact stress) does not make any contact with the rail. Allowing this contact stress to occur can lead to cataclysmic failure of either the wheel or the rail. If failure does not occur, excessive wear is a sure result. Care has to be taken that both the crowned wheel and the surface of the curved track it rolls on are firm enough to not distort.

Scope structures

It's easy to imagine a 1 to 2 meter alt-az telescope as a scaled up Dobsonian, but rather than scaling up, let's look at scaling down – down from the world's largest scopes that is. If a telescope structure is strong and stiff enough for an 8 to 30 meter scope, it should work beautifully for a 1.5 meter scope.

For instance, look at the use of truss elements throughout the Thirty Meter Telescope (TMT) telescope design. Mirror cells are an excellent area for a multi-truss structure, as is a non-flexible ground ring support and a stiff rocker structure.

Figure 15: Computer model of the Thirty Meter Telescope gives an idea of what the final product may look like.

Taking cues from non-optical telescopes is helpful as well. Looking at the truss structures that hold the receivers above the main dish in radio telescopes shows that three or four truss elements can hold a prime focus instrument package precisely in place. This could be a perfect solution for the low optical quality fast cadence, high precision photometry scope that doesn't require high quality optics and doesn't need to worry about the possible extra diffraction caused by these truss structures where they cross into the light path.

The long and the short of it is that 1 to 2 meter alt-az scope design is wide open to using design influences from the world's largest optical and non-optical scopes. It's tempting to dismiss these designs as too complicated for amateurs to build, but the traditional wooden Dob reaches its limit right around 1 meter. Building successful alt-az amateur scopes larger than 1 meter will require a different approach, especially if they're meant to be transportable.

While the genesis of professional telescope design requires exacting engineering processes, amateurs can approach the design process somewhat differently. Fast prototyping can be more efficient than a specification process, so simply trying something new can be highly successful, especially when playing around with structures originally engineered by professionals. That's not to say that a careful design analysis isn't useful, only that when trying something new, just building it will often provide useful information and empirical testing more quickly.

No doubt, some exciting amateur telescopes will come from creative use of professional telescope design concepts. Combined with direct drive motors, a typical amateur alt-az scope 10 years from now may more closely resemble the 10 meter Keck telescopes or the 8 meter Gemini telescopes – and in many cases will be considerably larger than the average Dobsonian of today.

Adaptive Optics (AO)

Greg Jones is constructing an adaptive optics (AO) system for his 20 inch Cassegrain by following the basic design of professional AO systems. Nearing completion, his system is straightforward and promises to produce a small corrected field of view. Although the corrected "isoplantic patch" is small, it is large enough for small objects, such as Pluto and its largest moon or close double stars, or to concentrate light from a star for high resolution spectroscopy or similar tasks.

For now amateur AO isn't promising a magic bullet to create Hubble-like sharp images, but Greg's efforts do point the way for its use on specific science missions with amateur scopes that are large enough to make the technology practical.

The Future

Construction on the Thirty Meter Telescope (TMT) scope will begin in the near future and the European 42 meter OverWhelmingly Large

(OWL) scope isn't far behind, so the size of the world's largest optical telescopes is currently undergoing a revolution.

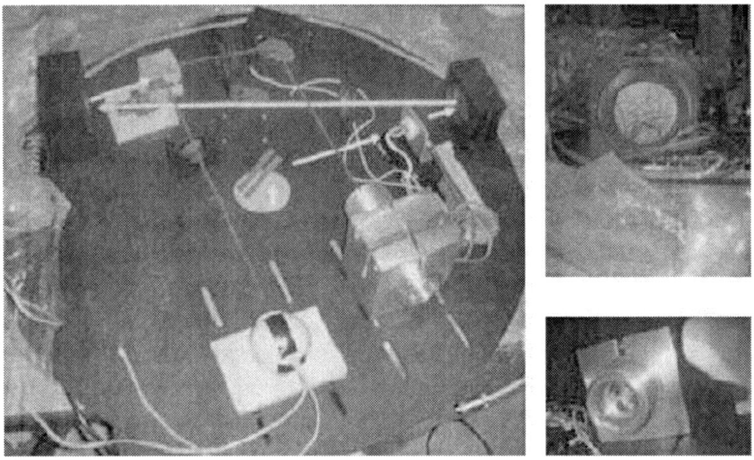

Figure 16: Greg Jones AO system he plans to use on his 20 inch Cassegrain telescope. He may be able to image Pluto and its largest moon, Charon.

Figure 17: Craig Breckenridge

Craig Breckenridge of Dynamic Structures, the company that will build the TMT, gave us a preview of the design. As thrilling as that was, even more interesting was discovering that Craig is just as much an amateur telescope maker and observer as he is a professional telescope engineer.

The 30 meter telescope and dome will be about the size of a football stadium and at first blush feels like a bit of science fiction. How can a telescope possibly be that big?

But getting used to a new idea or sense of scale can spark a paradigm shift and hopefully kick-start new ways of thinking. What's possible for amateurs to build? As new mirror technologies are developed, professionally inspired telescope structures, drives, and possibly AO will find their way into amateur telescopes considerably larger than 1 meter. The efforts of our informal group have produced some impressive results so far, but we've only scratched the surface.

Even so, let's imagine what all this might lead to in the next 10 years if our efforts continue to be successful. There are many variables at this point, but for fun let's say we meet our performance, price, and weight goals. A 1.5 meter (60 inch) alt-az telescope will not only be possible for amateurs to build but they will likely lead to even larger amateur instruments. These will probably be multi-purpose telescopes so they will be capable of collecting high quality data, produce beautiful astro images, and also be killer visual scopes.

But will amateurs really build 60 inch alt-az scopes that weigh only 300 pounds and are as easily transportable as a current 30 inch truss tube wood Dobsonian? If the optics become even remotely affordable the answer is most definitely yes.

Chapter 5

Visit to Gemini and Subaru 8 Meter Telescopes on Mauna Kea

Russ Genet[1,5], Dan Gray[2], Chris Carter[3], Michael Sheehan[3], Daigo Tomono[4], Hideki Takami[4], Cheryl Genet[5], and Jon Archambeau[3]

[1]California Polytechnic State University, [2]Sidereal Technology, [3]Gemini Observatory, [4]Subaru Telescope, [5]Cuesta College

The 5 meter Hale telescope on Mt. Palomar is the largest equatorially-mounted optical telescope ever built. A motor moving at constant speed in a single axis allows this giant telescope to smoothly track the stars. There are now some 18 larger telescopes. All of these larger telescopes employ compact, gravity-symmetrical altitude-azimuth (alt-az) mounts, which allow them to be built at much lower cost. The catch is that these telescopes require constantly-changing drive rates in three axes: altitude, azimuth, and field de-rotation. This was not possible before the advent of reliable digital computers. Recently, alt-az digital telescope control systems have become so inexpensive that the alt-az revolution can now go in the opposite direction, from larger to smaller telescopes.

Astronomers (both professional and amateur), engineers, students, and smaller telescope manufacturers are working together in an informal group to develop economical research-grade alt-az telescopes in the 0.5 to 2.0 meter range. Large alt-az telescopes have pioneered many of the techniques that can now be applied to their diminutive cousins at greatly reduced cost and weight, thanks to advances in electronics and materials. A key element of this smaller telescope, alt-az developmental program is discussing technical aspects of the large mountaintop telescopes with the engineers who design and maintain them. They genuinely enjoy considering the technical challenges of applying what they have learned on a much smaller scale with greatly reduced budgets.

Large and small research telescopes are complementary. The larger telescopes, which are few in number, concentrate on fainter objects or on discovery surveys, while the growing army of smaller telescopes concentrates on brighter objects as well as follow-up time-series observations of newly discovered objects. By sharing large and small telescope design knowledge, we hope to develop 0.5 to 2.0 meter alt-az research telescopes that emulate their big mountaintop cousins at costs that will allow many small colleges and even high schools to acquire such modern research instruments.

Figure 1: Meeting participants at the Gemini headquarters in Hilo, left to right: Chris Carter, Mike Sheehan, Daigo Tomono, Hideki Takami, Doug Simons, Russ Genet, and Dan Gray.

Figure 2: Cheryl, Russ, Dan, and Chris stopped for a leisurely lunch and altitude adjustment on the way to the summit. This facility at the 9000 foot level is a hotel and restaurant for all the astronomers.

Recently, two of the Alt-Az Initiative team members (Russ Genet and Dan Gray) met with Gemini engineers Chris Carter and Mike Sheehan and Subaru engineers Daigo Tomono and Hideki Takami in Hilo, Hawaii. A visit to the 8 meter Gemini and Subaru telescopes on Mauna Kea followed this meeting. Chris Carter guided Russ Genet and Dan Gray through Gemini, while Daigo Tomono guided them through Subaru. Cheryl Genet kindly took photographs throughout the Mauna Kea visit, and Jon Archambeau provided transportation and Gemini tour guidance.

Our goal was to observe and discuss various aspects of the Gemini and Subaru designs we might be able to transfer to smaller telescopes.

Figure 3: As we arrived at Gemini, the main shutter was facing us, flanked by two of the doors that cover the louvers. When open, the louvers allow a smooth airflow that reduces the effects of dome- and mirror-seeing.

Figure 4: Cheryl, Russ, Chris and Jon stand by Gemini's Cassegrain focus. One optical path is straight through to the bottom instrument behind Cheryl, while four other instruments are above.

The Gemini telescope's instruments are all located at its Cassegrain focus. An instrument selector can rapidly switch the optical path between the instruments. This allows queued programs to dynamically adjust to observing conditions. Also, having all the instruments at the bottom end of the optical tube assembly (OTA) results in a low elevation axis location, which places the primary mirror well up in the dome where air flow from generously proportioned shutters helps eliminate mirror-seeing.

Figure 5: Dan examines the flex plate and azimuth high torque motor. Dan's cap helped keep him warm. The interior is kept at the expected nighttime temperature.

Figure 6: The small band illuminated by the flashlight is the encoder tape. Our small alt-az telescope uses a similar tape, although it is significantly shorter!

The Gemini telescope uses one stage of friction drive reduction in azimuth. There are four pairs of two high-torque motors that drive a large-diameter azimuth disk. The motor pairs are fastened to substantial flex plates and pressed against the azimuth disk by hydraulic actuators. Position feedback is provided by a Heidenhain tape encoder and read heads, while velocity feedback is provided by tachometers. A key suggestion made by the Gemini and Subaru engineers during our meeting was that we add velocity feedback to our small telescope control systems.

Figure 7: The top end of Gemini's OTA. The 1 meter secondary is the small bright item in the middle of the secondary baffle, which was designed for infrared observations.

Figure 8: Dan (left) and Chris examine the control electronics for the secondary mirror. Both are control system experts and spoke at times in their strange language.

Figure 9: The azimuth cable wrap avoids cable drag. The scuff marks on the floor reveal the movement—somewhat reminiscent of baggage.

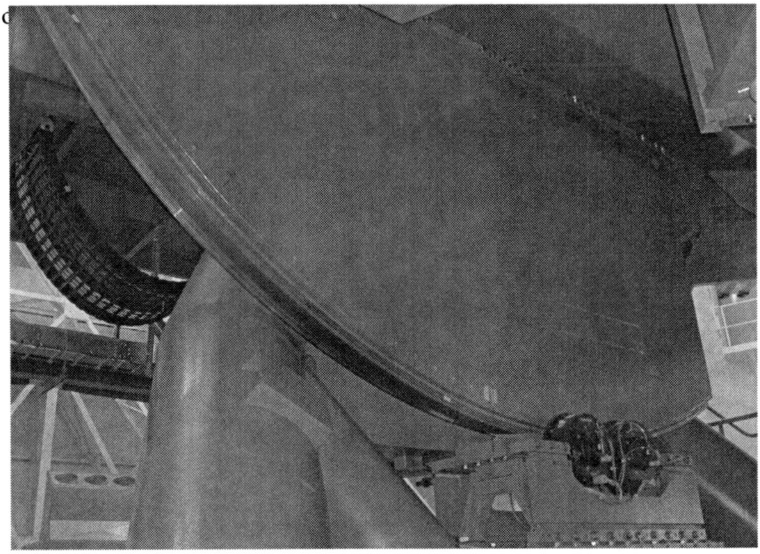

Figure 10: The elevation drive is similar to the azimuth drive, although smaller in scale. The end of the elevation sector can be seen at the lower right.

The top end of Gemini's OTA consists of a two-bay truss. Single-bay trusses are appropriate for smaller telescopes. Interestingly, although Gemini's top end, which supports its spiders, is a round ring, the Gemini engineers recommended that we use square top ends on our smaller alt-az telescopes. The spiders can then be put under considerable tension, placing the square under pure compression. This tightens up the entire top end of the telescope. But why doesn't the Gemini have a square top end? The straight top members would have been over 28 feet long, and buckling would have been a problem. Small has its merits!

Gemini's basic telescope control system is augmented with a tip-tilt secondary mirror and autoguider camera to keep stars centered in the instrument under use. The tip-tilt secondary handles the higher frequency disturbances caused by the wind, while the telescope control system takes care of tracking and lower frequency corrections passed off from the autoguider via a low-pass filter, thus reducing the range of corrections the tip-tilt secondary needs to make.

Needless to say, a 1 meter secondary mirror buzzing away at the top end of the telescope making corrections, not to mention providing "chopping" for infrared (IR) observations, would have made a very good telescope vibrator were it not for a reaction mass that is moved exactly opposite to the secondary mirror. For our much smaller alt-az telescopes we are planning on using a deviator (wobble) plate. Interestingly, the 3.6 meter Canada France Hawaii Telescope (CHFT), which is right next door to the Gemini telescope, uses a deviator plate for its higher frequency corrections.

Figure 11: As we left the Gemini for the short drive to Subaru, snow was being shoveled off the top of one of the shutters in preparation for the night's run. The snow removal team was dwarfed by the massive dome.

Figure 12: View of Haleakala on Maui from Subaru. Haleakala and Mauna Kea were separated by a sea of clouds which drifted below the summits. They are even lower during the night thanks to an inversion layer.

A laser can project a bright artificial star in the atmosphere's sodium layer about 60 miles above the telescope, very close to the object of interest. This artificial star is used to determine wavefront distortions with a Shack-Hartmann sensor. A high speed computer calculates corrections that are fed to a small deformable adaptive optics mirror at very high frequency to cancel out the distortions. Bright natural stars further away in the field cannot be used because of isoplanetism (uncorrelated motions), thus the need for the laser. Jon Archambeau trains and directs the visual, human aircraft spotters who keep a vigilant lookout during laser operation.

Daigo Tomono kindly led us through the Subaru telescope. Unlike Gemini, which has all its instruments at the Cassegrain focus, Subaru has a Cassegrain focus and two Nasmyth foci. Furthermore, the secondary mirror can be swapped out for a prime focus instrument. All these additional features require a very sturdy structure, so the Subaru telescope weighs more than the Gemini telescope.

Figure 13: Access to the Subaru telescope is via the tower on the left and an underground tunnel which leads to the control room in a separate building.

Figure 14: Russ stands under the Subaru telescope beside the Cassegrain focus instruments. This is just one of the four focus positions.

Both the Gemini and Subaru telescopes use thin meniscus primary mirrors. The mirror blanks were cast at Corning from ultra low expansion (ULE) fused silica. Passive supports alone would have been inadequate for these thin mirrors, so their figures are fine tuned with actuators which actively apply forces to the backs of the mirrors. The Subaru mirror has a diameter to thickness ratio of 60:1. What was of interest to us was that once the necessary mirror force corrections for various elevations and temperatures were determined for Subaru, they were used for months in a table look-up fashion before they needed to be refined. This may be important for our larger-aperture small alt-az telescopes. We might use a thin meniscus mirror in our 1.0 to 2.0 meter telescopes with active control, and hope we could make calibration measurements off line and just interpolate from a table during normal operation.

Figure 15: Subaru's primary mirror cell. The mirror was being uncovered as we watched, in preparation for the night's run. Its diameter is 60 times its thickness.

Figure 16: When the thin meniscus mirror is removed from the telescope and brought down to be recoated, it must be very carefully supported in this mirror transporter.

Figure 17: The entire Subaru building rotates on dozens of massive wheels. Russ gives perspective to their size.

Figure 18: The complex of walkways and machines below the Subaru telescope was an engineer's delight.

Like all large modern telescopes, both Gemini and Subaru are operated from warm rooms. Remote observations in real time are possible with these and many other large telescopes. Queue scheduling allows observations planned out well in advance to be switched in or out as conditions dictate. A resident astronomer usually monitors the queue observations. Many smaller observatories now also support warm rooms, remote observing, and queue observing.

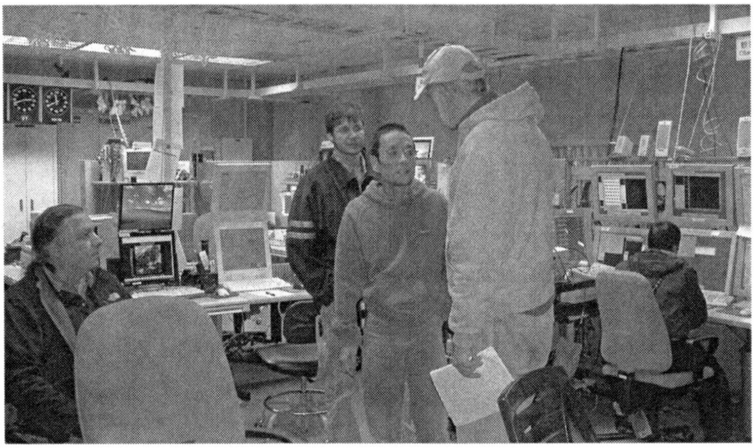

Figure 19: Dan and Jon listen in on the left, while Daigo and Russ discuss Subaru operations. The night operator on the right prepares the telescope for the night's run.

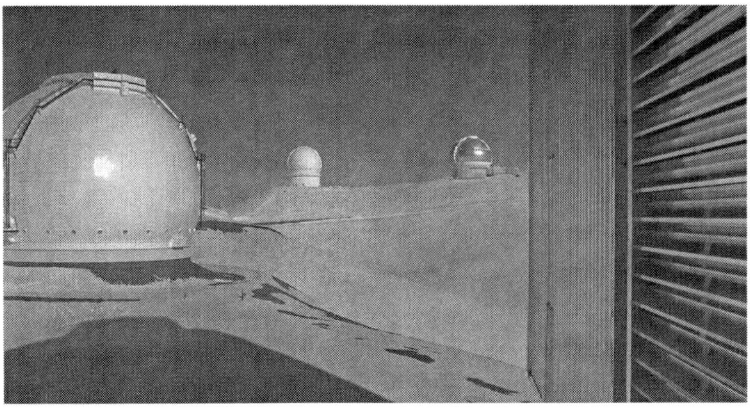

Figure 20: Looking from the Subaru telescope, we could see the nearest Keck telescope. The Canada France Hawaii telescope and Gemini are on the ridge behind.

As we left Mauna Kea, final preparations were being made for another night of observing at the top of the world. During our descent, we reflected on how some of the advances made by these mountain-top giants could be translated into economic improvements for our smaller, lowland telescopes.

Chapter 6

Return to Mauna Kea

Telescope Structures and Active Primary Mirrors

Russell M. Genet

California Polytechnic State University

Introduction

I returned to Hilo for discussions with the Gemini and Subaru engineers on March 3rd and 4th, 2008. Further discussions were held on structural dynamics and active primary mirror control, and the design of a lightweight 2 meter telescope was roughed out.

Figure 1: At Gemini Headquarters, from the left, Chris Carter (Gemini Controls), Daigo Tomono (Subaru Mechanical), Mike Sheehan (Gemini Mechanical), and Russ Genet (California Polytechnic State University).

Telescope Dynamics

Mike Sheehan (Gemini Mechanical Engineering) and I discussed telescope structures. Mike drew a force versus frequency diagram for wind disturbances on telescopes with the more or less steady forces below 0.2 Hz and the turbulent forces higher than 0.2 Hz, where resonances can create problems for control system stability. In general, if one makes the telescope stiff enough to avoid resonances in the higher frequency regime, the structure will be more than robust enough for the constant wind forces. Thus in structural design, one has to be primarily concerned with higher frequency resonances.

We also discussed tip tilting and rotational forces on the soil through the pier for systems not on bedrock. These movements are not sensed by the telescope's position encoders, of course, but show up as motions that must be removed by active star-tracking tip/tilt systems. Soil springiness may be a particularly difficult problem for small portable science/astrophotography telescopes. We discussed whether or not placing a number of very heavy weights on the bottom, non-moving portion of a portable telescope might improve this situation. Such weights might actually lower the overall telescope/soil resonant frequency, but make it less likely to be excited.

We then discussed dampening. The very stiff telescope structures required to avoid low frequency resonances generally end up having very little dampening, i.e. are high "Q." Mike demonstrated the vibration of two strips of aluminum. One vibrated for quite some time. The other, which had the same amount of aluminum in it, but was in two half thicknesses separated by a thin absorbing material, did not vibrate at all. The samples were supplied by CSA Engineering in Mt. View, California. Ty Safreno has mentioned this firm before. They specialize in solving (and avoiding) vibration issues.

A 2 Meter 1000 lb. Multiple Mirror Telescope

To inject some fun into our discussions, I asked Mike for some design ideas for a 2-meter equivalent telescope with four 1 meter, thin spherical mirrors. Our alt-az development group occasionally considers various options for extending small telescope apertures beyond 1 meter while still maintaining low costs and lightweight features. Such telescopes might be especially useful for time series photometry in the optical and near IR, and for fiber fed optical spectroscopy.

For the telescope we considered, the four spherical mirrors would individually be *f*/6, and in an aggregate *f*/3. The mirrors would be co-aligned and co-focused, but would not be co-phased, so their resolution would not exceed that of a single mirror (not much of an issue at typical lowland sites).

Dave Rowe has done some very preliminary optical design work that suggests that a 2 or 3 element spherical lens corrector could provide acceptable images (for typical lowland sites) over fields, for instance, for an SBIG ST-7 or ST-8 camera for a 2 meter equivalent telescope with a spherical primary. This would be adequate for time series photometry and fiber fed spectroscopy, but not wide field imaging, so would be considered a somewhat specialized telescope. Dave is also considering a Pressman-Carmichael Cassegrain design with an oblate convex secondary for spherical correction and refractive elements for coma correction. This would provide a much wider, reasonably well corrected field of view, although the secondary would not be low cost.

Mike recommended a pure truss structure, suggesting that even for smaller telescopes such as this 2 meter, a truss structure would be superior to a monocoque plate structure. Advantage could be taken, Mike suggested, of the space between the four mirrors, assuming they were not jammed right on top of one another. The trusses could extend beyond the tops of the mirrors to support the overall mirror system.

Instead of a conventional optical telescope top end, Mike suggested a quadrapod, with the four trusses outside the converging light, and an additional fifth truss extending from a spider near the top of the telescope down through the space between the mirrors. As the top end of the OTA would be very light compared to the back end, the altitude axis would be near the top surface of the primary mirrors.

Mike thought that conventional axles and bearings in altitude might be preferable to large diameter elevation disks on both sides. He suggested, however, that we might consider using two direct drive motors in altitude, one on each side to avoid twisting the bottom section of the OTA.

In azimuth, we first considered using off-the-shelf slim line bearings, but if one were to avoid fork flexures, it would be good to bring the forces straight down from the altitude bearings to the azimuth rotation surface and hence to a cylindrical pier. This would require about an 8 foot diameter bearing, probably putting it out of the small telescope price range.

Instead, Mike suggested four wheels or sets of wheels rolling around on a flat plate (track) on the top of a cylindrical pier. The four wheel sets would be placed at the four corners of a square truss structure. Diagonals from the corners would meet in the middle at a centering bearing.

Although a line-contact roller would provide the most surface contact area, it would probably need to be slightly conical in shape and the wheels slightly tapered, both somewhat difficult to implement. We discussed, alternatively, having one or more "U" shaped wheels at each corner (if more than one, they would need to be on independent bearings).

We made some rough weight estimates. If the 1 meter mirrors were somewhat thin meniscus mirrors, say 25 mm thick, they would weigh about 120 lbs. each, and the four together would weight about 500 lbs. Mike thought that if the truss structure was made primarily from carbon fiber tubes, that the entire telescope, not including the mirrors, should weigh less than 500 lbs. This would give a total weight for a 2 meter equivalent telescope of less than 1000 lbs. It could probably be assembled by two persons without any cranes or hoists. The telescope would, roughly, be 8 feet across and 24 feet high.

Active Primary Mirrors

I met with Chris Carter (Gemini Controls Engineer) on the following day, and over lunch we discussed active support of primary mirrors and, later on, included Mike Sheehan and Daigo Tomono (Subaru Engineering) in the discussion. The 182 actuators on the Subaru actually hold almost the entire weight of the Subaru's primary mirror (except where it very lightly rests on three defining hard points). Thus the Subaru is magnetically levitated. Each actuator provides considerable force and has an individual feedback loop. Gemini's primary mirror, on the other hand, is supported primarily by air pressure, with a seal around the periphery of the mirror and the ability to change the air pressure versus altitude. Some 120 voice coils and permanent magnets supply the "tweaking" forces required to bring the mirror into its desired shape.

Figure 2: Chris Carter and Russ Genet in the Gemini electronics lab. Between them is the voice coil mechanism that adjusts the shape of the Gemini's 8 meter mirror. Some 120 similar coils are on the telescope. This one is a spare.

These voice coil / magnet assemblies were made by BEI Kimco. Chris' lab test setup can be seen in the figure above. At 5 volts and about 1 amp of current, I could no longer force any significant coil movement. Chris pointed out that we could easily make our own coils. Dave Rowe has located some low cost speakers that should work well, complete with voice coil, magnet, and centering spider. We would just toss the speaker cone and frame in the trash bin.

Chris recommended, and I think we all concurred, that it would, for smaller telescopes, be worthwhile to consider floating the mirror on conventional whiffletree supports and just tweaking the mirrors with the voice coil actuators at locations between the whiffletree points. The goal here would be to just correct the lower order Zernike terms, so this scheme would work best on mirrors of medium flexibility, and would only require a dozen or so actuators.

Figure 3: Close up view of the voice BEI Kimco voice coil and magnet. The magnet is inside the metal outer sleeve, and cannot be seen. At 5 volts and 1 amp, I could no longer move the coil in and out. Very much smaller (and lower cost) coils and magnets could be used for mirrors weighing pounds instead of tons.

Chris has considered how we might implement the control system for a dozen or so voice coil actuators. He suggested we consider using the SPI buss, which can be easily driven from the parallel port of a PC. Maxim Max 5352 12-bit D/A converters are about $12 each from DigiKey. The audio/operational amplifiers required are about the same price. If one wound one's own coils, the cost in electronics, magnets, and wires would be around $50 per channel. Perhaps a unit with a dozen or two channels, power supply, etc., could be built for under $2000.

All four of us discussed, at some length, whether or not implementation of an active primary mirror would require some sort of on-line wavefront sensor, such as a Shack-Hartmann sensor. Although Gemini does utilize an on-line wavefront sensor for primary mirror adjustments, Subaru does not. Subaru calibrates their system offline with a wavefront sensor on bright stars with 30 second integrations to average out atmospheric effects. (Bright to them is 14th magnitude—brighter than that they overflow their camera in 30 seconds.) However, they only need to do this at infrequent intervals, every six to eighteen months.

When they make a calibration run, they find the sensor positions that give the best mirror figure for every 15 degrees in altitude. During normal operation, a lookup table and interpolation provides the settings for the actuators.

We discussed whether or not, for a telescope with just a dozen or two actuators fine tuning a mirror supported by a whiffletree, one would even need a wavefront sensor for calibration runs, let alone regular operation. Our conclusion was that a wavefront sensor would probably not be required. Given some time for experimentation, a computer algorithm could be developed to automatically search for the best settings at each of several altitudes, observing a bright star at each altitude and making adjustments to minimize the image size. These settings could then be stored in a lookup table and interpolated for any specific altitude. This is not to suggest that one might not want to employ a wavefront sensor, only to suggest that it might not be necessary.

Acknowledgements

My thanks to Mike, Chris, and Daigo for taking the time to consider how the technology used on large alt-az telescopes might be applied to smaller aperture telescopes.

Alt-Az Initiative Workshops and Conferences

The first remote Internet access of automatic telescopes was made over two decades ago for telescopes at the Fairborn Observatory on Mt. Hopkins. Shortly thereafter came the publication of two 1989 books, *Robotic Observatories* (R. Genet and D. Hayes) and *Remote Access Automatic Telescopes* (D. Hayes and R. Genet, eds.), as well as a conference, Remote Access Automatic Telescopes (RAAT I). RATT II, pictured above, was held in 2007 at the historic Cloudcroft Inn high in the New Mexico Mountains overlooking White Sands. A major conference on Robotic Observatories will be held in Waimea on the Big Island of Hawaii at the very end of 2011/first of 2012.

Intense conversation at the Portland III workshop held in Dan Gray's factory in Portland where he makes control systems for large ocean going ships (Technical Marine Services) and also controls systems for telescopes (Sidereal Technology). This now annual event is relaxed and informal and occurs on a weekend near the end of July/first of August.

Alt-Az Initiative 2010 workshop held on the leeward side of Oahu at the Makaha Resort. Hot topics included lightweight mirrors, light bucket astronomy, low cost instrumentation for near IR photometry, spectroscopy, polarimetery, and undergraduate research. This was the second annual Hawaii Alt-Az Initiative meeting.Two more annual meetings are already planned: one on *Light Bucket Astronomy* and the other on *Robotic Observatories*. Both will be held in Waimea on the Big Island, the headquarters location for the Canada France Hawaii Observatory and Keck Observatory.

Two teams of mechanical engineering students at California Polytechnic State University (Cal Poly) in San Luis Obispo used finite element analysis (FEA) to design the OTA and fork for a pioneering 18 inch alt-az telescope that employed direct drives. Pictured are the participants at the Critical Design Review for this telescope.

Enjoying dinner after a busy day at the Small Telescopes and Astronomical Research (STAR) workshop held at Cal Poly in San Luis Obispo. This was the Initiative's second workshop, held just a couple of weeks after the first one in Portland, Oregon.

Attendees at the STAR workshop met on the campus of Cal Poly, San Luis Obispo. The Research Scholar in Residence program at Cal Poly kindly sponsored the workshop.

Tom Frey, a chemistry professor at Cal Poly and amateur astronomer, readies his 18 inch Obsession alt-az telescope for visual double star measurements with an astrometric eyepiece. Tom explained to Jerry Aanenson the advanced features of the Obsession telescope.

The participants of the 2nd Annual Pine Mountain Student Research Workshop relax after dinner. This annual event focuses on research that can be completed in a very busy long weekend, including project planning, observations, data analysis, and writing results. The workshop brings together established members of the astronomical community with school staff and students.

Dan Gray and Cheryl and Russ Genet (left to right) have lunch at the southernmost bar in the United States, located on the southern tip of the Big Island of Hawaii. They were making their way around the island toward Hilo and a meeting with Gemini and Subaru engineers and an insider's tour of the 8 meter telescopes on Mauna Kea.

Richard Berry (right) watches Arne Henden (middle) type notes into his computer at the first Hawaii Alt-Az Initiative conference, Galileo's Legacy, 2008.

Participants in the first Alt-Az Initiative Hawaii conference listened to some 30 PowerPoint talks and participated in a number of panel discussions. There was plenty of time for swimming and boogie boarding at the nearby Makaha beach—home of the famous annual big board surfing contest.

Doug Simons, left, led the panel discussion at Galileo's Legacy on tech transfer from large to small alt-az telescopes. Left to right is Doug, Dan Gray, Arne Henden, Russ Genet, Alan Holmes, and Richard Berry. As new technical approaches are developed for larger telescopes, and the costs of components drop as technology marches on, clever adaptations of big telescope "high tech" approaches can benefit smaller telescopes.

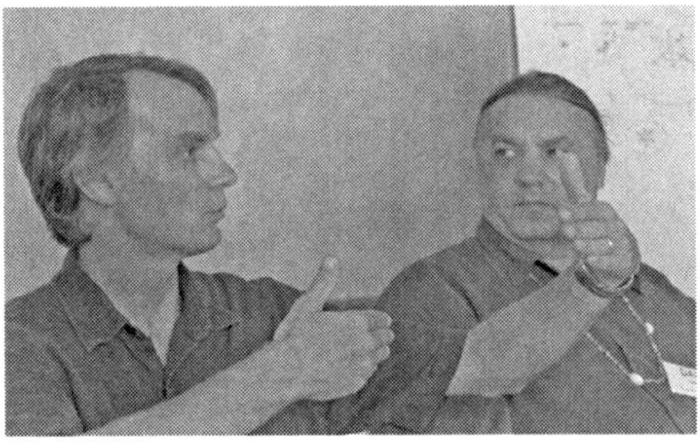

Doug Simons (left), the Director of the Gemini Observatory, explains how the active primary mirror technology used in large telescopes could be applied to smaller telescopes. Dan Gray (right), an expert on control systems, listens intently.

Russ Genet (left) and Alan Holmes. Alan, the President of Santa Barbara Instrument Group (SBIG) made commercially produced CCD cameras affordable for smaller observatories. The combination of CCD cameras, personal computers, and compact Schmidt-Cassegrain telescopes revolutionized small-telescope astronomy.

Dan Gray, President of both Technical Marine Services and Sidereal Technology. Dan became interested in the development of servomotor control systems after reading the 1985 book *Microcomputer Control of Telescopes* (M. Trueblood and R. Genet). Dan developed the electronics and software for the first low cost direct drive system for smaller telescopes.

Tom Johnson (right) and his son Greg Johnson at Galileo's Legacy. Tom Johnson gave a talk on Schmidt-Cassegrain Telescope optics and the clever way he figured out how to make the corrector plates at low cost, thereby launching Celestron. Greg worked on Celestron control systems and other projects.

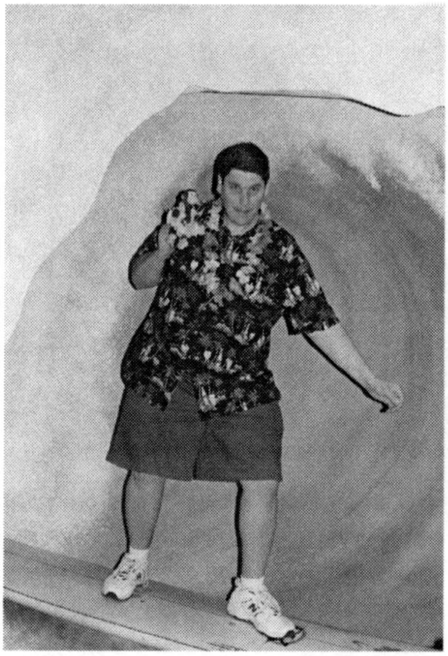

Andrew Aurigema surfs the big one at the Makaha Resort. Andrew developed a lightweight composite mirror with a thin glass front plate and foam glass backer made from Starstone (a product more commonly known as Earthstone). Andrew has a vacuum chamber, mirror grinding machine, and CNC machine for making up to 60 inch diameter mirrors at home in his airplane hanger.

One of the tables at the aloha dinner at the Makaha Resort. Some 30 amateur and professional astronomers from around the world gathered at this international conference to discuss various technical aspects of telescopes, instruments, and the scientific research one can accomplish with smaller telescopes.

An informal meeting of Initiative members at the Riverside Telescope Makers' Conference (RTMC) in 2009. Left to right: Dan Gray, Richard Berry, Joe Haberman, Tom Mathis, Russ Genet, Rick Hedrick, Floyd Fiechtner, Alan Keller, and Danyal Medley.

At RTMC in 2008, Initiative members met at a restaurant in Big Bear City, where Dave Rowe and Russ Genet made a gentlemen's bet on someone bringing a 1.5 meter portable telescope to RTMC. This bet morphed into the Rowe 1.5 meter challenge. Exactly a year later in the same restaurant, members gathered again (pictured here) to discuss progress. Left to right: Rich Hedrick, Floyd Fiechtner, Joe Haberman, Ed Ginan, Alan Keller, and Russ Genet.

Howard Royster (left) explains his clever alt-alt telescope control system that uses the sky as its encoder. A small auxiliary telescope and camera provide continuous position and tracking information to the control system's computer such that the telescope quickly figures out where it is and locks in and tracks on the desired celestial position.

Participants from the 2010 Hawaii Alt-Az Initiative workshop stand on the higher reaches (near the primary mirror) of the 8 meter Gemini North telescope. Left to right: Josh Walawender, Gary Cole, Clifford Livermore, Doug Arion, Mike Sheehan (Gemini's Chief Mechanical Engineer), and Russ Genet.

Keck II 10 meter telescope as seen from its backside. The hexagonal multiple mirrors are pointed away and to the left, while the space frame mirror support structure can be seen in the center of the picture taken by Russ Genet.

The Silversword (center) is an endangered species on the Big Island of Hawaii. The human species (behind) is, most likely, the reason the Silversword has become endangered (Josh Walawender, Joe Garlitz, Doug Arion, Russ Genet, and Gary Cole).

While the Mauna Kea Observatories are well known, the observatories on the other major volcanic peak on the Big Island, Mauna Loa, are not so well known. Driving to the observatory at 11,000 feet on the north slope of Mauna Loa, one passes through a number of recent lava flows. The tower behind Paul Fast, Russ Genet, and Josh Walawender (left to right) gathers air samples for continuous analysis by the National Oceanic and Atmospheric Administration. Their observations of the rising levels of tropospheric carbon dioxide, starting in 1957, first alerted the planet to the possibility of global warming.

These domes at Mauna Loa Observatory house two University of Hawaii Institute for Astronomy's totally robotic telescopes. Josh Walawender, a UH/IfA post-doc, automated these telescopes. The view is to the north, across the "saddle" to Mauna Kea, where the giant telescopes at the summit appear as small white dots.

Part II

TELESCOPE DEVELOPMENTS

Chapter 7

Alt-Az Telescopes for Research, Astrophotography, and Education

Russ Genet[1], Dan Gray[2], Howard Banich[3], Dave Rowe[4], Tom Smith[5], and Tom Krajci[6]

[1]California Polytechnic State University; [2]Sidereal Technology; [3]Rose City Astronomers; [4]Sierra Monolithics; [5]Dark Ridge Observatory; [6]Astrokolkhoz

This chapter presents the case that the combination of low-cost alt-az telescope control systems and affordable aerospace materials has reached a point where a revolutionary new class of light weight, highly capable alt-az telescopes is emerging. Similar to SCTs, they will not only be used visually, but for CCD-based scientific research and astrophotography. Similar to Dobs, they will have larger apertures than SCTs yet will be light in weight. Unlike visual-only Dobs but similar to their closely related giant mountaintop alt-az relatives, this emerging class of precisely controlled telescopes will handle a variety of instruments mounted on a field de-rotator with generous back focus.

Introduction

Thanks very much to Tom Johnson, Schmidt-Cassegrain telescopes (SCTs) revolutionized astronomy by making production telescopes compact, portable, and affordable. Computerized go-to capabilities and quantum-efficient, affordable CCD cameras—pioneered by Richard Berry's "Cookbook Camera" and also by SBIG and now also available from Apogee, Finger Lakes Instrumentation, QSI, Starlight Xpress, Yankee Robotics, and others—transformed SCTs into powerful instruments for research, astrophotography, and education. Beyond an aperture of 14 inches, however, the cost and weight of SCTs rises sharply.

Enter altitude-azimuth (alt-az) Dobsonian telescopes (Dobs) which, thanks to their transportability and relatively low cost, now dominate the class of visual only telescopes with apertures greater than 14 inches. Dobs, because they can be built with ordinary woodworking tools from readily available materials, inspired numerous amateur telescope makers and manufacturers into action.

Human vision is amazingly adaptive, easily compensating for image motion. Cameras, however, are unforgiving. Thus the use of the large aperture Dobs in research and astrophotography is limited by their lack of precise, variable rate, two axes, computerized tracking, and their need for field de-rotation. Dobs also generally lack the backfocus and weight capacities required to accommodate instrumentation.

For reasons of compactness and economy, all recent large mountaintop telescopes have been alt-az, not equatorial. The availability of computerized control systems to constantly calculate the ever changing drive rates in altitude, azimuth, and field rotation made this revolution possible. Although these telescopes weigh many tons, they are actually very lightweight, stiff structures when one considers the weight of their mirrors. Steel, pound for pound, is one of the stiffest of all materials, and on a cost per pound basis it has no peer. Only advanced materials such as carbon fiber are, pound for pound, stiffer than steel. Similarly, aluminum honeycomb panels, for their weight, resist bending better than solid sheet steel. Although in the past, such advanced materials were deemed too expensive to be used in telescopes, this situation is changing rapidly as costs continue to decrease and telescope makers become more familiar with the use of such materials.

The cost of alt-az telescope control systems has plummeted over the years. Initially, PDP-8 control computers and telescope control electronics cost tens of thousands of dollars and filled entire equipment racks. Today, Sidereal Technology makes a microcomputer-based alt-az telescope control system for about $1,000 that you can pack in your briefcase with room to spare. Meanwhile, the cost of aerospace materials has also plummeted as their use has moved beyond aircraft and spacecraft to outdoor signs and building exteriors. Given these two dramatic drops in cost, research class alt-az telescopes fabricated from lightweight materials are now economically viable, at least for modest to intermediate apertures where the cost of materials is not prohibitive.

The primary reason for developing lightweight alt-az telescopes with apertures larger than SCTs is to conduct research on, image, and view fainter objects. Dark skies are also helpful, but the costs of establishing and maintaining dark site observatories are considerable. Lightweight yet large aperture portable research grade telescopes avoid these costs. Many backyards are blessed with modestly dark skies. Observational pads accommodate lightweight telescopes which can be covered when not in use. During extended periods of inclement weather these telescopes can be safely stored in garages or basements. Alternatively, these alt-az telescopes can be permanently installed in unobtrusive roll-off roof shelters or even under domes.

Bright sky locations can also benefit from large aperture, precision alt-az telescopes. Narrow-band astrophotography produces good results by observing H-alpha, O III, and S II emissions, which are brighter than even their urban sky backgrounds. Such narrow band, photon-limited astrophotography, much like spectroscopy, benefits from raw aperture.

The emergence of this new class of lightweight, highly capable alt-az telescopes is being facilitated by innovative solutions to challenges in six vital areas:

(1) lightweight affordable optics
(2) back focus for autoguiders
(3) high natural frequency mechanical structures
(4) precision control systems and drives
(5) field de-rotation
(6) ease of construction and affordability

Each challenge area is discussed below.

Lightweight Affordable Optics

Significant astronomical research and education, not to mention public outreach, is still visual—eyeball to the telescope. A Newtonian optical configuration on an alt-az mount places both eyepieces and instruments at a comfortable working height for telescopes of modest to intermediate aperture. If, however, the sole purpose of a telescope is non-visual, a prime focus configuration can save weight and cost and is slightly more efficient in optical transmission. However, for the general purpose telescopes we are primarily considering, placing instruments at the Newtonian (secondary) focus is not only convenient, but it shortens the optical tube assembly (OTA). The small telescope community is blessed with a plethora of high quality, thin Pyrex, finished parabolic mirrors for Newtonian telescopes. They are available from Galaxy Optic, Hubble Optics, Kennedy Optics, Lockwood Optical, Pegasus Optics, Orion Optics, Zambuto Optics, and many others.

In this chapter we are concerned with somewhat larger than normal apertures, so we offer several optical guidelines. Focal ratio is the key determinant of the difficulty in figuring primary mirrors. A 20 to 30 inch *f*/4 mirror is relatively standard these days. An f/3 mirror in this size range is difficult to figure, but still doable by the best opticians, but one pays for it in cost, delivery time, surface accuracy, and tight collimation tolerances. If high optical performance is a requirement, then mirrors faster than f/3 should be avoided, thus setting a lower bound on the focal length of larger aperture mirrors in demanding situations.

Opinions vary as to an upper bound to focal lengths. Eyepiece access and observer safety are obvious issues. Also, as the OTA gets longer, its resistance to bending decreases by approximately the cube of its length. Thus, to retain stiffness, the structural members must grow in cross sectional area, which causes wind loading to increase. Furthermore, longer OTAs give these wind gusts an increased lever arm. Countering highly leveraged wind gusts with an active control system—most likely a must with any large imaging Newtonians out in the open—requires high torque drive motors. In general, shorter telescopes will always perform better in the wind than longer ones.

Higher torque motors won't solve all one's problems, however. Instruments at the Newtonian focus provide added leverage for movements of the entire telescope with respect to the ground, as well as motion and vibration at the mechanical interface between the altitude axis and the OTA. Thus as apertures increase, one should increasingly consider folding the optical path to bring the instruments nearer the intersection of the telescope's axes, a design change that also reduces the moment of inertia and hence increases the natural frequency.

If one is willing to settle for significantly longer effective focal lengths, classical or Ritchey Chrétien (RC) Cassegrain or corrected Dall Kirkham (CDK) optics can be used along with a tertiary flat to bring the optical path out on one side. The corrected Dall Kirkham optics with a spherical secondary mirror and spherical corrector lenses are not only economical to produce, but the spherical secondary, compared to Cassegrain or RC systems, is less sensitive to misalignment.

Although Newtonian telescopes with parabolic primaries provide acceptable visual images over a small field, coma is objectionable for even small CCD chips and is unacceptable for large format chips. For example, a 20 inch *f*/4 Newtonian with an uncorrected parabolic primary has 50 micron RMS diameter comatic spots 10 mm off axis. Fortunately, off-the-shelf Wynne coma correctors are available for parabolic systems, such as the two inch Baader MPCC and Televue Paracorr correctors, and the three and four inch Keller correctors.

Thin, monolithic, annealed Pyrex mirrors supported by whiffletrees have worked well for apertures up to 32 inches and beyond. However, other types of mirrors may weigh less, cool faster, or be easier to support. These include cast mirrors, cellular mirrors (Wangsness Optics), and fused sandwich mirrors (Hubble Optics). Furthermore, although parabolic primaries coupled with off-the-shelf Wynne coma correctors provide good images, a hyperbolic primary—when coupled with a matching Ross-type corrector—can provide excellent images with significant cost and mechanical advantages.

Figure 1: Tong Liu and a 20 inch Hubble-Optics fused sandwich mirror. Because cooling air can circulate between the relatively thin front and back plates, the mirror cools up to ten times faster than a solid mirror with an equivalent thickness. Hubble Optics has produced about 1,000 mirrors, with some 200 being sandwich mirrors.

Back Focus for Autoguiders

Long-exposure spectroscopy, astrophotography, and time-series photometry can all benefit from the precise tracking provided by autoguiders. A small pick-off prism or mirror on the periphery of the field directs the light from a guide star to a small autoguide camera or to an auxiliary CCD chip built into the main camera itself, as in many SBIG cameras. Deviations of the guide star from its desired position are sensed and appropriate corrections are provided by the telescope's control system or by a small, refractive "deviator plate" (SBIG and Starlight Xpress) tilted in two dimensions. Because deviator plates have low inertia, they can be moved much more rapidly than entire telescopes. This allows them to correct for mount errors and, perhaps, atmospheric turbulence if a sufficiently bright guide star is available.

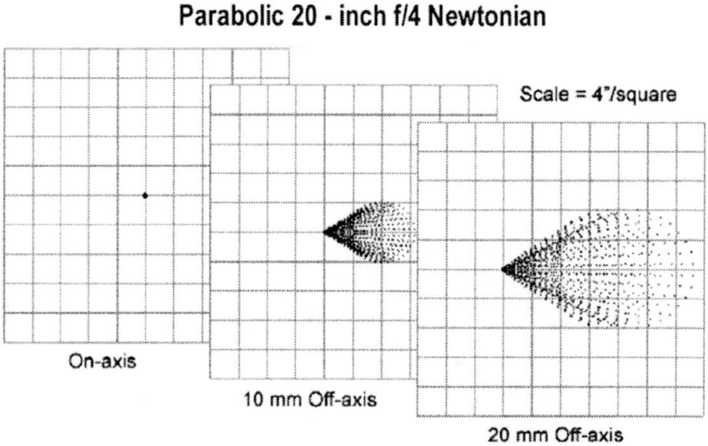

Figure 2: Ray trace of a standard 20 inch *f*/4 parabolic mirror without any corrective optics. Each square represents 4 arc seconds.

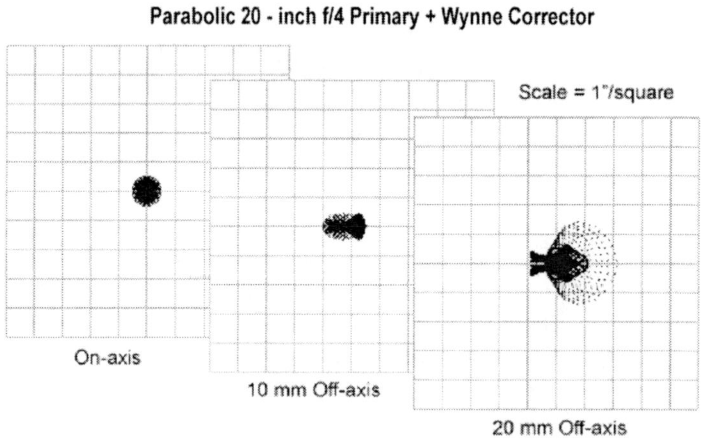

Figure 3: Ray trace of a parabolic 20 inch *f*/4 primary mirror with a three-element Wynne-type corrector over a flat field. The corrector is 6″ long, 4″ in diameter, and has a 95 mm back focal length. It does not vignette a 20 mm diameter image circle. Note that each square now represents only 1 arc second. The simulated wavelengths are 400 nm, 550 nm, and 800 nm.

However, there is a problem with Wynne correctors for parabolic mirrors mentioned previously. The back focus distance between the outer surface of their last corrector lens and the CCD camera's chip is, typically, only 55 mm. CCD cameras and their accompanying filter wheels take up most of this back focus, leaving insufficient room for an off-axis guider (OAG) pickoff prism or mirror, let alone a deviator plate. Fortunately, a Wynne coma corrector with greater back focus is available as a modified 4 inch Keller from Dream Telescopes (82 mm back focus).

Alternatively, there are four solutions compatible with 55 mm of back focus. First, if exposure times are kept short, autoguiding may not even be required. Images can be aligned and stacked using software such as RegiStax. Furthermore, control systems can be designed that track stars with considerable precision for long periods of time. Control systems can even be designed to sense and counter wind gusts, as will be described below. Thus an autoguider may not be required for longer exposures.

Second, one can employ an SBIG camera with an in-camera OAG CCD chip that fits within a 55 mm back focus. This works well for many applications, although one often needs to rotate the camera to find a suitably bright guide star—a task easily accomplished by alt-az telescopes equipped with instrument rotators. Such autoguider chips may not work well for narrow-band astrophotography, however, as there may be insufficient photons passing through a narrow filter for guiding.

Third, one can use a low profile camera such as the SBIG ST-402 with its built-in filters. With the ST-402 there is enough back focus within the 55 mm space for an OAG pickoff. Similarly, astrophotographers can use a color camera, such as those from Starlight Express, where the "filters" have been incorporated into the CCD chip itself.

Finally, one can forego an OAG and, instead, autoguide with a camera on a separate guide scope sturdily mounted to minimize differential flexure. Although the autoguider's field will rotate and the optical axes of the two telescopes might not coincide, these difficulties could be handled in software.

High Natural Frequency Mechanical Structures

Portable and roll-off-roof alt-az research telescopes must, obviously, be designed to operate out in the open. A good share of small telescope research involves a time series "movie" that is follow-up to "snapshot" observations made with larger survey telescopes. Thus science alt-az telescopes are workhorses. Observing cannot be suspended with the first significant breeze, even though instruments and baffles are exposed to wind gusts. Furthermore, a Newtonian configuration gives these gusts a long lever arm. One good, although expensive, solution is to place telescopes under permanent domes, but there are other solutions. A low aerodynamic cross section at the top ends of Newtonian telescopes, for instance, can help reduce the effects of gusts, as can an open truss design, although they are more susceptible to stray light.

Al Kelly points the apex of a V-shaped wind screen upwind to deflect gusts around his portable imaging alt-az telescope. Placing portable telescopes inside portable observatories reduces the effects of gusts, not to mention stray light. Commercially manufactured portable observatories are available from AstroGizmos, Kendricks, and Observa-Tech. Although portable observatories require transport and setup, they are low in cost and provide significant protection from both wind and light.

Whether operating out in the open or under a dome, how can the telescopes themselves be designed to be less susceptible to wind gusts? In the past, when control systems were mere clock drives, the immense rotational inertia of massive mounts resisted wind gusts. That approach still works but is no longer necessary. Today, modern mountaintop telescopes with their high resolution encoders and backlash free drives counter wind gusts with servo loops closed through high torque motors. When a wind gust pushes on the telescope, encoders sense this motion and servo motors generate appropriate counterbalancing torques constantly revised on sub-millisecond time scales. Thus active control systems provide "electronic stiffness," obviating the need for "inertial stiffness." Alt-az telescopes with active wind-countering control systems can have much lighter weight structures than would otherwise be possible. Besides costing less, such lightweight structures also have lower moments of inertia, allowing the use of lower torque, and less expensive drive motors. Furthermore, lighter structures deflect less due to the weight of the structures themselves.

Any system that is excited by a broadband energy source, such as wind gusts, tends to concentrate energy at its resonant frequency. Most of the energy transferred to telescopes by wind gusts is below 10 Hertz, therefore modern telescopes strive to have their lowest resonance (their natural frequency) greater than 10 Hertz. This allows control systems to directly counter wind gusts without encountering destabilizing resonances. The advantage of even higher natural frequencies is that control system bandwidths can be increased so telescopes settle faster and track with even smaller following errors.

A small mass suspended from a stiff spring will vibrate at a high frequency. Similarly, a low inertia (lightweight) stiff telescope will have a high natural frequency. Unlike yesteryear's passive, low natural frequency telescopes where high mass was an asset, now just the opposite, low mass, is a virtue with today's active, high natural frequency, gust countering telescopes. Besides being stiff and lightweight, modern alt-az telescopes keep their rotational inertias low by placing their OTA centers of gravity coincident with the intersection of the telescope axes, not offset and counterweighted as in some other telescope configurations.

Most materials are strongest in compression and tension and are weakest in bending and torsion. Truss geometries, which avoid bending and torsion, have high natural frequencies, while cantilevers have low natural frequencies. Although Dobs, especially those with truss tubes, use geometry to good effect, even the best birch plywood is not as stiff and is somewhat heavy compared to the advanced composite materials used in aircraft and spacecraft. Thus it is not surprising that lightweight aerospace materials which are, pound for pound, stiffer than steel, are ideal for use in smaller alt-az telescopes. The relatively small amount of such expensive material does not preclude its use.

Modern aerospace materials include carbon-fiber laminates as well as foam and honeycomb core panels. Geometric shapes formed from aluminum alloy extrusions can also be used to good effect. Honeycomb panels—which act as thousands of miniature I-beams running in two directions at once—are incredibly resistant to bending. Carbon fiber laminates, besides being one of the stiffest known materials for their weight, also have a near zero coefficient of thermal expansion, a property that can be used to good effect in truss tubes to reduce focus shift due to temperature change.

Precision Control Systems and Drives

Alt-az telescopes require constantly-varying motions in three axes (altitude, azimuth, and field rotation). Typically, a dedicated low-level control microcomputer, such as the one in Sidereal Technology's control system, adjusts these three motions at millisecond intervals. A PC provides both mid-level servo control and high-level supervision via a telescope control program such as Maxim DL, TheSky, or MPO Connections. Compatibility between the PC's servo and telescope control software can be assured if both adhere to the Astronomical Common Object Model (ASCOM) interface standard.

To actively counter wind gusts, not only must telescope structures be very stiff, with natural frequencies above 10 Hertz, but their drive systems must be both stiff and backlash free. Changes in torque and position must take effect within a few milliseconds. Encoders with sufficiently high resolution on both telescope axes allow direct closure of the altitude and azimuth servo loops during tracking. This enables control systems to actively counter wind gusts. The use of high resolution encoders also obviates the effects of friction roller slippage or gear errors in drive trains.

The cost of high resolution encoders has dropped dramatically in recent years. Top-of-the-line "Mercury II" encoders from Micro-E Systems, with amazing resolution, cost well under $2,000. Heidenhain recently (2010) came out with very high resolution encoder (25 bit/35 million ticks/revolution) for less than $1000. Renishaw and Micro E Systems produce encoder stick-on tapes that can be fastened to drive wheels and decoded with small read heads. Friction roller schemes have been successfully used to mechanically step up lower cost, medium resolution encoders such as the 230,000 counts per revolution encoder from Gurley which costs about $500.

Mountaintop alt-az telescopes often employ direct-drive torque motors. However, these large diameter torque motors—which are built right into the telescopes themselves—are very expensive, although Dave Rowe has recently devised a low cost direct drive motor (see later chapters in this volume). A lower cost approach is to mechanically step up the torque with a small friction roller pressed against a large diameter disk. A smaller, less expensive torque motor, such as the "pancake" DC motors on electric bicycles, can then be used.

If one is willing to forego the very considerable advantage of actively countering wind gusts, there are other, lower-cost control system options. The most basic just employs low-resolution encoders integral to gearhead servo motors. For slews, the controller simply commands the number of encoder "ticks" required to arrive at the desired destination. There is no independent check to see if the system actually arrived at its destination, which could be off if there was slippage in a friction-roller drive or error in a gear train.

Another approach utilizes low resolution encoders built into the servos, as above, but adds other low-resolution encoders on the telescope axes to position the telescope. Once positioned, typically with a precision of an arc minute or so, the rather coarse telescope axis encoders are ignored during tracking, and the highly geared-down servo motor encoders simply implement the constantly calculated, ever changing, drive rates.

Although the above scheme largely overcomes drive inaccuracies during initial positioning, these errors reappear once the system starts tracking without the benefit of telescope encoder "tick" updates which—with low-resolution encoders—occur too infrequently to provide smooth closed-loop servo tracking. Sidereal Technology recently implemented a "tick management" scheme that, as each low-resolution encoder tick occurs, "stamps" its location in its low-level microcomputer and then hands off this information to the control system software in the higher level PC which uses it to interpolate the telescope's actual position and adjust the tracking rate commands to the telescope's servo motors. Tick management has improved the tracking precision of lower-resolution encoders by a factor of ten on several systems.

Field Rotation

Image rotation in alt-az telescopes is countered by an instrument rotator, although if exposures are short enough, images can be rotated and stacked in software. Highly capable, Peltier-cooled CCD video cameras, such as MalinCam and StellaCam, have been used to good effect on alt-az telescopes without de-rotation. However, rotating and stacking consumes memory, increases reduction time, and reduces time on target with frequent image downloads. Furthermore, each sub image adds additional camera read noise which, cumulatively, can create problems for faint-object photometry, spectroscopy, or narrow-band astrophotography—all important applications of large aperture telescopes.

There are several stepper controlled instrument rotators available off the shelf (Meade, Optec, and RC Optical Systems). They are primarily used to rotate off-axis guiders (OAGs) to a suitably bright guide star, although the 3 inch Optec rotator has also been used for field de-rotation. RC Optical Systems has announced that a servo-controlled system specifically intended for alt-az imaging de-rotation will soon be available.

Although "dust donuts" and other near-field irregularities in filters and coma correctors rotate with the camera, and can thus be removed via normal flat fielding, far field irregularities and asymmetries vary with the position angle of the rotator and cannot be removed via flat fields taken from a single position angle. Thus telescope designs with non-symmetrical vignetting in their far fields should be avoided. Although mountaintop alt-az telescopes routinely obtain high precision photometric data with flats taken at a single position angle, some astronomers take a series of flats at different position angles and interpolate between them.

Ease of Construction and Affordability

In the aerospace industry, riveting and welding are being replaced with structural adhesives. The latest Boeing airliners are quite literally glued together, albeit not with any ordinary adhesive. The preferred gap-filling, stronger-than-steel, stick-to-anything structural adhesives are methacrylates and their ilk, with trade names such as Fusionbond (Heron Manufacturing). Fusionbond is easy to use and inexpensive. A 400 ml dual plunger is only $20. The two components are combined in a disposable static mixer when the plunger is depressed. Structural components need to be held in place while the adhesive sets—fully hardened in less than 24 hours. Full scale drawings, three feet wide and any length printed out at Kinko's, can be used as assembly templates, all rather similar to building model airplanes from balsa wood glued together with model plans as positioning templates.

Aerospace materials can be cut and shaped with ordinary woodworking tools and, as suggested above, fastened together with adhesives. Thus "aerospace" telescopes do not require machining or welding capabilities and can be built by students and amateurs, a source of "free" labor than can significantly reduce telescope costs.

Conclusion

We envision, just a few years from now, a virtual army of lightweight alt-az large-aperture telescopes. Built by amateurs, students, and commercial firms, they will be widely used for research, astrophotography, and education. Their superior light grasp, wide sharp fields, and precise go-to and tracking control will please scientists and delight both astrophotographers and visual observers. Their active, gust countering control systems will facilitate their operation in the field or in roll-off-roof observatories. Narrow band astrophotography, as well as the astrometry of faint asteroids, precision follow-up photometry of the many exciting objects now being uncovered by robotic surveys, and long-term spectroscopic studies of variable stars, will all benefit from the raw photon-gathering power of this revolutionary new breed of telescopes.

Chapter 8

An 18 Inch Direct Drive Alt-Az Telescope

Russ Genet[1], Dave Rowe[2], Dan Gray[3], Tong Liu[4], John Ridgely[1], Howard Banich[5], Cary Chelborad[6], Allan Keller[7], Billy Albertson[1], Wilson Chiu[1], Michelle Kirkup[1], Drew Murphy[1], Josh Schmidt[1], Matt Swanson[1], and Rob Urban[1]

[1]California Polytechnic State University, San Luis Obispo; [2]Sierra Monolithics; [3]Sidereal Technology; [4]Hubble Optics, Hong Kong, China; [5]Nike, Portland, OR; [6]Optical Structures; [7]PlaneWave Instruments

Introduction

Members of the Alt-Az Initiative group decided, in early January 2008, to develop and evaluate some of their ideas by designing and building an 18 inch Newtonian telescope. In this paper we provide a summary description of the design features of this telescope, beginning with the direct drive motors and control electronics.

Figure 1: Alt-az group members at the 18 inch Critical Design Review at California Polytechnic State University in mid March 2008.

Direct Drive Motors

The objective of direct drive motors is to dispense entirely with gears, belts, shafts, friction drives, or any other mechanical transmission, and replace them with large-diameter, high-torque stators and rotors directly mounted to the telescope itself. The telescope's bearings become the motor's bearings, while the telescope's structure becomes the motor's "frame," hence the term "frameless motor." Positional feedback is provided from high resolution encoders mounted directly on each telescope's axis, and a servo controller closes the loop around the encoders.

The primary advantage of direct drive is the elimination of the mechanical transmission and the compliance, backlash, periodic error, and other undesirable characteristics associated with them. Properly implemented, direct drive motors in conjunction with directly-coupled high resolution encoders and a stiff structure can, at least in theory, provide the stiffest, most responsive telescope control possible.

As suggested by Beley (2003) in his classic book, *The Design and Construction of Large Optical Telescopes*, "Direct drive motors are structurally simple and have no moving parts. They are intrinsically free from friction and stick-slip effects and need no maintenance."

Beley then points out that the disadvantage of direct drives is their relatively high cost. This is why they are not usually incorporated in smaller telescopes. Rowe, however, has circumvented this difficulty by designing, for the Alt-Az Initiative, a direct drive motor that uses readily available parts and materials (costing about $200) that is amazingly easy to construct.

Most currently available direct drive motors are radial flux motors with permanent magnets mounted on an iron cylinder forming the rotor, and radial coils wound over iron cores situated around the periphery forming the stator. While such a radial flux, direct drive motor is energy efficient, it is difficult and expensive to construct using the common techniques in current practice.

Rowe realized that power efficiency was not a critically important requirement for smaller telescopes because they generally move slowly (only infrequently slewing at higher speeds) and that an axial flux motor as opposed to radial flux motor would be much easier to build and integrate into the telescope, albeit with lower overall efficiency.

The motor is quite simple to build. For the rotor, 4N rare earth permanent magnets are evenly spaced and epoxied onto a soft iron annulus, arranged with alternating poles. (These magnets are very powerful, and one must not let two of them get too close together as they can easily crush a finger.) The stator consists of 3N coils epoxied to a non-ferrous aluminum or plastic annulus of the same size. The coils are wired in three phases. On the first test motor, the annuli had a 20 inch OD and an 18 inch ID. The rotor ring was made of soft steel while the stator ring was made of epoxy and plywood. The prototype motor achieved a torque constant of more than 10 N-m/amp during bench testing.

For the 18 inch telescope, the azimuth direct drive motor, encoder, and bearings have been integrated into a single, quite thin unit designed by Rowe. Two 18 inch diameter cast aluminum disks have been machined to provide V-grove bearing tracks on their periphery, support for the magnets and coils of the direct drive motors (inward from the bearing groves), and space toward the center for an 8 inch diameter Renishaw tape encoder disk and read head.

Figure 2: In the foreground is a complete azimuth assembly, and in the background a test coil assembly.

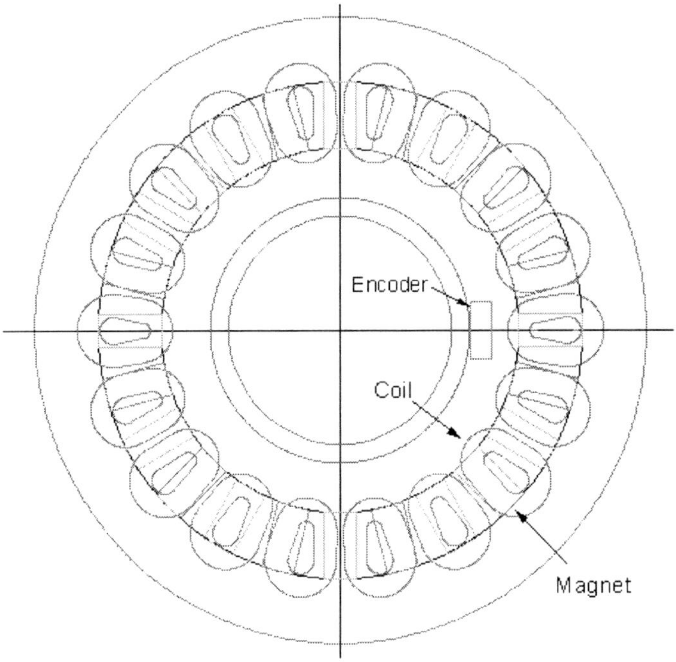

Figure 3: A simplified plan view of the direct drive motor, encoder and bearing assembly used for the azimuth axis.

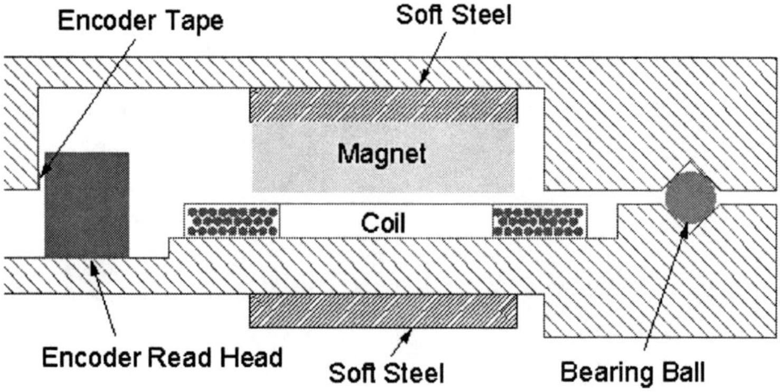

Figure 4: A simplified cross section of the direct drive motor, encoder, and bearing assembly for the azimuth axis.

Control Electronics

Direct drive motors can be operated in two quite different modes. By far the most common mode is a permanent magnet, brushless DC motor. Such brushless motors are widely used in air conditioners, refrigerators, and electric bicycles, not to mention new, top-end washing machines and driers. In these motors, commutating brushes have been replaced with Hall effect sensors and electronic switches (typically MOSFETs). These motors have no brushes to wear out.

Less frequently, when exact positioning or very high precision speed control is required, direct drive motors are operated inside a servo control loop. In this mode, the angular position of the motor is sensed by a high resolution encoder. This position information is used to provide, via trigonometric functions and precision analog-to-digital (A/D) converters, precise currents for the motor's multiphase windings. In this mode, the motor is effectively generating torque, commanded by the closed loop servo algorithm. This closed-loop control system keeps the motor essentially "locked" to the exact desired position whether the motor is rotating or standing still.

The control electronics for direct drive motors operated in the brushless DC motor mode are quite simple, and only cost a few dollars. The control electronics for direct drive, brushless DC motors operated in closed-loop feedback mode are much more complex, requiring high resolution encoders, a computer to run the feedback algorithms, high resolution analog-to-digital (A/D) converters, and high current switches.

Gray, at Sidereal Technology, has developed the electronics and firmware that will control two direct drive servo systems using three-phase brushless DC motors and high resolution encoders. In addition, this new control system will operate brushed DC servo motors and stepper motors. With the ability to read encoders at a 6 MHz rate, this new "Turbo Drive" controller accommodates both high resolution encoders and fast slew rates. A 5x7 inch, five layered board contains the complete telescope control system.

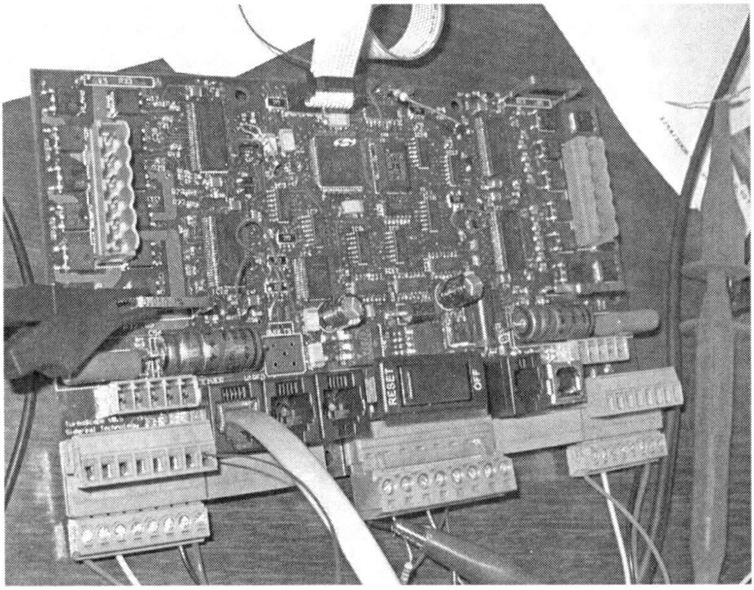

Figure 5: Sidereal Technology Turbo Drive control system.

OTA and Fork Structures

The telescope's primary structural components—the optical tube assembly (OTA) and fork—were designed and built by two student teams at California Polytechnic State University in San Luis Obispo. The OTA team consisted of Chiu, Kirkup, Schmitt, and Swanson, while the fork team consisted of Albertson, Murphy, and Urban. The OTA and fork team efforts were senior projects in Professor James Widmann's mechanical engineering class. The teams were advised by Professors Ridgely and Genet.

Both simplified mathematical analysis and computerized finite element analysis (FEA) were used to design these stiff structures. It was specified that the lowest resonant frequency of the mechanical system not be below 10 Hz, allowing the control system to effectively counter wind disturbances up to roughly this frequency without the control system's performance being degraded by interactions with structural resonances.

These goals were achieved in the design. In simulation by FEA, the lowest resonant frequency of the OTA (loaded with its optics and instruments) is approximately 40 Hz, while the lowest resonant frequency of the fork (loaded with the OTA with its optics and instruments) is greater than 10 Hz. Carbon fiber composites (CFCs) were utilized in the OTA's truss tubes and top end, and also in most of the fork to facilitate a stiff, lightweight structure.

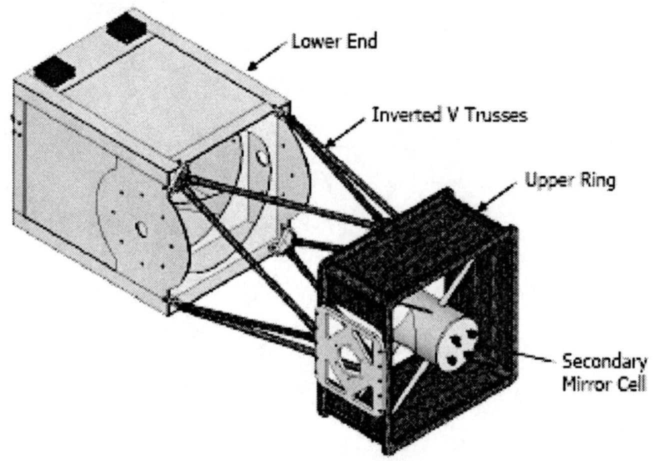

Figure 6: Isometric View of the OTA.

The OTA's square top end allows the spiders for the secondary assembly to be placed under considerable tension, helping to maintain stiffness in the entire top end. The square top also provides a flat surface for mounting the instrument rotator / focuser at the Newtonian focus.

The OTA's square lower section is made from aluminum alloy. The combination of aluminum and carbon fiber along the optical axis compensates, fairly closely, for temperature induced changes in the focal length of the primary mirror.

The fork was designed to directly transition the load from the two altitude bearings to the large diameter (18 inch) azimuth bearing. The fork features a polyurethane foam core, CFC exterior, reinforcing internal steel plates, and a steel azimuth bearing interface cylinder. A compartment in the bottom of the fork provides generous space for the telescope's control electronics, instrument power supplies, etc.

Figure 7: OTA team, left to right: Michelle Kirkup, Wilson Chiu, Matt Swanson, and Josh Schmitt.

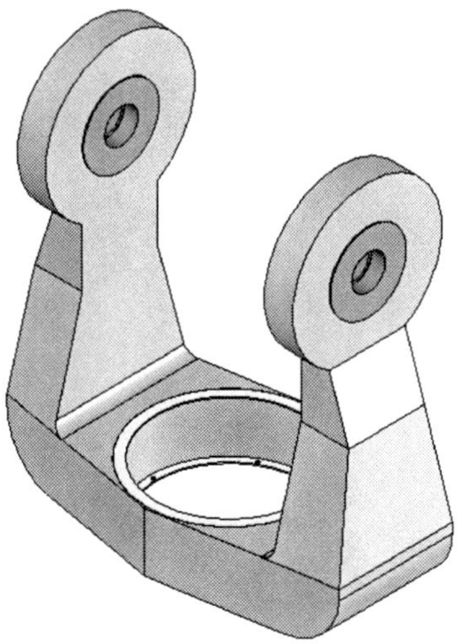

Figure 8: The carbon fiber fork with steel center ring.

Figure 9: Fork team, left to right: Drew Murphy, Billy Albertson, and Rob Urban.

Optics and Optical Support

The optical layout, including the baffles, was designed by Rowe. The primary mirror, designed and fabricated by Liu at Hubble Optics, is an f/4 parabola 18 inches in diameter. Two relatively thin disks of plate glass are separated by round glass spacer disks, and the entire structure has been fused together.

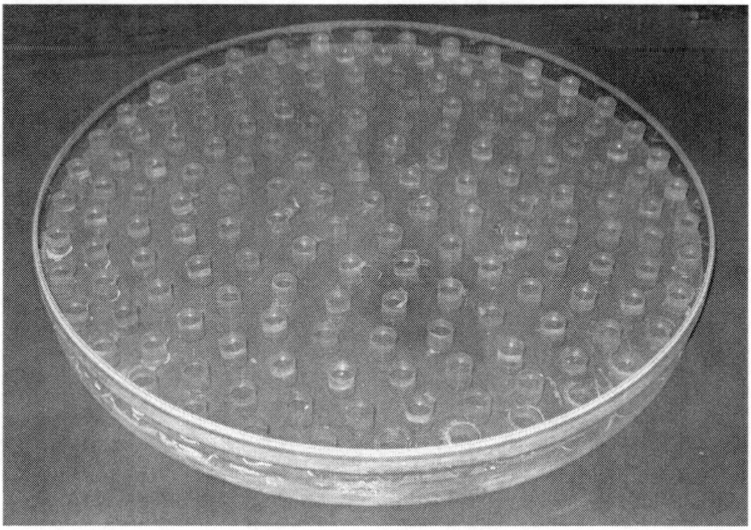

Figure 10: 18 inch *f*/4 Sandwich Mirror.

This sandwich structure has two advantages: (1) the two spaced out thin plates cool quickly as compared to an equivalent solid mirror, and (2) the sandwich mirror weighs less than a solid mirror of similar stiffness. This 18 inch, f/4 mirror has a flat back and two layers of glass separated by spacers.

While the coefficient of thermal expansion (CTE) of plate glass is higher than that of borosilicate, the changing primary focal length has been almost exactly compensated for by an appropriate combination of carbon fiber trusses and aluminum lower-end structure.

The 18 point primary mirror support was designed and fabricated by Banich. The parts were cut out by a numerically controlled water jet machine.

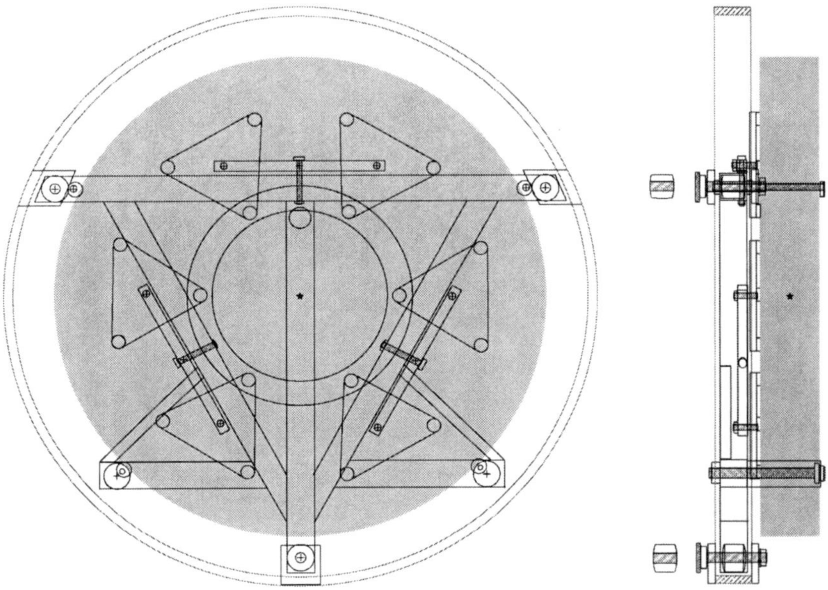

Figure 11: The 18 inch primary mirror cell, back and side views.

The secondary mirror, being designed and fabricated by Chelborad at Optical Structures, is a very lightweight, cast borosilicate design. The 5.5 inch minor axis provides the generous backfocus needed by the instrumental payload, which can include an instrument rotator, deviator plate, off-axis guider, filters, and CCD camera.

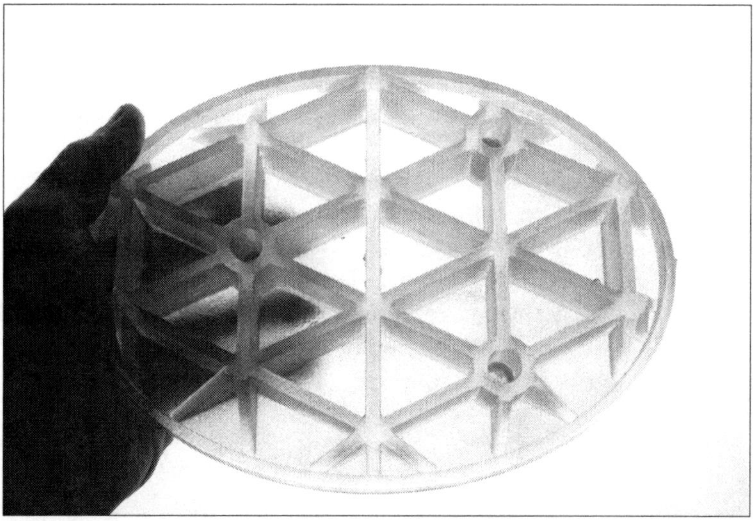

Figure 12: The lightweight cast borosilicate secondary mirror made by Optical Structures.

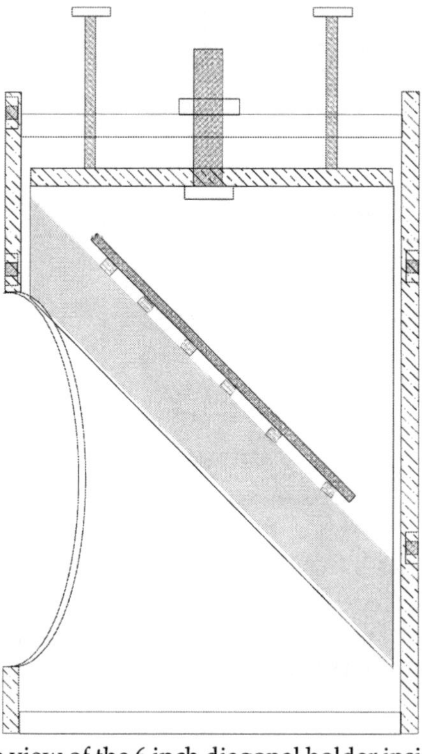

Figure 13: Side view of the 6 inch diagonal holder inside its baffle and spider hub.

The secondary optical assembly which surrounds and supports the secondary mirror was designed and fabricated by Banich. The secondary mirror support structure and secondary baffle have been integrated into a single structure designed to transmit forces from opposing, tightly tensioned spiders.

The center of gravity (CG) of the secondary optical assembly has been located to minimize torsional loads on the spider vanes. The spider vanes are tensioned to the corners of the square OTA top end (the four corners at the top of the top end "cage" and four at the bottom).

Conclusion

The combination of direct drive motors operated inside a servo loop closed around high resolution encoders, lightweight mirrors, and an FEA designed CFC/aluminum structure can provide an unusually light and stiff telescope that should perform well under gusty wind conditions out in the open or in a roll-off-roof observatory.

Acknowledgements

We thank the Mechanical Engineering Department at California Polytechnic State University for hosting the Critical Design Review for the 18 inch telescope in mid March 2008.

All the authors of this paper have generously contributed their time and funds to this project. We are pleased to acknowledge component contributions by the following firms: Hubble Optics (primary mirror), Optical Structures (secondary mirror), PlaneWave Instruments (truss tubes and clamps), and Sidereal Technology (control electronics).

Finally, we are pleased to acknowledge the financial donations by the following firms for the materials and components needed to fabricate the OTA and fork structures by the two California Polytechnic State University student teams. These donors included Dan Gray (Technical Marine Services), Charles Harper (Sierra Monolithics), Richard Hedrick (PlaneWave Instruments), Richard Kay (Impact Bearings), Ken Kissell (Kissell Associates), and George Roberts (Inca Inc.).

Reference

Beley, Pierre Y., ed. 2003. The Design and Construction of Large Optical Telescopes. New York: Springer.

Chapter 9

A Corrected Dall Kirkham Nasmyth-Focus Telescope

Dave Rowe[1], Rick Hedrick[2], Dan Gray[3], Joe Haberman[2], Bob Peasley[2], Allan Keller[2], Kevin Iott[2], and Russ Genet[4]

[1]Sierra Monolithics; [2]PlaneWave Instruments; [3]Sidereal Technology; [4]California Polytechnic State University

Introduction

This work describes the design and development of a 0.7 meter alt-az telescope, known as the CDK700, which bridges the gap between large, high-performance mountaintop observatories and the modest, but affordable, telescopes in use today by amateurs and small universities. It takes the high-performance, but previously high-cost, technology from its larger brothers, and, without compromise, brings it within the reach of a large group of users. In doing so, it sets a clear and viable course for future generations of telescopes in the range of 1 to 1.5 meters and provides an example of modern telescope technology that is well within the capabilities of the advanced amateur telescope maker.

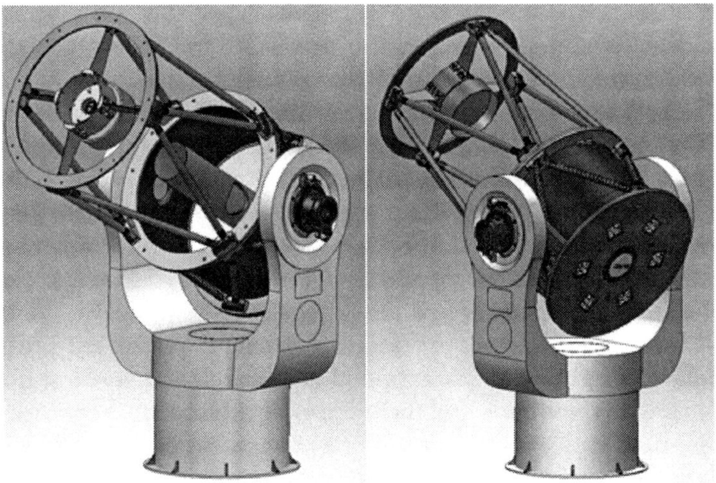

Figure 1: Two views of the CDK700 telescope CAD design. The combination of an alt-az mount, lightweight structure, and dual Nasmyth foci provide a compact, transportable, low cost, and easy to use research-grade telescope.

The principle design objectives for this telescope were:

- Wide-field, photo-visual optics for science, imaging, and visual observing.
- Ease of manufacture, for an excellent price-performance ratio.
- High-accuracy, full-sky pointing, and tracking for research and imaging.
- Field de-rotation with sufficient accuracy to allow long-exposure, wide-field imaging.
- Very high structural stiffness to resist wind pressure and gravitational loading.
- A high precision, wide-bandwidth motion control system to counter disturbances.
- Flexible, maintainable, computer-standards-compliant control software.
- A transportable and easy to install design.

Optical Design

The optical tube assembly (OTA) of any telescope is one of the principal drivers of the overall telescope design. The diameter, length, and moment of inertia of the OTA force many of the mechanical requirements of the mount, and therefore its weight and cost. The OTA diameter is determined by the aperture of the telescope, but the length and moment of inertia are determined by the optical design.

The first and most important decision was the selection of the aperture diameter. We felt that a price and performance gap existed in fully installed, high capability telescopes in the range of 0.7 to 1.5 meters. High performance professional telescopes in this aperture range are very expensive per unit of aperture area; amateur-built telescopes are rare and usually lack the necessary mechanical systems and positioning and tracking accuracy to allow long exposure imaging. We felt that the 0.7 meter aperture was an ideal size for many reasons, including the availability of mirror substrates and as an excellent starting point for developing the methods and techniques necessary for larger apertures.

The first optical configuration to be considered was a Newtonian parabolic configuration with a Wynne corrector. This combination has worked well for many telescopes. Low cost parabolic correctors are available from TeleVue, Baader, and Lumicon, and larger diameter Wynne correctors are available from Keller. This tried-and-true approach has much to commend it, and it is entirely appropriate for smaller apertures and lightweight instruments. However, most of these correctors have limited backfocus, which, in many configurations, precludes the use of both an off-axis guider and a filter carousel or other backfocus-hungry instrumental combinations.

We then considered a Newtonian configuration with a hyperbolic primary mirror and a two- or three-element corrector. Unlike the Wynne design, correctors for hyperbolic mirrors are specific to the focal length of the primary. Nonetheless, the corrected hyperbolic system can simultaneously provide generous backfocus and large, well corrected flat fields. In addition, the corrector lenses for these systems can be manufactured in large diameters at reasonable cost. Their generous backfocus facilitates the incorporation of a deviator plate, off-axis guider, filters, and other instrumentation as needed. The large field-of-view allows, simultaneously, both large format cameras and access to a wide selection of guide stars, necessary if one desires a bright guide star for high frequency deviator plate corrections. This is a good configuration for 0.5 meter class telescopes.

Figure2:

The Nasmyth optical configuration provides a convenient, fixed-height eyepiece for visual observations, as well as an instrument location at the altitude axis that does not require rebalancing, even with instruments of widely differing weights.

As the aperture increases, access to a Newtonian focus via increasingly tall ladders becomes unattractive even for experienced visual observers, not to mention educators with young students. Furthermore, placing large scientific instruments at the Newtonian focus increases the moment of inertia of the OTA and lowers its natural resonant frequency. Such instrument placement also adds to the wind cross-section at its most leveraged point.

For these reasons, we chose for the 0.7 meter telescope's basic optical configuration a Nasmyth focus with a tertiary mirror. Eyepieces and instruments are at the pivot point of the altitude axis, so even with large changes in instrument loads, no counterweights are required and the instrument's moment of inertia is minimized. A tertiary mirror, admittedly a significant complication, is required because a compact, transportable alt-az configuration is not well suited for access at the Cassegrain focus in this aperture range.

The Nasmyth focus is inappropriate for small aperture telescopes because the native focal plane will be located too close to the ground for convenient use. However, as the aperture is increased, a comfortable sitting height is eventually reached. This minimum natural height is achieved at approximately 0.7 meters of aperture, although smaller telescopes could use piers or other methods to raise the height of the focal plane. At larger apertures, such as 1.0 meters, there are comfortable standing-height eyepiece positions.

Three optical systems met the criteria for our Nasmyth-focus telescope: the classical Cassegrain, the Ritchey Chrétien, and the corrected Dall Kirkham Cassegrain. We chose the corrected Dall Kirkham (CDK) for its wide, well corrected flat field, generous backfocus, and its insensitivity to secondary mirror lateral misalignment (the secondary mirror is spherical). PlaneWave Instruments manufactures an OTA with CDK optics designed by one of the authors (Rowe); this design has proven to work well for demanding, wide field astrophotography.

CDK Optical Design

The optical layout of the 0.7 meter CDK is shown in the accompanying diagram. The tertiary mirror can rotate 180 degrees, making both Nasmyth focus locations available. We selected this optical configuration for the following reasons:

- It has a wide, flat, well-corrected field over a very broad wavelength range.
- The spherical secondary has no optical axis, making the telescope easier to collimate than other Cassegrain configurations. In addition, a sphere is easier and much less expensive to figure than the convex hyperboloid needed in other Cassegrain designs.
- The under-corrected primary is easier and less expensive to figure than its counterparts.

The CDK can be designed with a large back focus distance between the corrector and the focal plane, making it ideal for a Nasmyth configuration where the light must pass through the altitude bearings and field de-rotator.

Two very common Schott glasses, BK7 and SK11, are used in the two-element corrector. The prolate ellipsoidal primary mirror is 64% corrected from a sphere to a paraboloid. At *f*/6.45, the telescope is quite fast for a Cassegrain configuration, providing the user with excellent *etendue* (the product of the collecting area and field solid angle) when a large imaging sensor is used. This well corrected field supports the largest commercially available CCD sensors available today, and should continue to do so well into the future. The large, flat field is very useful for a variety of off-axis guider (OAG) configurations.

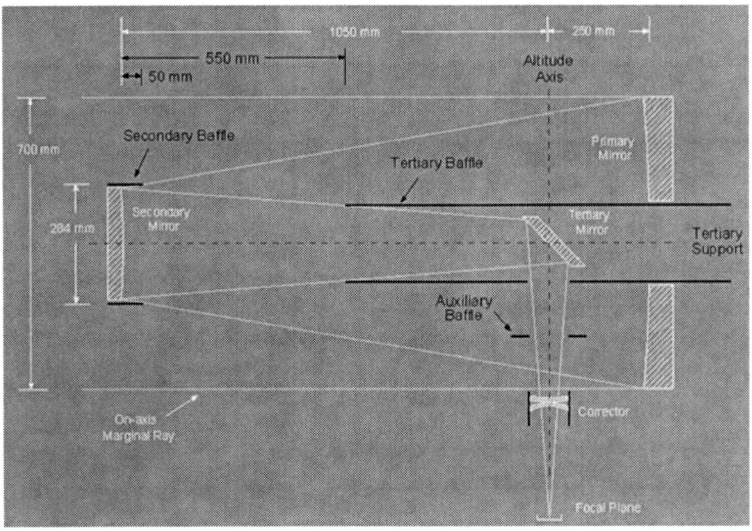

Figure 3: The optical layout of the 0.7 meter CDK telescope. The optical axis and altitude axis are coincident—often called the Nasmyth-focus configuration. The telescope is fully baffled over a 60 mm diameter image circle. The linear central obstruction is 41% (17% in area) and the effective focal ratio is f/6.45. The auxiliary disk baffle prevents undesired illumination of the focal plane from the opposite side of the telescope.

	Radius 1	Thickness	Radius 2	Aperture Diameter	Spacing
Primary	-4200 mm SC=-0.64			700 mm	1250 mm
Secondary Baffle				284 mm	50 mm
Secondary Mirror	-3000 mm SC=0			270 mm	550 mm
Tertiary Baffle				188 mm	500 mm
Tertiary Mirror	Flat			135 mm Minor Axis	372 mm
Lens 1 -- BK7	-1203 mm	12 mm	-356 mm	100 mm	1 mm
Lens2 -- SK11	143.3 mm	7 mm	-109.3 mm	100 mm	274.3 mm
Focal Surface	Flat				

Table 1: The optical prescription of the 0.7 meter CDK telescope. SC is the conic deformation constant. The primary is a prolate ellipsoid that is 64% corrected from a sphere to a paraboloid. The secondary mirror and all four lens surfaces are spherical. The secondary and tertiary baffle dimensions and locations are provided.

Clear Aperture	700 mm
Effective Focal Length	4516 mm
Effective Focal Ratio	f/6.45
Back Focus	274 mm
Central Obstruction	41%
Fully Baffled Field Diameter	60 mm
Polychromatic RMS Spot Diameter at Field Edge from Ray Trace (400 nm – 800 nm)	8 microns 0.36 arcsecond
Vignetting at Field Edge	20%

Table 2: The main performance characteristics of the CDK700 telescope.

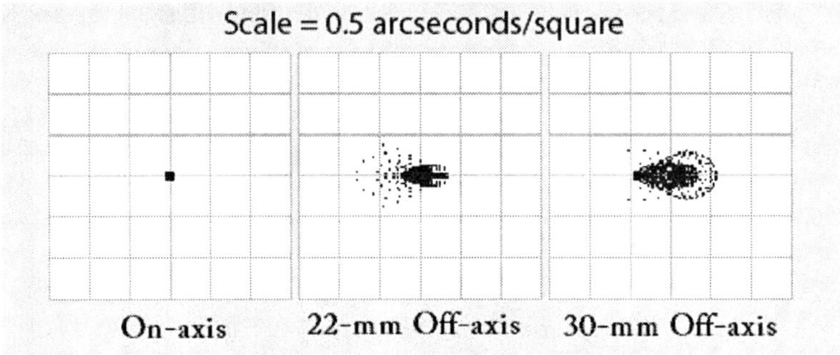

Figure 4: The polychromatic geometrical ray trace of the 0.7 meter CDK at three off-axis distances and three different wavelengths over a flat field. Each small grid square is 0.5 arc seconds. The wavelengths used in the ray trace are 400 nm, 550 nm, and 800 nm. The telescope is very well corrected over a broad band and wide, flat field, and will be seeing limited under almost all situations.

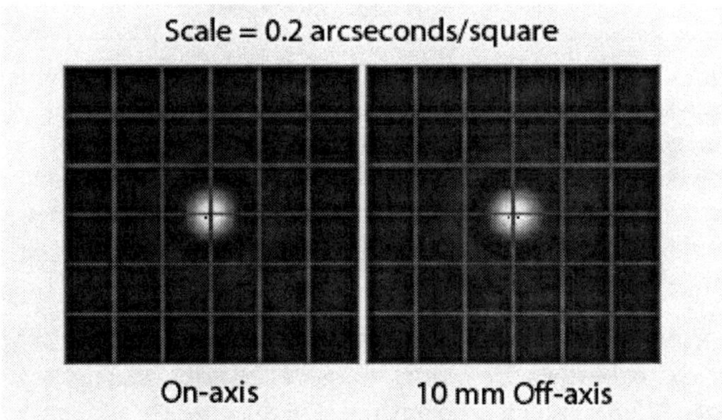

Figure 5: The diffraction performance of the CDK700 at 550 nm. Each small grid square is 0.2 arc seconds. The 41% central obstruction has been included in the simulation. The on-axis Strehl ratio is greater than 0.97. In other words, there is less than 3% degradation in the peak intensity of the full diffraction image versus the theoretical value. The Strehl ratio remains greater than 0.8 over a 20 mm field diameter, making it an excellent telescope for visual use.

Telescope Motion and Control

For small to intermediate telescopes operating within a dome or outside under mild conditions, simple open-loop stepper control systems—or their servo equivalent with low resolution encoders on the back of gearhead servomotors—are often adequate when used to drive telescopes through worm gears, belts, or friction drives. In these cases, pure telescope mass and the stiffness provided by the gearing can help resist the occasional wind gust. Even then, backlash and gear train compliance can easily ruin an imaging session under moderate wind conditions. It doesn't take much wind disturbance to cause significant trailing during an exposure.

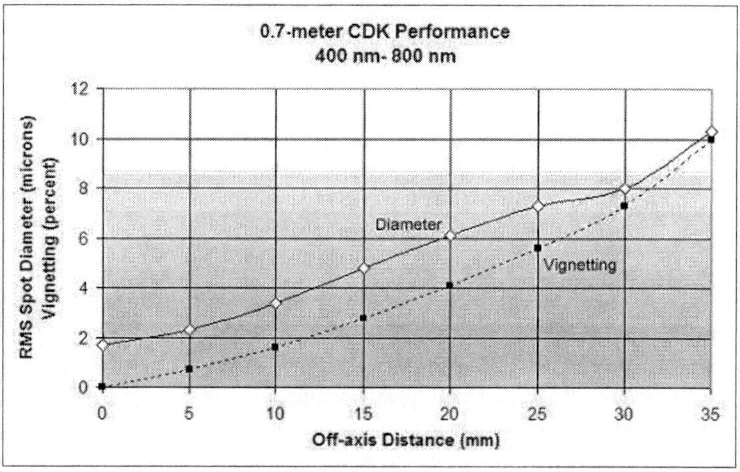

Figure 6: The overall performance of the telescope as a function of the off-axis distance. Vignetting is in percent and the polychromatic geometric RMS image diameter is given in microns. The performance of the telescope is excellent over a 70 mm circle, and remains usable over an 80 mm diameter. The simulated wavelength range is 400 to 800 nm. It should be noted that one arc second is 20 microns at the image plane for this telescope.

An easily transportable telescope, almost by definition, cannot fight external disturbances by being massive. It can, however, fight wind gusts by utilizing "electronic stiffness" provided by a fast servo control loop. Such lightweight telescopes, if they have stiff structures and low moments of inertia, allow wind gusts to be countered—one gust at a time—on millisecond time scales by the servos. This requires very stiff and low inertia OTAs and mounts, ultra-high resolution encoders on both telescope axes, stiff and accurate bearings, and responsive, stiff drives free of backlash and compliance.

Since gear backlash and compliance is effectively impossible to counter in a high bandwidth servo loop, one is left with two options: (1) one or two stages of friction reduction driven by a moderate-torque motor, or (2) a direct drive system where a high torque motor is mounted on each axis. It should be noted that both of these approaches have been used to good effect on large alt-az telescopes, and both should work well on smaller systems. It is, however, worth quoting Pierre Bely (2003) from his classic *The Design and Construction of Large Optical Telescopes:*

> In contrast to all other drive mechanisms, in which the force is concentrated on a pinion or wheel, direct drives distribute the thrust along the structure, thus minimizing localized deformation and maximizing the structural stiffness. This results in the highest drive stiffness possible.... Another key advantage of direct drives is their insensitivity to mechanical misalignment, dramatically reducing telescope installation time.

While challenging, we elected to use direct drive motors and ultra-high resolution encoders directly coupled to the axis and controlled by fast (wide bandwidth) servo electronics. We feel that the reduction in mechanical complexity, alignment, and maintenance afforded by direct drive is well worth the cost of the large diameter motors.

Since our telescope uses tightly coupled motors and encoders, the servo software becomes the heart of the telescope's motion control. We have chosen the current version of Sidereal Technology's control system. The software is ASCOM compliant, so it works well with high-level software such as Maxim DL and TheSky. This control system supports the operation of the ultra-high resolution encoders and AC synchronous brushless motors.

Direct Drive Motors

Large diameter, frameless, direct drive motors are available off-the-shelf. They are, however, somewhat expensive, typically costing several thousand dollars per motor. They are designed to handle large loads at high RPM with excellent efficiency. For those who wish to construct cost effective telescopes on a limited budget, these off-the-shelf motors are overkill for the very low speed, modest torque requirement motors on telescopes. Fortunately, the permanent magnets and coils required to construct a simple, axial flux direct drive motor are quite inexpensive.

The motors must supply sufficient torque at reasonable current to overcome wind gusts and static imbalance while accelerating the telescope axis at the highest desired rate. The Nasmyth-focus configuration completely eliminates altitude axis imbalance associated with instruments and eyepieces, easing the direct drive torque requirement.

Rotational Inertia	20 kg-m^2
Maximum Desired Acceleration	0.25 radians / sec^2
Torque Needed for Acceleration	5 N-m
Torque Needed to Overcome Wind Gusts	8 N-m
Torque Due to Imbalance	2 N-m
Torque to Overcome Static Bearing Friction	5 N-m
Peak Torque Needed from Motor, Including Margin	30 N-m

Table 3: Torque Requirements for the CDK700 telescope.

Motor drive electronics are relatively inexpensive for peak currents up to 11 amps, and telescope control systems can be powered by two 12 volt batteries in series or an inexpensive 24 volt power supply, so we optimized the direct drive motors for a minimum supply voltage of 20 volts. For more demanding applications or larger telescopes, the controller is designed to handle up to 55 volt supplies and 10 amps of current to each motor. Under the normal usage conditions of light winds, sidereal tracking, and good balance, the torque needed is very low and thus the DC current draw will also be low. This effectively eliminates the need for expensive, high-efficiency motors.

Radial-flux motors have the highest efficiency and motor constants, but they also require machined or stamped cores and machined rotors and stators. An alternative, lower cost (but somewhat lower efficiency and higher weight) approach is to employ an axial-flux configuration with a large diameter rotor and stator. The large available diameter is used to the designer's advantage, providing more torque per unit force from the coils. The CDK700 has ample room, so no penalty is incurred for a large-diameter motor.

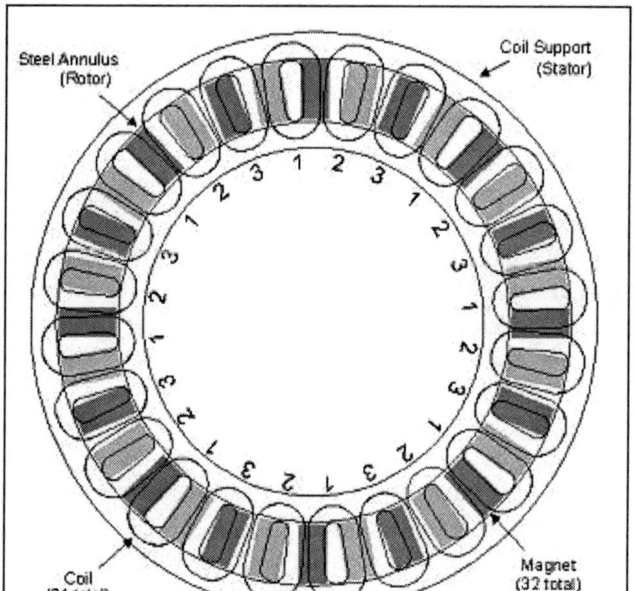

Figure 7: The plan view of the motor. The poles of the magnets alternate between North and South. The coils are connected in series in three phases, where the phases are given by the numbers. The outside diameter of the coil support is approximately 20 inches (0.5 meters). The NdFeB magnets are 0.5″ x 1.0″ x 2.0″.

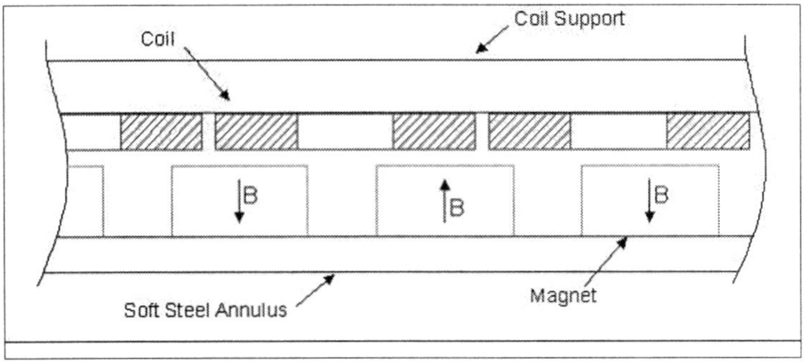

Figure 8: A portion of the cross section of the axial-flux motor. The coils are 0.3″ thick and the gap is 0.1″. For the highest flux through the coils, the gap should be kept as small as possible. The soft steel annulus provides a return path for the magnetic flux. In the version of the motor shown, the coil support is non-magnetic. Higher flux can be achieved if the coil support is also soft steel. This provides a significant pre-load on the bearing, which may or may not be desirable, but must be considered.

Magnets	(32) Type N42 NdFeB, 0.5" X 1.0" X 2.0"
Coils	(24) 22 AWG 100-turn, L=175 cm/coil
Average Magnetic Field	0.65 Tesla
Torque Constant	7 N-m/amp
Winding Resistance	1.8 ohms per phase
Minimum Motor Voltage	20 Volts
Maximum Current	5.5 Amps
Maximum Torque	37.5 N-m

Table 4: Direct drive motor performance parameters.

AC Synchronous Motor Driver

Most modern direct drive motors are three-phase brushless, with current supplied to the phases via solid state linear amplifiers or pulse width modulated (PWM) switches. Low resolution Hall-effect sensors often supply the position information needed by the controller to correctly phase these switches. Washing machines, air conditioning compressors, and many other low cost commercial appliances now use brushless DC motors. However, this simple motor control scheme is not adequate for our application.

For telescopes, the motors must be operated in an AC synchronous, servo-control torque mode. The position of the telescope is sensed with ultra-high-precision encoders on the two telescope axes, the position error is calculated, and the desired torque feedback is computed. The three phases of the motor are then energized based on this torque. One of us (Gray) has modified his Sidereal Technology control system to drive brushless AC synchronous motors.

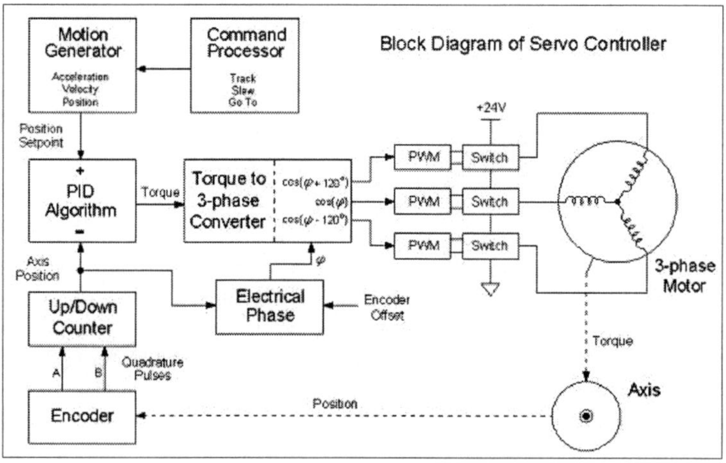

Figure 9: Block diagram of the servo controller. The three-phase, Y-connected motor applies torque directly to the axis. An ultra high resolution encoder creates quadrature incremental pulses to the controller, used to determine the axis position. The controller compares the desired axis position to the measured position, creating an error signal. This error is the input to a proportional, integral, derivative (PID) algorithm which calculates the torque to apply to the axis. Pulse width modulation (PWM) is used to control the current to the motor, a very high efficiency technique.

We have chosen to use 10 bit PWM current switches instead of linear amplifiers for reasons of cost, simplicity, and power efficiency. Although PWM drives can cause undesirable jitter at the telescope's axes if the resolution is too low, analysis and the measurement of a prototype drive have shown that 10 bit PWM is sufficient for this application, and supports servo bandwidths of at least 10 Hz.

The other difficulty with high current switching is the noise conducted on signal, power, and ground lines, and also the electromagnetic interference introduced into nearby sensitive components, such as encoder read heads or CCD cameras. We designed the motor driver board with edge control, so that the high frequency switching components are greatly attenuated before leaving the controller. In addition, the drive board uses excellent grounding and shielding to minimize conducted and radiated interference (EMI).

Encoders

Positioning errors due to wind gusts, bearing cogging, and other external torques are removed via the servo control system. The error is sensed by comparing the desired position with the actual position, and removed by the application of a counteracting torque. To do so, the telescope must first move very slightly, and this slight movement must be sensed before the correcting torque can be applied by the control system. One or two encoder resolution units (ticks) are sensed as the correction is being applied, and the typical RMS error at the axis is on the order of one tick. Thus, for excellent tracking accuracy and low jitter, an encoder resolution of approximately 0.1 arc seconds is required (about 12 million ticks/axis revolution).

We considered two basic approaches to obtaining the requisite encoder resolution. The first was to mechanically step up, via a large friction disk drive, a medium resolution encoder such as the Gurley 230,000 ticks/rotation device. One of us (Gray) has tried this encoder on his 14 inch alt-az telescope. One problem with the friction drive gearing, which is very difficult to resolve, is the periodic error induced by the run-out of the gearing and encoder. Other issues with this approach are accuracy and reliability (slipping and dirt) and the requirement for a large mechanical step up.

The second approach is to place high resolution encoders directly on the telescope's axes. A good encoder for this is the Mercury II made by MicroE Systems. It uses a high degree of interpolation to achieve the required resolution.

Somewhat lower in cost are striped tapes with their associated read heads, such as those made by Heidenhain and Renishaw. Adhesive-backed stainless steel tapes with closely spaced lines can be adhered to the outer edge of a machined disk and read by an optical read head. The disk can have a large diameter (up to 10 inches in our case), naturally providing a high angular resolution.

One difficulty of using a tape in azimuth is the break in the tape, since the telescope cannot maintain precise tracking when it passes over the gap. There are at least two solutions. One is to purchase a continuous tape pre-mounted on a ring (more expensive, and requires exact placement on a machined shoulder). The other is to allow the telescope's OTA to rotate in altitude 180 degrees instead of just 90 degrees, allowing the OTA to rotate from horizon-to-horizon, completely passing through the forks. With this capability and a simple software algorithm, the telescope can be pre-positioned such that any object can be tracked across the sky without encountering the azimuth break, thus avoiding the alt-az equivalent of a GEM's meridian flip.

For simplicity, greatest accuracy, and full flexibility, we chose to use a continuous ring encoder from Renishaw, having 100 nm resolution, a diameter of 200 mm, and an angular resolution of approximately 0.1 arc seconds.

The Azimuth Assembly

The azimuth motion assembly, comprised of the azimuth bearing, direct drive motor, and on-axis encoder, is fabricated as a single unit from a top and bottom plate machined from low-stress cast aluminum. The 20″ diameter V-groove bearing is lapped for smoothness and accuracy, and when loaded with the weight of the telescope, provides a bearing that is very stiff against forces that would cause tip, tilt, or radial motion. Stiffness in the radial direction is very important when using an on-axis encoder because motion in this direction is falsely sensed by the encoder as a rotation.

The integrated assembly is mechanically very stiff in the direction of rotation, with very low compliance between the motor and the encoder. This places the mechanical poles in the servo loop at a very high frequency, which in turn allows a high closed-loop servo bandwidth.

Figure 10: The azimuth assembly stator. Twenty-four coils are mounted to the aluminum frame and are wired into 3 phases in a series-parallel configuration. The machined V-groove azimuth bearing race has been lapped to make it very smooth and uniform.

The stator, which holds the 24 motor coils, is shown in Figure 10 (opposite page). The magnetic flux through the coils is returned via a steel ring that is placed on the other side of the stator. The coils are each wound on a non-magnetic core. The rotor, which holds the 32 magnets, is shown in Figure 11 (below). The magnets are mounted on a steel annulus that supplies the magnetic return path. Figure 12 (below) shows the Renishaw encoder ring and read head mounted to the azimuth assembly.

Figure 11: This photograph shows a section of the azimuth assembly rotor. The full rotor has 32 NdFeB magnets with alternating poles. The magnets are mounted onto a steel ring that provides a flux return path. The V-groove bearing race is filled with stainless steel bearing balls.

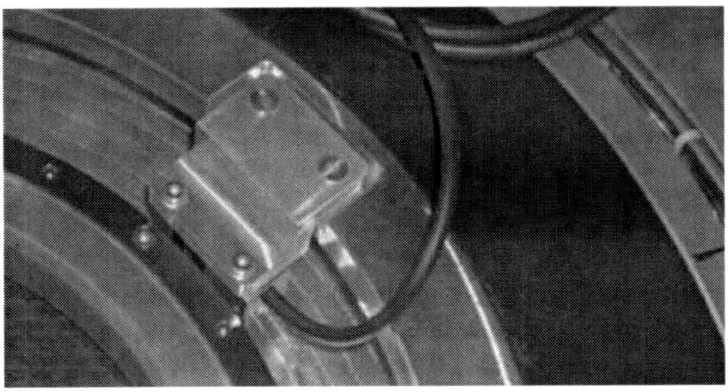

Figure 12: This photograph shows a section of the assembled azimuth motion unit. The Renishaw ring encoder is at the lower left and the read head is mounted to a bracket near the center of this image. The wiring channel for the coils can be seen at the top right. The read head is connected to the controller through the shielded cable.

Instrument Rotator and Focuser

For imaging, photometry, and spectroscopy, the focal plane of an alt-az telescope requires derotation. For our purposes it is sufficient to simply calculate the desired angular position as a function of time and drive the instrument platform to that position. The CDK700 uses a simple geared servo motor for driving the rotator.

The instrument rotator can be used to pre-position an off-axis guider (OAG) pickoff mirror or prism on a guide star. By providing a large, well-corrected field of view (FOV), the telescope optics supports both a large format main camera and a large pickoff region for a sizeable guide camera.

Guide-star position information is used for low to medium rate corrections via the telescope's drive system. If the guide star is bright enough, the integration time can be reduced to a fraction of a second, and high-frequency corrections can be fed to a low-inertia deviator plate to counter the effects of wind gusts and drive irregularities not handled by the telescope's basic control system. Our optical design has sufficient back focus to allow for an autoguider pickoff, a deviator plate, and a filter carrousel in the optical path, if desired.

The motorized focuser is built into the same assembly as the rotator. Three linear bearings and three drive screws are used to make an extremely stiff linear motion stage for very low radial compliance against the heavy, asymmetric loads that are typical of modern cameras, filter wheels, and OAGs.

Figure 13 (opposite page) shows the top view of the focuser/rotator. The worm gear drives the rotating stage. In Figure 14 (opposite page), the three large cylinders are the housings for the linear bearings, and the gear head motor is shown coupled to the three drive screws through a timing belt and pulley system.

OTA and Mirror Cell

The CDK700 requires a high ratio of stiffness to weight. Along with the usual considerations of precise and repeatable motion, weight and stiffness are primary considerations in the structural design. The requirement for light weight is more stringent than that of a typical mountaintop alt-az telescope, because most large telescopes are under a dome that gives them significant protection against the wind—typically reducing wind gusts by an order of magnitude. The CDK700, on the other hand, is designed for roll-off observatory operation out in the open where telescopes are exposed to the wind.

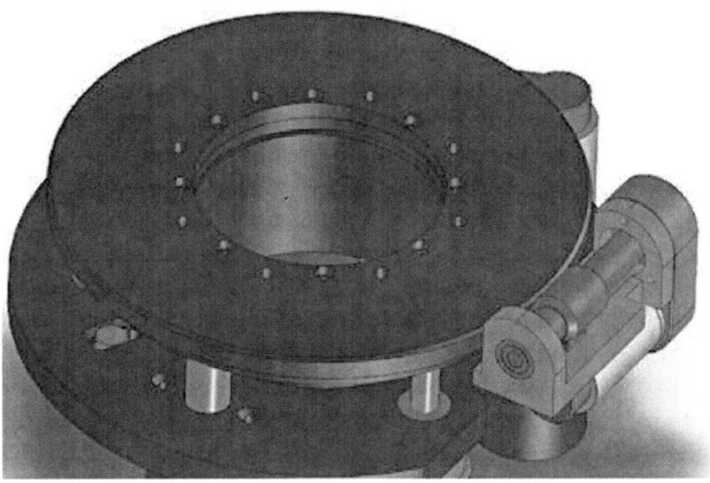

Figure 13: CAD drawing of the Field Derotator Focuser Assembly (FDFA). The field rotation is accomplished by a motor coupled to a worm, shown at the right. In use, the camera, filter wheel, OAG, and other optical hardware are bolted to the top plate, which is supported and allowed to rotate on a thin ball bearing. This plate also acts as the worm wheel. One of the three linear bearings for the focuser can be seen at the bottom center.

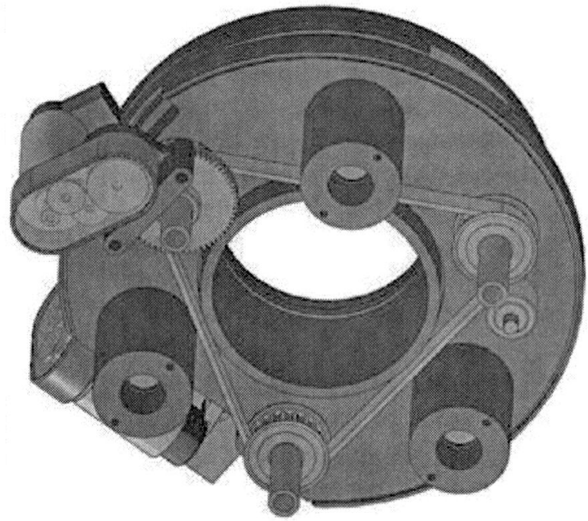

Figure 14: Bottom view of the FDFA. A gear head motor, shown at the upper left, drives a spur gear that in turn drives a timing belt and three pulleys. Each pulley drives a precision thread, moving the focuser axially. The three housings for the linear bearings can also be seen in this view.

In order to be site transportable without high costs, the telescope needs to be lightweight. Yet to achieve precise pointing and tracking, the telescope structure must also be very stiff. The stiffness requirement is manifest in two ways. First, the structure must not bend enough to significantly degrade image quality when subjected to disturbances such as wind gusts. Deflections of the optical tube and fork cannot be sensed and compensated for by the control system unless autoguiding is being performed by sensing guide star position, and it was our goal to have the telescope precisely track for 5 minutes without autoguiding. Second, the structure must be stiff enough to force all mechanical vibration frequencies to be higher than the servo loop bandwidth, otherwise servo instabilities could result. In summary, the design goals of the structure were: (1) that it will hold the telescope pointing angle to within 1 arc second RMS for wind gusts from 0 to 10 MPH, (2) all vibration frequencies are greater than 20 Hz, and (3) that it will allow unguided exposures of 5 minutes, with 1 arc second accuracy.

The tolerances for the structure can be divided into two categories: those which cause repeatable errors and those which cause non-repeatable errors. We can relax our tolerances with respect to the former somewhat, as the tracking software is programmed to compensate for repeatable errors such as the flexure of the optical tube as a function of altitude. Non-repeatable errors, such as the slipping of bolted joints or roughness and rumble in bearings, must be designed out of the system, and received close attention during the design process.

The truss tubes are made from carbon fiber composite, which has the highest stiffness-to-weight ratio of any readily available material, and also is close to zero expansion over temperature, an important property to maintain focus over changing temperatures. The OTA rings and bottom plate are machined from aluminum plate, and forks are made from welded steel plate. A unique ball joint clamp is used to fasten the truss poles to the aluminum rings and backplate. All surfaces are either power coated or anodized for full protection against the environment.

From many years of experience with telescopes, we have found the primary mirror cell to be one of the most difficult and important structures within the OTA. It must be very stiff against axial forces to prevent focus from shifting as a function of elevation. It must also float the mirror in the axial direction with very little stiction, yet decouple the radial forces and motion from the axial forces and motion in order to prevent astigmatism introduced by gravity. These difficult requirements are satisfied in the CDK700 by its unique mirror cell, shown in Figures 15 and 16. An 18 point floatation cell is coupled axially to the mirror through stainless steel balls that are free to rotate, thus preventing significant side to side forces on the back of the mirror that can cause warping of the optical surface. The radial forces are taken up by two balls that ride on a stainless steel ring which is glued to the inside of the core hole. These balls are also free to rotate, allowing axial motion without significant force or stiction in the axial direction. The two balls are positioned on the plane through the center of mass of the mirror, eliminating gravity-induced torques, also very important for the prevention of astigmatism.

The use of balls for the floatation joints makes the design very compact, light weight, and easy to manufacture.

Figure 15: CAD drawing of the primary mirror cell. An 18 point flotation cell is used to provide axial support for the mirror. On this cell the mirror is allowed to move in the plane of the mirror (radially) with very low friction, preventing distortion of the mirror from stiction. The radial loads are taken up by two ball bearings that ride on a stainless steel annulus glued to the core hole of the mirror. This arrangement decouples the radial and axial loads due to gravity.

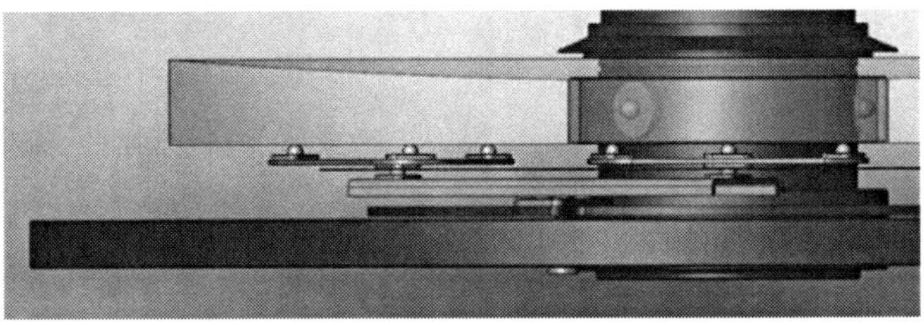

Figure 16: A side view of the primary mirror cell. This CAD drawing shows the mirror riding on ball bearings that are allowed to rotate with low friction. Visible are the two radial-support ball bearings in the core hole.

Instrumentation and Observing

The CDK700 is a multi-use telescope and it is designed to accommodate a variety of instruments. In typical applications we envision that one of the Nasmyth ports will be utilized for CCD cameras and eyepieces, while the other port will host larger instruments such as spectrographs or will be outfitted for visual use.

Specialized instruments are often quite bulky and heavy—a major reason the CDK700 has a Nasmyth optical configuration—but their complexity and requirement for precise adjustment preclude them from being repeatedly installed and removed from the telescope, hence the benefit of two Nasmyth ports.

Newtonian optics have long been the mainstay of amateur Dobsonians, but the height of the eyepiece for large aperture telescopes discourages many observers. In contrast, the CDK700 folded Cassegrain design offer an eyepiece position that allows seated observing at the Nasmyth focus. Figure 17 shows the convenient focus location of the finished telescope.

Figure 17: A photograph of the assembled CDK700 with one of the authors (Allan). The convenient and static focal plane height makes this instrument ideal for visual use, and allows large and heavy instruments to be mounted on either side of the dual Nasmyth foci.

Acknowledgements

Many innovations in the CDK700 are the direct outgrowth of years of collaboration with the Alt-Az Initiative as described by Russ Genet in the cited reference. The CalPoly 18, also referenced, was the first implementation of direct drive motors and encoders in a small Alt-Az telescope.

Author's Note

David Rowe is Chief Technology Officer and Co-founder of Sierra Monolithics and can be reached at dave@starryridge.com

References

Bely, P. 2003. *The Design and Construction of Large Optical Telescopes*. New York: Springer.

Genet, R. et al. 2008. An 18-inch direct drive alt-az telescope. In *The Society for Astronomical Sciences 27th Annual Symposium on Telescope Science*.

Genet, R. 2009. The alt-az initiative for lightweight research telescopes: scientific research programs and engineering developments. At the American Astronomical Society Meeting #214, #409.12. *Bulletin of the American Astronomical Society, 41*, 673.

Chapter 10

SpicaEyes

A Unique Line of Metal Telescopes

Tom Osypowski

Equatorial Platforms

Introduction

During the last five years, our company, Equatorial Platforms, has developed a unique line of metal telescopes, called *SpicaEyes*. The *SpicaEyes* telescopes are a Newtonian optical system mounted in an alt-az framework. They are offered in two drive configurations:

Figure 1: (Left) *Spica Eyes* telescope with integrated equatorial platform; (Right) *SpicaEyes* telescope with advanced SlipStream go-to drive.

The salient features of both systems will be described later, but first it would be appropriate to list the elements they have in common.

SpicaEyes Construction, Design, and Materials

Materials

Most commercially made alt-az telescopes in the 16 to 32 inch range are Dobsonian-type structures made out of wood. This construction design has the advantages of low cost and easy workability with ordinary tools. Disadvantages include *lack of durability* (the wood deteriorates with weather and needs upkeep) and *heat retention* (the mirror is at the bottom of a wooden box where heat can be trapped unless actively exhausted). Wood is also subject to many expansion/contraction cycles, which can strain joints and loosen fasteners.

These issues are dealt with in the *SpicaEyes* telescopes by using rectangular aluminum tubing for the structural material of the rocker and mirror box. This tubing is available in sizes such as 1″ x 4″ x 1/8″ wall and 1.5″ x 6″ x 1/8″ wall. It can be cut and welded together to form ultra-strong and rigid low-profile boxes. These metal boxes shed heat rapidly and have no tendency to trap hot air. To cool down the mirror quickly, all that is needed is one or two fans (depending on the size of the mirror) in the rear of the cell, exhausting out the back. A rugged powder-coat finish of the aluminum frame provides an extremely durable and tough finish that is easily maintained.

f/ratio

Another defining characteristic of the *SpicaEyes* telescopes is their size. They are short, with typical *f*/ratios in the sub-*f*/4 range (*f*/3.3 – 3.8). In general, the shorter the telescope, the stiffer it can be made. Also, with the Newtonian focus at the top end, a short telescope means a short ladder, or perhaps no ladder at all. Until recently, sub-*f*/4 parabolic mirrors were not generally available. Now several small commercial mirror-making firms are offering high-quality mirrors with deep curves. These companies include Kennedy Optics, Lockwood Custom Optics, and Lightholder Optics. All of them have websites that are easily accessible with a simple Internet search.

While a short f/ratio makes for a desirable short telescope, some issues do come up that need to be addressed. One is the increased coma inherent with short-focus mirrors. Thankfully, there are excellent coma correctors on the market that will adequately deal with this problem, such as ParaCorr. Also, a short-focus telescope requires more exacting mirror collimation for peak performance. A metal structure, if properly made, has very little flexure, and modern collimation tools make the job of achieving precise alignment an easy task.

Mirror Support

Short-focus thin mirrors demand a properly designed mirror support system to prevent flexure of the optic in use. The mirror cells installed in the *SpicaEyes* telescopes have been designed to do the job. The frame is constructed of welded steel tubing and is black powder-coated for durability. The back support for the mirror is a standard 9, 18 or 27 point flotation assembly, calculated by the well-known program, PLOP. The flotation triangles, laser-cut from stainless steel, carry Teflon pads that support the mirror without binding or stiction.

The edge support for the mirror is at least as important as the back support, particularly when the telescope is aimed lower in the sky. Traditionally, a mirror sling has been used for edge support in Dobsonian-type telescopes, either using a belt of some kind (nylon or Kevlar) or a cable. This system works well, but it must be implemented carefully for best results. The sling must be centered exactly on the balance point of the edge and not allowed to shift laterally. The mirror is free to move sideways in the cell with a sling—a big disadvantage when using an equatorial platform, which will tilt the whole telescope sideways while going through its tracking run, and thus cause the mirror itself to move out of alignment.

Given these limitations of the sling, the *SpicaEyes* telescopes utilize a dual rocker-arm system, with four points supporting the edge of the mirror, as seen in Figure 2. This works as well as a sling and has the big advantage of constraining the mirror and preventing sideways slippage.

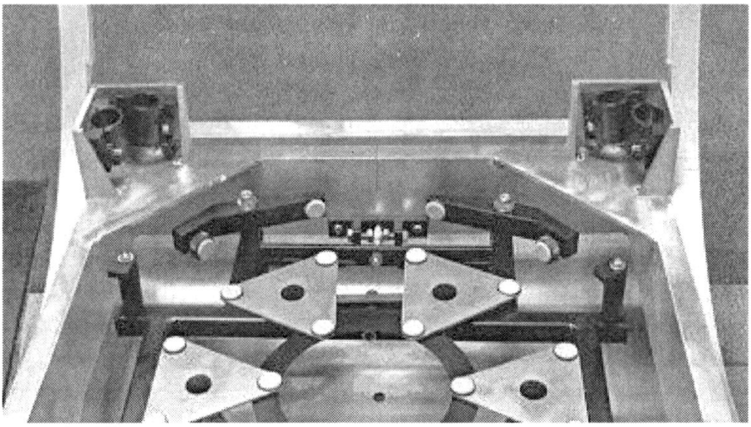

Figure 2: Mirror Support System.

For mirror collimation, most commercial cells provide three knobs in back which actually move the support triangles in and out, thus physically moving the glass. This system works all right with a sling edge support, but would be detrimental with a rocker-arm support, as the placement of the pads on the edge of the mirror would change laterally as the mirror is collimated. So, in our design, the whole cell is tilted by two knobs on the top end, with the bottom end swiveling on a ball joint. This allows the mirror to rest against its back and edge supports without movement, and the placement of the collimation knobs is much more convenient to reach, especially when the scope is pointed near the zenith.

Figure 3: A 16 inch mirror cell.

The picture in Figure 3 shows this collimation arrangement, with the two large brass knobs near the top of the cell and the swivel joint at middle bottom. The cell is also very low profile, which puts the mirror back a little further in the mirror box and thus helps with the balance. Also visible are some other features of the cell structure, such as the fan and the rechargeable battery array. The control box for the fan has an on/off switch, fan speed control knob, and extra jacks to tap power for dew heaters or digital setting circles (DSCs), such as those available from Sky Commander or Argo Navis. The mirror box has tight-fitting covers for both the back and front, which protect the mirror during transport and storage from dust and insect invasion.

Upper Cage

The upper cage of the *SpicaEyes* telescopes is a rugged but lightweight assembly. It is constructed in a traditional fashion with two rings separated by aluminum tubing. (While a single ring structure would weigh a bit less, it seemed to us that such a design compromises rigidity, especially when using the heavier eyepieces and cameras available today.) The rings of the cage are water-jet cut from 1/8″ plate aluminum, with 1.5″ diameter tubing separating them and with smaller square tubing in between the round stock. So, the cage has four round tubes that hold the rings apart and supply a mounting place for the spider vanes. In between these round tubes are smaller square tubes that help with the support of the rings, since the rings themselves are just 1/8″ thick aluminum. A shroud of ABS plastic material, attached to the tubing, nests inside the rings for added rigidity and protection against stray light. The design with the vertical tubes is quite stiff and resistant to warping and bending. The focuser board also adds sheer stiffness, and the ABS plastic seals the deal as far as resistance to distortion of the cage is concerned. A stiff board carries the renowned Feathertouch 2-speed focuser.

Figure 4: *SpicaEyes* upper cage assembly.

An available option is a rotating cage assembly. This feature introduces a third "interface" ring that attaches to the truss tubes before the cage is put on. The interface ring has four 30-degree slots that mate with shoulder bolts mounted to the bottom cage ring, providing rotation of the cage through a 30-degree arc for optimal placement of the eyepiece. The interface ring also makes mounting the cage a cinch, since it is much easier to initially fasten one ring rather than the whole cage to the truss hardware.

Figure 5: Eight truss tubes on the *SpicaEyes* telescope connect to the lower cage assembly with hand knobs.

Truss Tubes

The eight truss tubes on the *SpicaEyes* telescopes are arranged in 4 pairs for easy assembly. On the mirror cell end, each tube fits into a specially machined holder that clamps onto the tube with a hand knob. On the top cage end, the pairs of tubes come together with a hinge arrangement that fastens with a single knob to the cage hardware. The whole system is easy to assemble and results in an extremely stiff structure. The truss tubes themselves are encased in shrink-wrap tubing, giving them a friendly feel and a dull, non-reflecting surface.

Drive Formats

The foregoing discussion has outlined the common features of all *SpicaEyes* telescopes. Two different drive formats are available for the *SpicaEyes* telescopes: the integrated equatorial platform and the SlipStream go-to drive.

Equatorial Platform

The equatorial platform format can give any alt-az (Dobsonian style) telescope excellent tracking ability. Indeed, manufacturing such platforms for commercial and homemade Dobsonians of all sizes has been the mainstay of our business for two decades. So it was natural that our first *SpicaEyes* metal telescopes should also be equipped with platforms.

An equatorial platform has been described as a kind of sawed-off horseshoe or split-ring mount. Just two short arcs of the horseshoe or split-ring are utilized, and the rest of the structure is made flat to carry the telescope. The wooden platform seen in Figure 6 neatly illustrates this description, though the situation here is a bit more complicated, as the sectors shown are not part of a circle inclined at the latitude angle, as with a split-ring mount, but instead they are sectors of a cone, with the axis of the cone being the polar axis of the platform. Still, it all amounts to the same final effect: with this compact structure sitting low to the ground, you are getting a true equatorial mount capable of solidly supporting a large telescope and giving it precise tracking for visual, imaging, and research endeavors. Because just an arc of the full ring is used, the platform has a tracking time limited by the length of its arc. Typically, the platforms we manufacture offer a tracking time of 80 minutes, with a quick pull of a handle to get another 80 minutes.

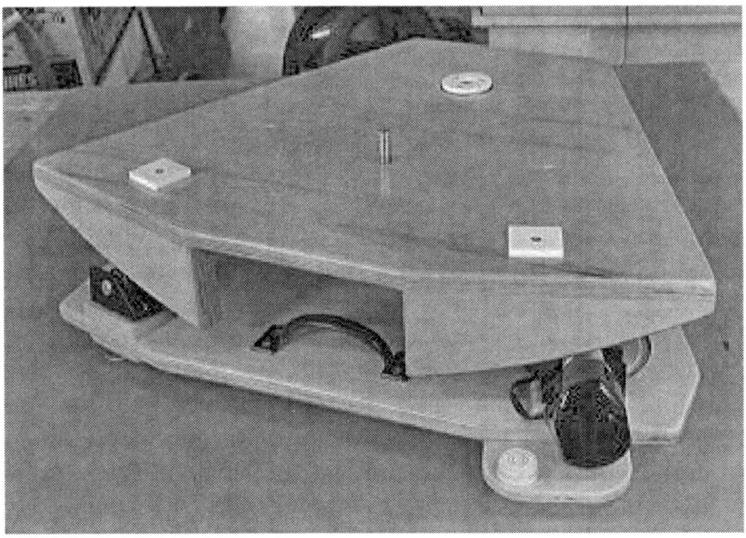

Figure 6: Wooden equatorial platform.

Generally, the platform acts as a new groundboard for the Dobsonian telescope. To that end, the top of the platform has a central pivot and three Teflon pads for the rocker box of the scope to sit on and swivel around. This arrangement keeps everything as low and stable as possible, since fewer parts are being stacked on top of each other. The aluminum platforms (see Figure 7) used with the *SpicaEyes* telescopes utilize this same configuration, even though the platform itself is just a framework instead of a solid structure.

Figure 7: Aluminum *SpicaEyes* platform.

An equatorial platform is a useful and versatile tracking device, and no modification of the telescope is required for it to operate properly. It opens up new areas of observing opportunity, allowing high-power visual examination and CCD imaging. With the dual-axis platforms, even autoguiding can be implemented for longer exposures, since the platform is a true equatorial mount with no field rotation typically, shorter exposures of 15 to 30 seconds need no guiding at all.

To sum up: combining an equatorial platform with the *SpicaEyes* telescope structure yields an instrument of exceptional quality, durability, and usefulness. Besides excellent tracking ability, the scope also includes digital setting circles (either Sky Commander or Argo Navis), which make finding objects easy and quick. It's a "go-to" feature, with your hands—instead of noisy motors—supplying the go-to power.

SlipStream Go-To Drive

The SlipStream is an alternative to our platform drive. These new *SpicaEyes* telescopes combine solid all-aluminum construction with a new alt-az go-to drive system that integrates a slip clutch on both axes for full-time manual and motorized slew control.

You can find objects fast by using the included Argo Navis computer. The telescope will automatically slew, or you can move it by hand. Once you find your target, the telescope will track it for you, keeping it centered in the field of view. The integrated SlipStream clutches allow you to move the scope at any time by hand or with the motorized slew control, with no levers to disengage.

The oversized altitude bearings sit on stainless steel shafts rotating in pillow-block ball bearings. Both axes are driven by servo motors through a stand-alone computer mounted to the telescope. The adjustable slip clutches are fully enclosed and protected in a metal housing. Thumbscrews allow one to adjust the drag of the clutches as needed.

A wireless hand control gives three-speed slew control on both axes. The fast slew will spin the scope completely around in less than a minute, useful for going across large stretches of the sky. The medium slew will nicely center objects in the field of view at low and medium powers, and the slow slew will do the same at high power. All three slew speeds can be individually calibrated to the user's requirements.

Figure 8: 28" *f*/3.66 *SpicaEyes* SlipStream telescope with go-to drive.

Summary

Both styles of *SpicaEyes* telescopes offer an extremely durable and rigid alternative to the wooden Dobsonian-type telescope structure. The *SpicaEyes* with integrated equatorial platform format offers true equatorial tracking with no field rotation. This gives the user the opportunity to do long-exposure CCD imaging with an autoguiding setup. Also, other areas of observing that are best done with an equatorial mount can be done with these scopes. Measurement of double stars comes to mind as a particularly suitable project for a *SpicaEyes* on a platform.

The *SpicaEyes* with SlipStream drive format offers full motorized go-to that allows the user to find any object in the Argo Navis database. After the telescope automatically slews to the object, it then tracks it precisely. And, unlike other go-to alt-az setups on the market, there is no need to lock down the drive to get it to track or to disconnect the drive to move the scope by hand. Instead, built-in slip clutches are included that allow the user to grab and move the scope at any time without losing positional accuracy or tracking ability. Also, if one wishes, the cordless hand control can be used for motorized slewing, centering, and guiding. The *SpicaEyes* SlipStream drive, being an alt-az format, will exhibit field rotation; still, very effective CCD imaging can be done if the exposures are kept short enough (typically under 30 seconds). A set of these exposures can then be stacked to achieve the equivalent longer exposure. There is also available an optional field de-rotator feature at extra cost, which allows extended single exposures.

More information on these *SpicaEyes* telescopes can be found on our website: www.equatorialplatforms.com. The same website also provides a wealth of information about our custom line of equatorial platforms for those who wish to use that type of mount to give their present Dobsonian telescope tracking ability.

Author's Note:

Tom Osypowski
Equatorial Platforms
530-274-9113
tomosy@nccn.net
www.equatorialplatforms.com

Chapter 11

Telescope Portability

Moving a Big Telescope Doesn't Have to Be Difficult

Howard Banich

Rose City Astronomers

Portability and Set Up

Designing easy portability into a telescope can be vital to its ultimate success. Whether it's transported a few times a year or every clear night hardly matters because when it's time to move a scope, portability and ease of set up become its most important attributes.

Practically speaking, portability and easy setup are defined by needing no more than two people using a pickup truck, van, or trailer for transportation and needing about thirty minutes to set up the scope. More than this and the term portability should be replaced with transportability.

The term "transportable" usually means a scope that can be moved with difficulty, which too easily becomes a permanent installation, or worse – a telescope that's seldom used. If an instrument is too much trouble to move, it will generally stay put. In this sense the concept of "transportable" is something to avoid when designing a telescope that's meant to be moved regularly. As a related thought, designing a telescope larger than 1 meter as if it were going to be moved would greatly reduce transport costs should it ever need to be relocated.

Strictly for convenience, a one person operation is desired so the observer has only their own schedule to consider, although with scopes over 1 meter aperture this will probably require some kind of motorized lifting or pulling.

A consideration as important as the size of the primary mirror is making sure the method of portability is compatible with your vehicle and the amount of effort you can happily give to each observing trip. Matching the design of the scope to the vehicle that will transport it is a crucial step in determining the best approach to take toward portability. How much room is there? How high off the ground is the cargo area? How big is the entry door? Will you use muscle power, ramps, a winch, a lift gate—or a combination?

Equally important is having a firm grip on your personal boundary between cheerful effort and grudging toil—no one wants a telescope that gathers dust because it's too much trouble to transport, set up, and tear down. Considerations of personal strength are a part of this calculation, but just being able to physically move something doesn't mean it's enjoyable to do so. Stronger observers will do well to design the largest, heaviest piece well below their limits even if they enjoy the challenge. Time will only make it more difficult and less enjoyable to move a heavy telescope, so design for the long term. With these variables in mind, here are three ways to design portability into a telescope:

Break down the telescope into small, lightweight parts
If they're easily carried and readily assembled, the maximum weight for a single unit would be between 50 and 100 pounds. Although this implies that a large scope could be transported in a relatively small vehicle, it also requires the most assembly and disassembly time. In addition, the structure of the connection points would need to stand up to repeated use.

Users who are well versed in putting together and taking apart the scope could make this a viable option, but it could become the least attractive way to deal with a scope that's frequently transported if there are lots of small connections, or the main units weigh over 100 pounds. However, this is an excellent option when designed well, and a good example of a two person set up is a 28 inch *f*/4.5 computerized string telescope designed and built by Dan Gray. This telescope is transported in a small trailer. www.siderealtechnology.com/28inch.

Figure 1: Dan Gray's 28" *f*/4.5 alt-az string telescope. Two people place the ground ring on the ground, which is then leveled. They then stack the flex rocker on the ground ring and the mirror frame on the flex rocker. The three tension poles of the string truss are then installed between the mirror frame and cage ring. The entire procedure takes about 10 minutes.

Transport the fully assembled telescope in one piece

This option requires a large trailer with an excellent suspension, but is certainly achievable. In this approach, the OTA would likely be pointed horizontally for transportation, which could place uncomfortably large edge stresses and vibration on mirrors and their cells. However, this has been done up to 1 meter. A variant of this idea would be to design the optics and their cells to be easily removed and reinstalled so they can be safely transported horizontally.

The major benefit is that set up and tear down time is minimized, but moving the entire scope is still a time consuming process because getting it in and out of a trailer is slow and careful work, even with motorized lifting. The chances for structural damage during transport could be unacceptably high, unless the entire scope is well secured in the trailer and isolated from the constant shaking and bumping from the road.

Figure 2: Steve and Bruce Swayze's 1 meter *f*/5 Dobsonian, transported in one piece in a custom trailer. A hand cranked winch was used to get the scope in and out of the trailer; it took one person to operate the winch and a helper to make sure the scope was rolling straight down the ramps. From trailer to sitting leveled on the ground took about 30 minutes.

Keep the largest and heaviest parts together and move this assembly on deployable wheels

In this scenario, only the upper cage and truss tubes would be assembled and disassembled, just like large truss tube Dobsonians. This approach was first developed for large amateur truss tube Dobsonians by Ivar Hamberg in 1982 and popularized by Dave Kriege's Obsession Dobsonians in 1988.

Deployment would either need: one strong person moving the alt-az base and mirror cell/frame/box like a big wheelbarrow; two regular people pushing this assembly on four casters; or a winch or Tommy Gate lift to mechanize the movement.

One of the benefits of this method is that the optics are moved while horizontal and well protected by their respective cells. Noting that most truss tube Dobsonians up to 1 meter have been moved this way, a feasible weight limit is perhaps 500 pounds for two people. Using a wheelbarrow-like approach has a manageable upper limit of about 300 pounds for one strong person.

A heavier scope would need the mechanized lifting power of a winch and/or a Tommy Gate lift. This approach opens possibilities to the construction of truly portable 1.5 meter and larger scopes, and with careful design consideration, given to the realities of portability, there's every chance for practical success.

Figure 3: Dan Bakken's 41.2 inch string alt-az Newtonian being manually pushed into its transport trailer. With the aid of a winch this could become a one person operation.

Figure 4: The author and his 28″ *f*/4 alt-az computerized one-person portable telescope. Complete set up from the van transport vehicle (seen in the right background) using ramps takes 20 minutes. The ground board, rocker frame, and mirror box always stay assembled as one unit and are moved by inserting wheels and wheelbarrow handles in two slots on either side of the rocker frame. The cage is secured to the top of the mirror frame for transport, leaving only the tubes to be moved separately.

A short time lapse video of a 28 inch *f*/4 computerized alt-az telescope specifically designed for one person portability can be seen here: sites.google.com/site/teardownvideo.

Chapter 12

One Meter Portable Alt-Az Telescopes

Russell Genet[1], George Roberts[2], Reed Estrada[3], and Chris Estrada[4]

[1] California Polytechnic State University, [2] Inca Inc.,
[3] Northrop Grumman Corp., [4] Allen Hancock College

Introduction

Howard Banich (2010) defines telescope portability and easy setup—practically speaking—as "needing no more than two people using a pickup truck, van, or trailer for transportation and needing about thirty minutes to set up the scope." Any more than this, he suggests, "the term probability should be replaced with transportability." Transportability, Howard admonishes, "is something to avoid when designing a telescope that's meant to be moved regularly."

Truly portable telescopes can be readily moved to dark sites. Many astronomers—amateur, student, and professional—do not live at dark sites nor can they afford to maintain a remote dark-site observatory. Also, many amateur astronomers can't afford either a dome or roll-off-roof observatory and need to easily move their backyard telescopes into their garages at night or during inclement weather. Similarly, astronomy instructors and students at small schools need to move their telescopes, when not in use, from outdoors into an equipment storage room. Some observations, such as asteroid occultations, actually require a portable telescope.

Almost all small (14 inch or less) telescopes are at least nominally portable—many of them equatorially mounted. Dobsonian telescopes, thanks to their compact alt-az mounts, are portable to considerably larger apertures than equatorial mounts, with 24, 28, and even 32 inch aperture portable Dobs becoming increasingly common.

We describe, below, our design of two similar 1 meter portable telescopes, with emphasis on the alternatives we considered during their design and the rationales for the choices we made, one of the most important being scalability in size from 1 meter up to 1.5 or even 2.0 meter apertures while still maintaining portability. Two other priorities for us besides scalability were light weight and low cost.

Two different versions of our 1 meter telescope are being built—one out of wood and the other from aluminum. The wooden version is being built not only as an easily modified prototype to pave the way for the harder-to-change aluminum version, but as a useful telescope in its own right. The wooden prototype is being built by Russ, Reed, and Chris. The aluminum version is being built by George and his associates at Inca as a pre-production prototype. These telescopes were still under construction as this was written, so we only show pictures of their construction to date (April 2010).

Portability

Our first design task was to select an overall approach to portability from three possibilities suggested by Banich (2010). First considered was rolling an entire completely assembled telescope on wheels in and out of a trailer. While this is an entirely reasonable approach in many situations, we rejected it in our case for two reasons:

(1) We were wary about transporting mirrors in a vertical as opposed to horizontal position. We felt the mirrors would be more securely supported if they were horizontal.(2) We did not feel this approach to transportation would scale up well beyond 1 meter.

Next we considered breaking the telescope down into a number of components, each of which could be handled by at most two persons. Banich (2010) suggests weights should be less than 100 lbs per component, with less than 50 lbs being desirable. Again, while an eminently reasonable approach for many situations, we also rejected this approach in our case for two reasons:

(1) We were uncertain whether or not a fairly complex, computer controlled telescope that had been broken down into small subassemblies could be reassembled in less than 30 minutes.

(2) Some of the components of telescopes larger than 1 meter (such as primary mirrors) could exceed the weight limit of 100 lbs, let alone the desirable limit of 50 lbs.

Thus we ended up accepting Banich's third approach, the one he prefers for large telescopes: keeping most all of the components together (especially the heavy ones) as a single wheeled unit that rolls on and off the transport vehicle. Only the top and relatively lightweight portions of the telescope are assembled.

We initially chose a sizeable, enclosed U-Haul type of trailer as the "typical" transport vehicle that would set our design constraints. After some analysis, however, we realized we could "shoehorn" a 1 meter telescope into the rear of a mid-sized sport utility vehicle. We chose a Jeep Cherokee (with its rear seat folded down) as a stand-in for the "typical" mid-sized SUV transport vehicle. We reasoned that if we could fit a 1 meter telescope into the confines of this Jeep, this exercise would help prepare us for fitting a 1.5 meter telescope into a U-Haul type enclosed trailer, or a 2 meter telescope onto a low, lightweight, roadable trailer.

Figure 1: Early on in the construction of the wooden prototype, a check was made to see if it would fit in the back of the Jeep Cherokee transport vehicle. The mirror box is on the top, the rocker box in the middle (without its altitude trunions), and the azimuth base on the bottom with its nose and steerable nose wheel pointed toward the camera. Three loading tracks are required: the middle track for the nose wheel, and the two outer tracks for the mains.

Azimuth Base

To roll the telescope about, we initially chose a wheel configuration similar to a shopping cart—two fixed and two castered wheels. However, Mel Bartels soon informed us—in no uncertain terms—that on uneven, sloped terrain, a heavy telescope with this wheel configuration would go where it wanted to go, not necessarily where we wanted it to go. So we switched to a configuration similar to that of small airplanes, namely two widely spaced main wheels slightly behind the center of gravity, and one smaller, steerable nose wheel well ahead of the center of gravity, with a tow bar temporarily attached to the nose wheel to provide both motive power and steering. Two of us (Russ and Reed) routinely move aircraft weighing 1000 lbs or greater (more than twice what the telescopes would weigh) by ourselves with this tricycle wheel configuration, albeit usually on more or less level pavement or grass. Following further unequivocal advice from Mel, Howard Banich, and others, we chose large diameter (10 inch diameter mains and 8 inch nose), wide, pneumatic wheels, the better to negotiate rough, uneven terrain as compared with smaller diameter or hard-surface wheels.

Figure 2: The nose wheel and its removable tow bar for the aluminum preproduction prototype telescope. The removable pin shown between the two is used to fasten them together. A chain keeps the pin from being lost in the dark.

We purchased a small, manually operated winch with a safety ratchet and cloth strap belt (instead of a more difficult to handle steel cable) to roll the telescopes up and down a three-tracked loading ramp (one track for each wheel). A slightly tricky part of the design was providing enough space under the azimuth base to clear the "hump" (the edge of the vehicle as one rolls the telescope from the ramp into the vehicle) without, at the same time, having the telescope be so high that its top leading edge would hit the ceiling of the tow vehicle. The wheels also had to be close enough to the front and rear edges of the azimuth base to prevent the leading or trailing edges of the base from gouging into the ground or track.

The wooden prototype has an alternative second method for moving the telescope about on the ground or in and out of the Jeep or other vehicle. Two large "wheel barrow" handles are inserted into brackets on each side of the azimuth base with the handles sticking out beyond the nose wheel. The two main wheels then become the "wheel barrow's" wheels. One advantage of this wheelbarrow mode is that one only needs two loading tracks, not three. A disadvantage is that it probably will not scale well up to 1.5 or 2.0 meter portable telescopes (or to older, less muscular astronomers).

The maximum length of the azimuth base (and the entire telescope) was, of course, set by the length of the Jeep's rear compartment. The spacing between the outside edges of the telescope's main wheels was set by the inside distance between the rear wheel wells of the Jeep. In turn, the azimuth base itself needed to fit between these two main wheels. However, as the Jeep was wider above its wheel wells than below (as is the case for many SUVs, pickup trucks, and trailers), the altitude assembly could be somewhat wider than the base assembly, and this was a good thing as otherwise we would not have been able to fit a 1 meter telescope into a Jeep Cherokee.

Figure 3: Reed Estrada (left) fastens down a standard 24 inch Inca bearing to the azimuth base while Chris Estrada looks on. Russ took the picture—his tilt-off roof observatory can be seen in the background. The azimuth base has two main wheels and one steerable nose wheel (left). A tow bar fastens to the nose wheel so the telescope can be moved about manually, much like a small airplane. For the wooden prototype telescope, we used a standard Inca bearing, while the more precise aluminum telescope uses a special Inca bearing manufactured to more exacting tolerances.

Unlike alt-az telescopes which have room for azimuth "tracks" significantly larger in diameter than their primary mirror, our azimuth track had to be smaller than the mirror to fit between the wheel wells of the transport vehicle. This obviously violates Mel Bartels' "Pyramid Rule" which suggests that, from top to bottom, telescopes should grow continually wider. Since our azimuth rotational base was somewhat narrow, we chose to use an azimuth bearing as opposed to azimuth wheels rolling about a track. Such azimuth bearings, which can be very stiff for their diameter, are widely used in various optical and radio telescopes, as well as radar systems, etc.

Figure 4: Aluminum pre-production prototype azimuth base. The octagonal tubular frame supports the azimuth bearing. Fastener inserts in aluminum frame members are stainless steel; all taping of aluminum was avoided wherever possible.

The base assembly supports the entire telescope. The aluminum pre-production prototype incorporates three 1 inch diameter stainless steel leveling screws with 3 inch diameter stainless steel pads. When lowered to the ground, these assemblies enable the operator to level the telescope and remove any contact between the pneumatic wheels and the ground, thus providing a stable platform for the telescope.

Inca Azimuth Bearing

Inca Corporation specializes in manufacturing automation equipment to move flat screen TVs in and out of concealments, doors that are remotely controlled, and numerous other robotic and semi-robotic mechanisms. In the course of developing these items, Inca needed economical swivel bearings—large area bearings that Inca initially imported. Over time, however, the bearing quality deteriorated to the point where they were unusable for swivel televisions and projectors because they could not carry the rated loads without distortion, affecting the sensing of the index location of the swiveled components. To avoid being at the mercy of an unreliable supplier, Inca developed their own large and inexpensive area bearings. Considerable experimentation eventually resulted in a reasonably strong, lightweight, and above all inexpensive bearing that did the job they asked of it. Inca now manufactures bearings that are 5.75", 8.5", 10.0", 12.0" 17.0" and 24.0" in diameter, and uses them in their machines.

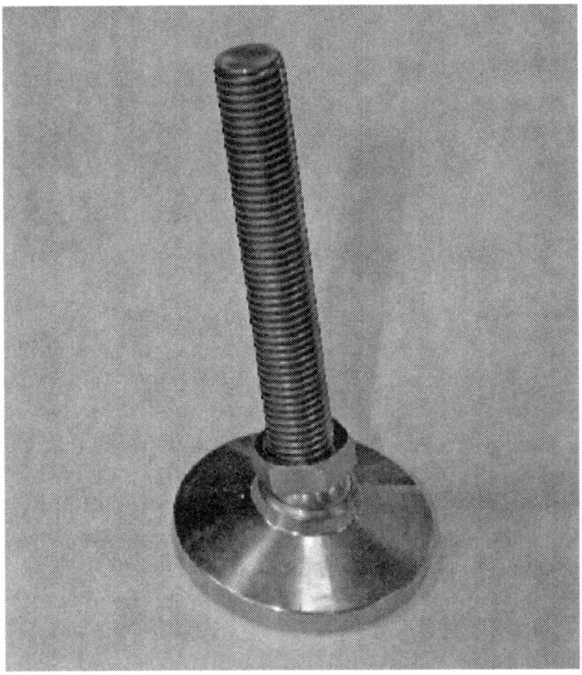

Figure 5: Three sturdy swiveled-foot adjustments are used to level the telescope and to elevate the telescope so that most all the weight is taken off of the wheels.

When the 1 meter telescope project came up, the 24 inch Inca bearing was a candidate for the azimuth bearing because it was lightweight, strong, thin, and could carry five times the load required for the 1 meter telescope. After some simple tests, the decision to use the 24" bearing was made. Since we had not designed the bearing for use in a telescope, we examined our tooling to see if we could improve the bearing's performance. We could have taken another path and gone to a much harder material than what we normally use, but that would have really increased the costs of fabrication without a corresponding improvement in life of the bearing or its performance. Hence we decided to proceed with a relatively inexpensive bearing that seemed likely it was up to the job. We could replace it in the future if necessary with another bearing. One could replace the Inca bearing more than 10 times before one began to approach the cost of a single hardened large area thin bearing.

Figure 6: George Roberts, president of Inca (left), checks the specially designed 24 inch Inca azimuth bearing for telescopes with Marco Estrada, the senior research and development master machinist at Inca (right). A stiffening ring has been added under the bearing as well as a stiffening annulus on the inside of the inner race.

Because the azimuth bearing is thin and the thinnest race is mounted to the base frame of the telescope, we deemed it appropriate to make the aluminum base adapter plate from 6061T6 and machine it such that it supports the inner race from side loads which would be generated by driving the telescope with a belt drive on the large and thicker outer race. The base plate adapter was lapped to fit the inner race of the 24 inch bearing to 0.001 inch. This "press fit" will transfer side loads immediately into the telescope base without distorting the swivel race, hence keeping the force to move the telescope in tracking or swiveling more constant and linear, even when the slewing changes the center of gravity of the telescope system.

Other considerations that were taken to improve the performance of the swivel system were the metallurgical characteristics of the components. To better protect the races, molecular hard anodizing was imposed to harden the surface for the ball bearings to roll on in the races. In addition, all aluminum surfaces were fabricated out of 6061T6 aluminum alloy for rigidity.

Figure 7: Inca special azimuth bearing and adapter plates installed on the aluminum pre-production prototype azimuth base.

Altitude Assembly

We initially considered a fork altitude assembly. While a good approach for many telescopes, we quickly concluded that it would not fit in the Jeep transport vehicle without removal of both the mirror box (the heaviest component) and the fork arms. Also, it didn't appear to be an approach that would scale up gracefully beyond 1 meter. We settled on the popular Dobsonian "rocker" design with two trunions riding on four wheels—two of the wheels on a common shaft driving the trunions, and the other two idler wheels also connected together but by an un-driven shaft. It might be noted that the second shaft could have been driven by a second motor, although only one of the two motors could have been encoded.

Compared with a fork design, the rocker design allows the primary mirror to ride much lower in the telescope—essentially near the "bottom" of the rocker when the telescope is pointed at the zenith (its usual position during transport). This "low slung mirror" feature was key to fitting the telescope within the allotted vertical space of the transport vehicle.

The problem with the rocker trunions, however, is that they extend up high on one side of the telescope, well beyond the mirror box itself. This problem, however, had already been neatly solved by the designers of ultra-portable Dobs with "fold over" top trunion segments. In our case there wasn't ceiling clearance room for these segments to literally fold over, so we simply made them removable.

Figure 8: Chris and Reed Estrada lower the altitude assembly of the wooden prototype onto the azimuth base before fastening the two together with screws through the standard Inca bearing.

Initial OTA weight and balance estimates set a range of possible trunion diameters, and we chose one that kept the telescope within the height restrictions of the Jeep while, at the same time, provided altitude access from the zenith almost down to the horizon. The spacing between the altitude wheels was such that the wheels would contact the outer ends of the bottom (fixed) trunion segments with the telescope pointed at the zenith. Thus the unavoidable small discontinuities between the fixed and removable trunion sections occur at the zenith, the one position in the sky not observable by alt-az telescopes.

Figure 9: The bottom plate on the altitude assembly fastened to the azimuth base on the Inca pre-production prototype telescope.

Figure 10: Marco Estrada (left) is welding together the main portion of the altitude assembly for the aluminum prototype telescope as Max Weisenburg watches.

Initial Drive Considerations

We initially chose direct drives for our telescopes. The advantages of direct drives—which have no gears, belts, etc.—are their stiffness and total freedom from backlash (Beley 2003). The coils and magnets of the motor are literally fastened to the telescope itself. The telescope's structure and bearings act as the "frame" and bearings for the "motor." Not surprisingly, another term for direct drive motors is frameless motors. Two excellent examples of large direct drive telescopes are the 2.4 meter telescope at the Magdalena Ridge Observatory and the 8 meter Subaru telescope on Mauna Kea.

Direct drive motors are normally quite expensive, but Dave Rowe developed an axial flux direct drive system that allowed the motors themselves to be built at a surprisingly low cost. This axial flux direct drive motor and its matching control electronics and software developed by Dan Gray has been incorporated in the CDK 700 telescope built by PlaneWave Instruments (Rowe et al. 2010).

In the end, however, we chose not to use direct drive due to the cost of the necessarily high resolution, on-axis encoders. Although such encoders are really not very expensive—considering the other usual telescope costs (under $1500 each)—they were somewhat beyond our meager budget. In researching high resolution encoders, we did find that Heidenhain makes a 25 bit (33 million counts per revolution) encoder that costs under $500. This encoder, however, utilizes an interface protocol, EnDat2.2, which requires interface development before it can be used with the control system electronics/software we had selected.

We then considered both friction drives and belt drives and combinations of these two. Friction drives have been used with success in many large telescopes, such as the 8 meter Gemini telescopes, as well as many smaller telescopes. Either gravity or spring forces can provide contact between a small diameter drive roller and a larger diameter circular driven surface. Because the area of the surfaces in contact is small and the pressures are high, careful attention must be paid to both alignment and the choice of materials if drive slippage or other difficulties are to be avoided. Properly executed, friction drives can be very stiff and backlash free.

Altitude Drive

In altitude, we chose, as our primary stage of reduction, to use friction drive. The two large diameter trunions and their four rollers, mentioned above, provided a ready-made friction drive. It is not surprising that many alt-az telescopes, large and small, make this choice. To minimize slippage and contact forces, we employed relatively large diameter drive wheels.

The wooden prototype telescope uses two 3 inch diameter drive wheels in altitude. These wheels are connected together by a ½ inch diameter steel shaft. The aluminum pre-production prototype is more sophisticated. The altitude bearings ride on custom machined sheaves, fabricated from 1010/1020 CRS steel bar. Each sheaf is 8.00 inches in diameter with a root diameter of 6.500 inches and a root width of 0.760" + 0.010" -0." The four sheaves which hold the altitude bearings are held in place with 0.75" OD drive shafts which are mounted to 2.5" OD x 16g CRS torque tubes, supported by 0.75" eccentric locking pillow block bearings. This configuration minimizes torsion distortion between the driven end of the axle and the drive end. The altitude bearing is fabricated from 0.25" thick 6061T6 aluminum plate and is sandwiched between two 0.25" thick stainless steel "tires" which ride on the roller drive sheaves.

Overcoming Significant Backlash

Further reduction in altitude is provided, in many smaller alt-az telescopes, by a gearhead servomotor (or stepper motor) that drives the friction roller via a belt and pulley. This approach can be straightforward and low cost. One of our requirements, however, was to have an essentially backlash free drive system. We were concerned that wind gusts on these sizeable telescopes operating out in the open could move the telescope back and forth within the backlash "dead zone." We were also concerned with the adverse effects any significant backlash could have on the many small but precise reversal motions required in the chopping or scanning modes used in some of the on-axis aperture photometry applications we had in mind. Most gearhead servomotors have between 1 and 3 degrees of backlash in their output gears. Even planetary gears have significant backlash. Harmonic drives are an exception, although they are not low cost.

One approach to eliminating backlash in gearhead servo motors is to simply dispense with their gears entirely. With reduced reduction, however, one needs either to add in another stage of non-gear reduction or, alternatively, to utilize a servomotor with a higher torque constant equipped with a higher resolution encoder to make up for the lost reduction. Some investigation revealed that high torque constant motors were available at quite reasonable prices (HT Servomotors), and that advances in encoder technology have led to a compact 10,000 line (40 K tics/revolution) Heidenhain encoder that could be installed directly on the back of many servomotors. In quantities of 5 or more, this encoder only costs $83.

The toothed pulleys normally used in conjunction with timing belts are purposely designed to have backlash. As is the case with gears, some looseness of fit enhances smooth running at high speed and reduces wear and noise. Compared to most applications, however, the rotational speed of telescopes is limited. Most of the time telescopes barely creep along at about 1 RPD (revolution per day). Thus smoothness and wear are not issues, but backlash is. A manufacturer was found, BrecOFlex, which makes zero backlash toothed pulleys that "squeeze" the sides of the teeth on a belt when it is under tension. The belts in regular pulleys are purposely designed to "bottom out" well before they significantly squeeze the sides.

The combination of a high torque constant, un-geared servomotor with a fairly high resolution motor encoder, and a zero backlash pulley/belt should allow us to achieve sufficient reduction in altitude with just two stages, one friction, and the other a belt/pulley.

Azimuth Drive

In azimuth, one does not have large diameter trunions and rollers already in place, not to mention convenient gravitational force to maintain a good friction contact between them. No free lunch in azimuth! This is not to suggest, however, that one cannot machine a large diameter azimuth disc and force a roller into friction contact with this disk via spring forces or, alternatively, wrap a belt around a large diameter azimuth disk and drive the belt with a small diameter pulley. We reasoned that the outer surface of the 24 inch diameter Inca bearing could serve double duty as an altitude drive disk if it were appropriately machined and the bearing was sufficiently stiffened against radial forces.

With this in mind, we had to decide whether we wanted to push on the outer surface of the Inca bearing with a spring-loaded friction roller or pull on it with a belt. The primary advantage of pushing with a friction drive was that it could be stiffer. The disadvantage, as mentioned earlier, could be high contact forces and potential alignment or slippage difficulties. The advantage of pulling with a belt was a somewhat easier design and, perhaps more trouble-free operation. The belt's disadvantage was that it might be less stiff. This was a close, difficult call, but we decided, after some stiffness analysis by Dave Rowe suggested that a belt drive would probably be adequate, in favor of a belt drive mainly because it was simpler. We reasoned that if it did not work, we could upgrade to a friction drive.

Control System

One of us (Genet) has co-authored two books on telescope control (Truebood and Genet 1985 and 1995), and was familiar with the difficulty of developing control system electronics and, especially, writing and debugging its attendant software. Thus we never considered developing our own control system, as such a development would have taken many years. Instead, a survey was made of the available control systems and one was selected based on its maturity, sophistication, user friendliness, support, compliance with interface standards (ASCOM), thoroughness of documentation, and the size of and likely assistance from its installed user base.

After careful consideration we chose a Sidereal Technology (SiTech) control system designed and built by Dan Gray in Portland, Oregon. This sophisticated system has automated telescope modeling, star field recognition, and many other advanced features. Several hundred telescopes of all sizes and shapes, including both amateur and professional, use this control system. It was purposely designed to accommodate a wide range of both equatorial and al-az telescopes—including alt-az telescopes with instrument rotators. The SiTech control system is fully ASCOM compliant, and is extensively documented in two very thick manuals written by Dan Gray (2010). The SiTech Yahoo discussion group is very active, and experienced SiTech members do not hesitate to help neophytes.

Mirror Box and Mirror Support System

The mirror box holds the altitude trunions on two sides, supports the mirror cell, and forms the base for the quadrapod truss. The mirror box was kept as short as possible to allow the main portion of the telescope to fit into the height-challenged transport vehicle. A mirror cover keeps out dirt and moisture when the telescope is not in use.

Figure 11: The 1 meter DOTI spherical optical mirror in the wooden prototype's mirror box. Reflected in the mirror are the two removable upper trunion sections and the removable brace between them. Also reflected is a nearly full moon.

The mirror support system for the meniscus mirror is a conventional 18 point wiffletree for axial support and steel cable sling for radial support. Provisions have been made to appropriately hold the mirror in place during transport. The wiffletree mount support is fabricated from steel for the wooden prototype, and from 0.25″ wall 2″x 2″ aluminum tubing for the aluminum pre-production prototype.

Figure 12: The 18 point wiffletree mirror support system for the wooden prototype telescope. The triangles are aluminum, while the rectangular support is steel tubing. A circular plastic annulus in the center positions the triangles. The steel cable radial support has been draped over the triangles.

Truss

Both telescopes have removable truss structures that hold the instrument clusters above the mirror box at an appropriate distance. We chose a quadrapod because our intended use did not involve visual observations, so physical access to the focal plane was not a requirement. Also, as our intended use was for on-axis light bucket astronomy with the wooden telescope with the spherical mirror, and CCD photometry with the aluminum telescope, the additional optical interference caused by having four truss members in the light path as compared to four thin spider vanes was not a major concern.

To allow individual quadrapod truss members to have an easily transported length, and to stiffen the truss, we went with a two bay quadrapod truss. Our truss design is very similar to many of the telescopes at the Fairborn Observatory, the VERITAS Cherenkov radiation telescopes, and a number of radio telescopes.

Our initial design for the wooden prototype truss used square aluminum tubing that fit into square aluminum receptacles. Assembling this truss, due to the limited degrees of freedom at each joint during assembly, proved to be difficult and time consuming—jeopardizing our claim to portability (the Banich half-hour time limit). So we switched to the ball and socket joints made PlaneWave Instruments for their telescopes. The sockets are opened widely during assembly, allowing easy fitting. Once assembled, the sockets are clamped tight and the truss becomes a rigid structure.

Figure 13: The aluminum prototype mirror box as seen from the bottom during construction. George Roberts (left) watches senior designer Max Weisenberg check the dimensions.

Figure 14: Reed aims the wooden prototype at the moon. The permanent and removable trunions are clearly visible, as is the quadrapod truss structure itself. The disk on the left is a lightweight mirror box cover that keeps dust off of the mirror when the telescope is not in use.

Spherical Meniscus Mirror and Corrector

Each of the 1 meter telescopes has a different optical configuration. Both telescopes have an instrument rotator/focuser designed and built by Andy Saulietis, a refractive corrector, and an instrument package at prime focus. The wooden telescope utilizes a 1 meter, *f*/4, spherical meniscus mirror. This mirror was made by DOTI Optics from ¾ inch thick soda lime plate glass slumped in a kiln over a mold. DOTI Optics makes many similar mirrors for use in flight simulators, Cherenkov radiation telescopes, and other applications that require low cost spherical mirrors. The mirror weighs 70 lbs. We chose this mirror because we wanted to explore the use of lightweight spherical mirrors for light bucket astronomy applications (Genet and Holenstein 2010). A 2 meter equivalent aperture multiple mirror telescope could use four of these mirrors.

Figure 15: The 1 meter mirror for the photometric telescope is only ¾ of an inch thick and weighs just 70 lbs. This spherical mirror was donated by Mike Itz at DOTI Optical. DOTI makes these mirrors for use in flight simulators, Cherenkov radiation telescopes, etc. Spherical mirrors are, of course, relatively easy to figure. The 1 meter telescope uses two $50 off-the-shelf Edmund Optics lenses to correct the spherical aberration—at least sufficiently for photometric observations. Tong Liu at Hubble Optics designed the corrector, optimizing it for the field of view of an SBIG ST-402 CCD camera.

Joe Haberman at PlaneWave Instruments ran the initial optical tests on the DOTI 1 meter spherical mirror. The mirror was tested while still in its shipping box (cover removed). Both Ronchi and knife edge tests were made at the mirror's radius of curvature, some 26 feet from the mirror. Two sets of tests were made. Between the two sets, the mirror was rotated by 90 degrees in its box in order to estimate which portion of the observed astigmatism was due to imperfections in the mirror itself, and which was due to gravitational warping (potato chipping). Joe estimated the mirror was showing about four waves of distortion, with roughly half of this due to the mirror sagging unsupported in its shipping container. We were very pleased to have, when it was properly supported, a roughly two wave mirror—better than required or expected.

Joe Haberman at PlaneWave Instruments took the first cut at designing a two-element refractive corrector. His design used 75 mm lenses with four custom spherical surfaces. Dave Rowe took this design one step further by showing that nearly the same results could be obtained with lenses only 50 mm in diameter, but suggested that only custom-made lenses would work. Tong Liu at Hubble Optics rose to the "only custom-made lenses" challenge and the very next day came up with a design that utilized two off-the-shelf 50 mm lenses from Edmund Optics costing less than $50 each—one plano-convex and the other plano-concave.

Tong's design minimized the on-axis spot size. We asked Tong if, instead of minimizing the on-axis spot size, could he minimize the spot size at the edge of the field that would be seen by an SBIG ST-407 /ST-7 camera? The rationale behind this request was that the on-axis spot size was much better than required by any anticipated on-axis only application, while the off-axis spot size was larger than desired for "fuzzy" image simultaneous photometry of variable, comparison, and check stars. Tong's final design reduced the on-axis spot size (compared with uncorrected spherical aberration) by a factor of 4. In terms of the reduction of area, this reduction was a factor of 16.

Parabolic Starstone Composite Mirror and Corrector

Compared to the wooden telescope, the primary mirror for the aluminum telescope is faster, *f*/3 as opposed to *f*/4, and parabolic as opposed to spherical. Although the corrector and instruments in the aluminum telescope remain at prime focus, there is no reason they could not have been at a Newtonian or Nasmyth Cassegrain focus if human access to the focal plane had been desired.

The 1 meter parabolic *f*/3 mirror is an experimental mirror being made by Andrew Aurigema from a slumped ¼ inch thick meniscus glass front plate that is kiln-fused to a foam glass supporting structure. The foam glass is Starstone (Earthstone as it is more popularly known) made from recycled colored glass (primarily beer bottles) with a small ceramic component. The development of these mirrors has been described by Aurigema, Davis, and Genet (2010).

Figure 14: The thin top glass plate for the 1 meter foam glass composite mirror being made by Andrew Aurigema sits on top of his automated mirror grinding machine. Andrew displayed some of his small experimental mirrors on top of the 1 meter mirror.

Uncorrected, an *f*/3 optical system would have significant coma toward the edge of large format CCD cameras. Thus a coma corrector is called for. We intend to use the fast, f/3 capable Tele Vu Paracorr recently developed by Al Nagler.

Moving Beyond Prototypes and 1 Meter Aperture

Telescopes with apertures larger than 1 meter would still be portable by Banich's definition if they could be transported with ordinary vehicles and set up by two persons in a half-hour or less. Although a 1.5 meter telescope with its 5 foot diameter mirror would not roll into the back of most mid-sized SUVs, it would fit in the back of some of the larger vans and pickup trucks. A lightweight 2.0 meter telescope could be designed to be narrow and lightweight enough that it could be transported on low, lightweight trailer towed by an ordinary car, SUV, or pickup truck.

If it could be setup in half an hour, it too would qualify as portable. What Inca has learned making a 1 meter telescope should enable it to scale it up to inexpensive yet functional telescopes that will help bring into being the next generation of portable 1 to 2 meter telescopes.

Acknowledgements

We are pleased to acknowledge the many contributions made by others to these 1 meter portable telescopes. The spherical meniscus mirror was donated by Mike Itz and his associates at DOTI Optical. A parabolic 1 meter parobolic mirror is being made by Andrew Aurigema at OTF Design using a large piece of Starstone foam glass provided by George Morandin at Earthstone. Joe Haberman at PlaneWave instruments kindly measured the optical performance of the DOTI spherical mirror. Joe Haberman, Dave Rowe (Sierra Monolithics), and Tong Liu (Hubble Optics) all worked on the design of the refractive spherical corrector. Dan Gray kindly contributed the latest version of his very capable Sidereal Technology control system. Finally, Rich Hedrick and his associates at PlaneWave Instruments contributed the ball and socket joints that rescued our truss assembly.

References

Aurigema, A., Davis, D., and Genet, R. 2010. Glass foam composite mirror developments. In *The Alt-Az Initiative: Telescope, Mirror, and Instrument Developments*, eds R. Genet, J. Johnson, and V. Wallen. Santa Margarita, CA: Collins Foundation Press.

Banich, Howard. 2010. Telescope portability: moving a big telescope doesn't have to be difficult. In *The Alt-Az Initiative: Telescope, Mirror, and Instrument Developments*, eds. R. Genet, J. Johnson, and V. Wallen. Santa Margarita, CA: Collins Foundation Press.

Beley, Pierre. 2003. *The Design and Construction of Large Optical Telescopes.* New York: Springer.

Genet, Russ and Holenstein, Bruce. 2010. Alt-az light bucket astronomy. In *The Alt-Az Initiative: Telescope, Mirror, and Instrument Developments*, eds. R. Genet, J. Johnson, and V. Wallen. Santa Margarita, CA: Collins Foundation Press.

Gray, Dan. 2009. *Sidereal Technology Control System, Vol. 1: Setup Manual, Vol. 2: Operations Manual.* Portland, OR: Sidereal Technology.

Holenstein, B., Mitchell, R., and Koch, R. Figures of merit for light bucket mirrors. In *The Alt-Az Initiative: Telescope, Mirror, and Instrument Developments*, eds. R. Genet, J. Johnson, and V. Wallen. Santa Margarita, CA: Collins Foundation Press.

Trueblood, M. and Genet, R. 1985. *Microcomputer Control of Telescopes*. Richmond, VA: Willmann-Bell.

Trueblood, M. and Genet, R. 1995. *The Telescope Control Handbook*. Richmond, VA: Willmann-Bell.

Rowe, D., Hedrick, R., Gray, D., Haberman, J., Peasley, B., Keller, A., Iott, K., and Genet, R. 2010. A corrected Dall Kirkham nasmyth-focus telescope. In *The Alt-Az Initiative: Telescope, Mirror, and Instrument Developments*, eds. R. Genet, J. Johnson, and V. Wallen. Santa Margarita, CA: Collins Foundation Press.

Chapter 13

Sensible Telescope Enclosure Design

Craig Breckenridge

Breckenridge Astronomical LTD

Introduction

While we all do not hesitate to think of professional telescopes situated inside their enclosures as the norm, it is worthwhile to consider what the actual relationship is between the telescope and its surroundings, and how the sensible design of an observatory should take this into account. I will explore this symbiosis and present some generalizations based upon my over 15 years of experience as a member of one of the leading enclosure design teams in the world. The philosophy of practical observatory design is applicable to any telescope enclosure, and even the amateur with little funds can benefit greatly from some simple forethought. I'll present some ideas in layman's terms that anyone can utilize to some degree, and while the complete integration of all these ideas may be beyond the resources of some, ingenuity can provide comparable systems once the intent is understood. The things to remember with enclosure design are what the end goals really are and how they can be sensibly achieved.

Function

What does an enclosure actually do? If we examine what the function of an enclosure really is, we can begin to formulate a practical approach to the design and construction of a suitable observatory. While many think of an enclosure as just a method of protecting the telescope for security reasons, there are some other features that really should be considered. First and foremost among these features is the ability of the enclosure to protect the telescope, not from would be thieves, but rather from the environmental conditions that it experiences throughout the daytime. In order for a telescope to perform at its best, we should be striving to minimize the amount of time we spend each evening bringing it into equilibrium with our nighttime observing conditions. Of course, this doesn't mean we should forget about the thieves!

Another function an intelligent observatory design should include is how the enclosure protects the telescope during its operation in the observing session. Does the enclosure prevent wind from buffeting the telescope while it is in use? Does the design allow the separation of pier and enclosure to prevent the transmission of vibration? These important questions must be addressed so the enclosure does not contribute to poor observing conditions.

An additional feature that must be considered is how the observatory is going to be used. Will the astronomers sit beside the telescope during operation or will they be at a remote location? Sometimes the remote location is simply another room but more often we are seeing operation of observatories across many miles, even on different continents. The operation of a remote observatory has some additional considerations that need to be taken into account.

Lastly, one must always include some budget analysis. The complete automation of an observatory may be the end goal, but this could be costly if it is approached without careful planning. This is another place where one's ability to see beyond the standard design can have a great effect. There are many paths to follow to one's goal, but by being inventive, the enclosure does not need to deplete all your resources.

The Enclosure As Protector

As I stated above, the primary function of an intelligent enclosure should be to provide protection for the telescope during the hours it sits waiting for its nighttime vigil. I am of the opinion that this includes keeping the telescope at near ambient conditions for the evening's foray into the wonders above. In order to accomplish this, an enclosure needs to be able to insulate its interior from higher daytime temperatures. There are several ways to perform this feat and each has been employed to great success on professional observatories.

Speaking of the enclosures with which I am familiar, the Canada France Hawaii (CFHT) enclosure, the twin Keck enclosures, and both Gemini North and South utilize an insulated interior, an exterior finish that reflects most of the solar energy away, and an interstitial space that provides some measure of active ventilation. There are differences between these examples but the basic principle is the same. In all cases, the insulation on the interior of the enclosure is sprayed polyurethane over an oil-tempered hardboard. The foam is coated with a fire retardant coating, but government regulations put in place since the construction of these enclosures now require a fire retardant within the foam itself. The hardboard is installed directly on the interior chords of the structural rib skeleton and together they make an airtight interior surface. The exterior of these enclosures is a welded steel skin that is attached to the exterior chord of the same structural rib system. This steel skin provides adequate protection from the elements, and as it is welded, it also provides the weather tight conditions required for the harsh climate at the summit of Mauna Kea. The interstitial space between the interior insulation and the exterior skin on these enclosures can be passively vented or actively vented by the use of fans at the upper levels of the dome. The CFHT and Keck enclosures are painted an observatory white that has proven to have low emissivity. The Gemini enclosures are coated with an aluminum solids reflective coating that has similar emissive features. All of these telescope enclosures function quite well and I only see one drawback: the thermal mass of the steel on the exterior of the enclosure provides a heat sink that will radiate its energy to the night sky for several hours.

The Subaru Telescope enclosure uses a slightly different approach. This enclosure is clad in bare aluminum that reflects a large portion of the thermal radiation from the sun. The Subaru enclosure is insulated on the interior side of this skin with the spray-on foam and fire retardant. All the structural steel within the Subaru enclosure is thermally isolated through extensive use of Fabreeka pads and washers. The interior of the Subaru enclosure contains a series of wind tunnels that can be controlled to vent the interior of the enclosure. Subaru's cylindrical design was developed in wind tunnel modeling and is intended to make the best use of the natural environmental conditions at its location. Again, the large thermal mass surrounding the exterior of the enclosure could be a concern.

Figure 1: The exterior of the Subaru enclosure on Mauna Kea, demonstrating its cylindrical shape.

In the summer of 2008 I had the very good fortune to visit the Large Binocular Telescope (LBT) on top of Mt. Graham in Arizona. While I am not a fan of the shape of the enclosure (I prefer the esthetics of a conventional dome), I was very pleased to see a recent development in insulating technologies being put to good use. In fact, the enclosure has a feature which I think represents the goal we should all strive for when considering enclosure design: the insulated surface is on the exterior of the enclosure and the structural steel framework is protected from the sun's rays. This means instead of having a thermal mass on the exterior of a hundred tons or more, we have only the smaller mass of the insulating panel's metal coating. The panel system used on the LBT is a composite panel with exterior steel skins and a Polyurethane foam interior. The panels are attached to the structural steel by self-tapping screws and sealed with butyl caulking at both the molded-in labyrinth seam between panels and the straight seams at the ends of the panels. These panels are manufactured by several companies worldwide and are used increasingly in conventional construction. The incorporation of such panels into the intelligent design of an enclosure is something I highly recommend. Careful study of the deficiencies of the joint design will benefit the enclosure designer.

Figure 2: The exterior of the Large Binocular Telescope enclosure on Mt. Graham, Arizona.

All of the enclosures I've mentioned have some measure of air conditioning that may be employed throughout the daytime. The ability to keep the telescope thermally stable and near the expected ambient observing temperature is something that makes all the forethought in thermal planning of the enclosure worthwhile. Development of the thermal management system will vary from the professional observatory to the amateur, but by giving this careful consideration, economical systems and procedures may be developed.

The Enclosure During Observation

Thermal protection of the telescope during the day is only one of the many functions an enclosure must perform. Another consideration is how the enclosure improves the environment during observations. One of the factors that affects this is the enclosure size relative to the telescope's length from primary to secondary mirror. By choosing this wisely, the basic geometry for the enclosure is developed. Make your enclosure too small and the telescope will be subject to wind buffeting. Make it too large and the interior space becomes too large to effectively cool during the day. For very large observatories, the stratification of temperature gradients becomes a very serious problem.

I find a simple rule of thumb for enclosure size relative to the telescope is to take an additional 1/3 of the primary/secondary length. For a telescope with a length of say 4 meters, I would expect the dome to 'feel right' around 6 meters radius. Simply laying out the telescope and checking will show this has the appearance of being about the correct size esthetically. Confirmation that this is correct requires proper Computational Fluid Dynamics (CFD) modeling and analysis to determine that the air flow in and around the enclosure will provide the maximum benefit, but these are often beyond the means of smaller projects. If the budget allows, full thermodynamic modeling should take place since many of the features I am suggesting need to be carefully coordinated with the enclosure design.

Figure 3: The Gemini South enclosure on Cerro Pachon, Chile, demonstrating the deployed windscreen.

This rule of 1/3 longer should provide some measure of protection from wind buffeting for the top end of the telescope. An additional feature that many observatories have is a secondary windscreen that can be raised and lowered when the shutters are open to minimize the amount of open slot. When observing at higher elevation angles, the windscreen is raised; when observing lower elevations the windscreen is lowered accordingly. This minimization of the slot assists in protecting the telescope from wind buffeting which, as we all know, is one of the leading sources of poor observations.

One thing that quickly becomes apparent when developing the initial geometry of your enclosure is the difference between an equatorial telescope and an alt-az telescope. Taking my example of a telescope with a 4 meter length from primary to secondary mirror as an example, consider that both the alt-az and equatorial telescopes would have the axis of rotation probably about 1/5 of this distance toward the secondary or about 800mm in front of the primary. Allowing for the telescope to sweep from zenith to horizon, pointing on an alt-az configuration, would require that the elevation axis be about 2.5 metres from the base supporting structure. An equatorial telescope would require about 2 metres for the telescope to sweep from zenith to horizon in declination. To this must be added the supporting structure for the RA axis. Conservatively this would add another 2.5 metres or so. When you compare these admittedly rough numbers to two telescopes with similar lengths, they require a height for the location of the center of rotation of 2.5m for the alt-az and 4m (taken at roughly 45 degrees latitude) for the equitorial telescope. Since we would prefer the axis of rotation for the telescope to be the centre of the enclosure diameter for geometric and esthetic reasons, the floor for the alt-az telescope would be a minimum of 1.5m closer to the axis of rotation than the equitorial telescope. The size of the enclosure is also affected by the location of the RA and azimuth rotational axis locations. Simply laying out the amount of space needed 'behind' an equitorial telescope will show that the dome size must be increased so the dome will remain clear of the pier of the telescope.

This brings us to another important feature of sensible enclosure design: the minimization of vibration transmitted to the telescope due to operation of the enclosure. The simplest method for achieving this is to design the telescope pier so that it is isolated from the enclosure foundation. On small observatories I have designed in the past, I have accomplished this by having vibration suppression material cast into the floor between the observation floor (usually connected to the observatory foundation and subject to vibration of both the operation of the enclosure and footsteps) and the telescope pier. I have always strived to have the pier foundations as solidly secured as possible. On some observatories I have been able to have the pier joined directly to bedrock through the use of chemical anchors and rebar, while allowing the observatory itself to 'float' on a layer of sand above. Careful design of this separation of enclosure and pier is imperative regardless of class of instrument; vibration is the worst condition to have to overcome with modification once an observatory has been built.

Another feature to try and incorporate in the enclosure design is the ability to assist in the creation of a laminar airflow across the primary mirror during observation. I am aware of many amateurs who have installed small fans to blow across the primary mirror during observing, as this improves the viewing. What we're really doing when we create this effect is limiting the ability of the primary mirror to have thermal 'hot spots' across its surface. By keeping the mirror thermally uniform, we have less chance of aberrations due to mirror deformity (always provided the mirror is properly supported of course). In recent years many observatories have been designed with the ability to provide cross ventilation by having a series of vents at the level of the primary mirror. The Gemini observatories are perhaps one of the best examples of this with their two vertically bi-parting sides able to be opened fully or partially depending on observing conditions. A benefit to this design is the ability to 'flush' the enclosure with the rapid exchange of interior and exterior air when preparing to observe. Commencing such flushing in advance of observations greatly facilitates having the telescope at the ambient observing temperature. The ability to control the airflow across the primary during observation can greatly improve a telescope's performance and bears serious consideration.

Figure 4: The Gemini North enclosure, Mauna Kea, showing the side vents opened.

The Enclosure Operations Room

The manner in which the enclosure will be controlled contributes on many levels to its efficient design. Primarily these have to do with equipment choice when the observatory is being designed and do not have much economic impact on the observatory itself in the overall scale of things. If the observatory will be operated from within the confines of its own space, it is possible to have simple switches to control the lights, fans, dome rotation, and telescope; however if it is to be controlled remotely (and this can simply be from an adjacent building), all of these controls need to be addressable with this in mind. In recent years the development of robotic observatories has progressed to the point that it is possible to control an observatory safely from the far side of the planet. I have participated for the last three years in the development and operation of two remote observatories. One of these was in a stand-alone dome of its own and the other was with several other telescopes in a roll-off roof enclosure. While I do not claim to be the developer for either of these observatories, I did participate in the development of the operational procedures for our two telescopes. Both of these observatories used off-the-shelf software and the ability of the Internet to operate. It is the careful selection of these key pieces that must be planned so that software conflicts do not develop. Knowing in advance if remote operation is going to be utilized is very important to everything from mount design to web cam position.

In both of the remote observatories I have participated in, we utilized a web cam mounted on the telescope to see where it was actually pointing. We found this beneficial several times when the enclosure would get out of sync with the telescope; we'd be imaging the inside of the observatory dome and wondering why we weren't getting any images! We were able to monitor the opening of the shutter with this low light camera as well, and on several occasions were able to determine that the shutter had malfunctioned while opening or closing. This was when having someone on site was a great benefit. We also had positioned a web cam on the enclosure to look at the telescope itself. This camera proved its worth when a cable came undone one night while imaging. The cable in question was our CCD cable and we had lost control of the camera. Again this was easily rectified once the problem was identified. Both of these examples serve to illustrate one drawback of remote observation: if something goes wrong, a plan of how to fix the problem must be conceived of in advance. This is particularly important with functions that are critical to the survival of the telescope. Failure of the shutter to close during a thunderstorm, for example, could be very expensive to repair.

Control of the enclosure rotation, shutter, telescope, lighting, fans, cameras, filter wheel, web cams, and local computer can all be accomplished over the Internet, provided the correct equipment is chosen in advance. We utilized the computer on site to control everything and used a remote desktop application to view the remote screen. All remote functions were controlled by that computer, including the compressing and transmission of data collected, which was timed for non-observing hours. We utilized Telnet to control the power and could even reboot the computer remotely if required.

In any observatory design, regardless of initial use, I feel it is wise to be prepared for the future. That said, I think it appropriate that any enclosure design should include the provision for remote operation.

The Enclosure as a Cost

Lastly (or firstly) the intelligent design of an enclosure must include the cost. There is no point in designing an enclosure that knowingly exceeds your available funds. Once your basic design is settled upon, take a look at the actual costs to purchase all your materials and construct the enclosure. Keep in mind things such as transport of the materials, remote site location, and labor experience. All affect your bottom line. Once your observatory is constructed, how much will it cost to operate? Keeping an observatory functioning for years to come is not something that can be taken lightly, as the costs do not diminish over the years; they grow. It has often been said that the most expensive part of a professional observatory is not the cost of construction or the optics but the day-to-day operational costs. This is especially true if the observatory is remotely located.

I have found it beneficial to perform cost analysis of my designs as I go, and to continue to develop the estimated cost as the observatory design progresses. I keep a contingency in mind as I go, as there is always something that comes up which you have not planned for. Once construction starts, keeping track of the actual costs against the estimated ones lets you know if you will encounter funding problems ahead. I guess I'm a product of my generation, as it has been noted that the baby boomers are planners. I like to know as much in advance as possible in order to plan out how to deal with problems when they arise.

It is simple enough to develop your cost estimate. Like everything else in life, address your enclosure design in parts. Divide the project into manageable sections that can be easily isolated for analysis. You'll soon find that the logical manner in which to break your project down is the same as the logical manner in which to design the segments.

This assists not only your budget considerations but makes it easier to monitor progress on your design overall. This is applicable to both professional observatories or to simple portable windbreaks to protect your portable telescope when observing.

A Final Word or Two

I have tried in the few paragraphs preceding this final section to give an overview of the nature of the beast that is the telescopic enclosure. I find great satisfaction in developing something that is both sensible and economic. I have many ideas for enclosure design that many might consider off the wall, which I think can be developed into practical applications of technologies now at hand. Just think what it must have been like for those original designers who struggled with engineering calculations that we can now perform almost at the push of a button. While the basics of telescopes and observatories are still the same, it is the refinement of the inherent principles that will lead to great ideas. I look at observatory design as a whole, as both the enclosure and the telescope must be considered to act together as one. The examination and thoughtful study of what we're really trying to achieve will lead us to sound, sensible solutions.

Chapter 14

MAGIC

The World's Largest Cherenkov Telescope

Michele Doro

University of Padova, Italy

Introduction

The Major Atmospheric Gamma-ray Imaging Cherenkov telescope (MAGIC), operating since 2003 in the Canary Island of La Palma, is currently the world's largest Imaging Atmospheric Cherenkov Telescope (IACT). An IACT takes advantage of the electromagnetic shower of particles initiated by very high energy cosmic gamma rays impinging into the high atmosphere. The collection of Cherenkov photons produced by the electromagnetic shower permits the reconstruction of the primary gamma ray energy and its direction.

Gamma ray emission is found in the non-thermal emission from galactic and extragalactic violent environments, such as pulsar wind nebulae, jets of particles from active galactic nuclei, the magnetosphere of pulsars, and gamma ray bursts. MAGIC consists of two telescopes with parabolas of 17 meters in diameter that allow attainment of the lowest energy threshold of all IACTs at only 25 GeV, thus covering the open energy gap with gamma ray satellite experiments. MAGIC is performing fundamental discoveries in astrophysics, such as discovering gamma ray emission from a black hole at five billion light-years away, revealing a more transparent universe than expected, and exploring the pulsed emission of gamma rays from the Crab pulsar—unveiling important results on the acceleration of cosmic rays in this extreme object.

In this document, a detailed view of the technical features of MAGIC is presented, with particular attention to its mirrors. Some scientific highlights are also presented.

Gamma Ray Astronomy as a Probe for Cosmic Rays and New Physics

Gamma rays (GRs) of astrophysical origin are produced as a result of non-thermal emission from extremely active regions in the universe. They are usually found in association with energetic parent particles that trigger their emission, mainly through inverse Compton scattering on low energy seed photons and synchrotron emission. The earth's atmosphere is continuously bombarded by such energetic particles (called *cosmic rays*) and by GRs. Cosmic rays (CRs) are mainly composed of protons and helium with secondary traces of heavier nuclei. They have a double provenance. Up to energies of 10^{15} eV, they are believed to be accelerated within our Galaxy. At larger energies, up to the most powerful CR observed at 10^{21} eV,[1] they originate in extragalactic regions (Gaisser 1990). Figure 1 shows the spectrum of CRs. Their flux ranges over more than 30 orders of magnitude, from several thousand particles per square meter per second, to few particles per century! Despite their discovery by V. Hess almost one century ago (Hess 1912), scientists could not answer two fundamental questions about their nature: how are CRs accelerated to such dramatic energies? and where are they accelerated from?

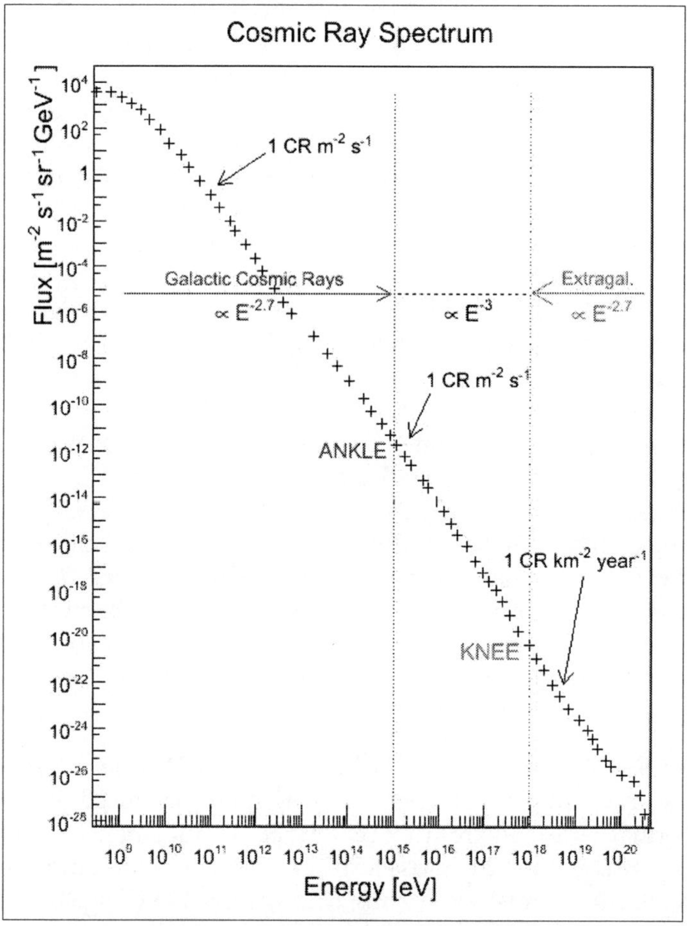

Figure 1: The spectrum of cosmic rays.

[1] 10^{21}eV is the kinetic energy of a tennis ball flying at 100 km/h!

These puzzling questions may soon be solved. New light has been recently shed by the Auger experiment in Argentina. Auger is composed of thousands of water tanks distributed over 3,000 km,[2] detecting fluorescent light from ultra-high energy (UHE) CRs (above 10^{15} eV) crossing the earth's atmosphere. They found a correlation between UHE CRs and active galactic nuclei (AGNs), i.e. galaxies hosting supermassive black holes ($M>10^9$ solar masses) at their center, which are believed to power extremely energetic jets of particles and radiation that can be propelled out of the galaxy for several kilo parsecs (Abraham et al. 2007). Below 10^{18} eV, CRs show a quite large degree of isotropy as a result of multiple interactions with interstellar and intergalactic magnetic fields. This constitutes one of the major difficulties in the study of CRs, whose origins cannot be directly traced.

We now turn to GR astronomy. GRs are the perfect probe to study the sites of CR acceleration, because GRs travel undeflected by magnetic fields. GRs are studied by satellite experiments, which measure their energy through calorimeters and from the ground. Satellite experiments suffer from typical weight and size limitations of the satellite payload and therefore have limited detection areas of the order of one square meter. The spectrum of GRs of astrophysical origin typically resembles that of CRs, with rapidly falling fluxes as the energy increases, and satellite experiments do not have enough event statistics above a few tens of GeV. On the other hand, ground-based experiments only indirectly detect GRs through the secondary products of their annihilation in the atmosphere, which is completely opaque to them. We will see below that this has the interesting advantage of multiplying the number of photons and increasing the collection area to around 50,000 m^2, thus making it possible to observe GRs above the GeV with sufficient statistics. While satellite experiments were historically the first to detect GRs (SAS-2 and COS-B) in the 1970's, it was only in 1989, after almost 50 years of attempts, that the first GR source was detected at ground level. The Whipple telescope (Weekes 1989) observed a signal from a supernova remnant, the cloud of ejected material following a supernova event. They observed the Crab Nebula, destined to become the standard candle for GR astronomy for its bright and steady emission. Since then, many detector technological advances permitted the extension of the catalog of GR sources with a hundred new members, while a thousand are expected to be discovered in the near future. This research field is very active, and important scientific discoveries could be around the corner. For a recent review of the prospects of the IACT technique, we refer the interested reader to the recent contribution from Buckley and others (Buckley et al. 2007).

The Observational Technique

The earth's atmosphere is opaque to GRs, which interact with the electric field of molecules in the top atmosphere (20 to 30 km altitude) and annihilate into an electron-positron (e+e-) pair. A cascade of millions of particles is initiated: the electron-positron pair radiates GRs via bremmstrahlung radiation, which again produces a second e+e- pair, and so on. The shower develops until the mean e+e- energy is below the threshold for ionization (around 83 MeV in air) and then it rapidly dies off. Its growth is accompanied by the emission of a second form of radiation: inside the shower, most of the e+ and e- are above the threshold for Cherenkov radiation. This radiation is emitted by the atmospheric particles whenever a charged particle is traveling at velocity larger than the velocity of light v>c/n (n is the air refractive index), because a coherent front wave is generated. The light is emitted within a cone of aperture θ=1/n which slightly widens as the shower develops as shown in Figure 2. The main emission is localized at around 12-8 km altitude, according to the primary GR energy, where the number of e+e- is maximum.

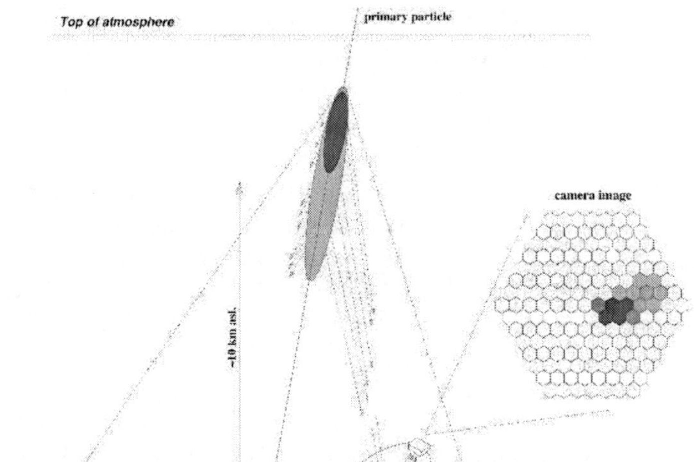

Figure 2: Scheme of detection of Cherenkov light from atmospheric showers. The Cherenkov emission produced by electromagnetic showers initiated by very high energy gamma rays are focused onto a multi-pixel camera (Courtesy of M. Lopez-Moya).

The Cherenkov light illuminates a nearly uniform circular spot of about 120 meter radius at the MAGIC altitude of 2200 meters. If a telescope is located inside this Cherenkov pool, the light hitting the reflector can be concentrated and focused onto a multi-pixel camera composed of a large number of photon devices, normally photomultipliers (PMTs). The effective collection area is therefore virtually the entire Cherenkov light pool, which has an extension of 50,000 m^2. The Cherenkov light cone is transposed into an ellipsoidal shaped image in the telescope focal plane. An image reconstruction algorithm (Hillas 1985) allows the determination of the major parameters of the primary particle, such as its energy and direction, as well as its likelihood to be a GR rather than a background event. The main background is constituted by the far more frequent atmospheric showers induced by CRs (roughly 1 GR for every 1000 CRs). Hadrons interact in the top atmosphere and produce principally pions and kaons. Neutral pions immediately decay into a pair of GRs that trigger a sub-electromagnetic shower, which can mimic a pure GR shower. Other sources of background are the Cherenkov light from muons produced in hadronic showers, the electromagnetic showers induced by cosmic electrons, and the diffuse night sky background (NSB) light. The background discrimination is performed through image discrimination and comparison with Monte Carlo simulations.

The MAGIC Telescopes

The MAGIC experiment is composed of two telescopes operated stereoscopically (Figure 3). They are maintained by a collaboration of around 150 physicists from 23 institutes and 7 countries, headed by Germany, Italy, and Spain.[2] The first telescope dubbed MAGIC I has been fully operational since fall 2003 and is currently in its fourth-year observation cycle. MAGIC has currently the world's largest mirror dish of 17 m diameter that allows the attainment of the lowest energy threshold of all current IACTs. There is, in fact, a strong physical motivation to perform observations in the energy band between 10 and 100 GeV, left open by past generations of satellites (limited to less than 10 GeV by insufficient statistics) and ground-based GR experiments (limited to more than 100 GeV by sensitivity). In the case of MAGIC, the large reflector area allows the reconstruction of primary GRs above 25 GeV (depending on zenith angle and analysis modes) with an integral sensitivity above 100 GeV of 10^{-11} photons cm^{-2} s^{-1} in 50 hours of observation.

---

[2] For further detail, refer to the experiment webpage, www.magic.mppmu.mpg.de

Figure 3: The MAGIC telescopes system. On the left, MAGIC I, operating since 2003. On the right, MAGIC II started operation in 2009. Courtesy R. Wagner.

MAGIC is located on the Canary Island of La Palma (28.75 N, 17.89 W) at an altitude of 2225 meters. The observation conditions on site are among the best in the world (Lombardi et al. 2008), even if occasional strong winds and rains, winter snowfalls, and high humidity demand a huge technical effort to prevent damage and aging. The size of the telescopes does not allow for dome protection as with optical telescopes. MAGIC I developed several technological advances:

- the world's largest mirror dish;
- a lightweight mirror dish constructed from carbon fiber reinforced plastics (CFRP);
- first use of diamond-turned lightweight all-aluminum sandwich mirrors;
- first use of an active mirror control (AMC) to counteract small deformations of the mirror during positioning and tracking;
- first use of low gain hemispherical PMTs with diffuse lacquer coating and special light catchers resulting in increased quantum efficiency and allowing prolonged operation during modest moon shine (Paneque et al. 2004);
- use of the transmission of the fast analog PMT signals over 160 m by means of optical fibers to a distant counting house;
- use of a 2 GHz FADC digitization of the fast PMT pulses.

A detailed report of the design ideas can be found in Barrio et al. (1998).

General Design Considerations

Despite its similarity to large optical telescopes, there are quite a few different constraints that very much influenced the design of MAGIC in particular and IACTs in general. In the case of optical astronomical observations, one observes stellar light, which originates basically at infinity. In case of ground-based GR astronomy, one detects Cherenkov light emitted by an extended air shower with a diffuse transverse extension and a typical length of several meters. Due to the stochastic showering process, the light emission region will have an inherent limitation in precision (improving normally with the energy). While in optical telescopes the mirror precision can be pushed to reach the diffraction limit, in IACTs the observation of a near extended object allows for a mirror of lower optical quality, which may be composed of many small mirror facets. In term of point spread function (PSF) requirements, IACT mirrors can have a factor 100–500 worse PSF compared to that of optical telescopes. The PSF of the single mirror facets is typically around 0.02° and that of the overall mirror 0.05°.

Again, while the tracking precision must be high for optical telescopes, the tracking and stability requirements for IACTs are much less demanding. Modern IACTs achieve a pointing resolution from single showers of 0.1–0.01° (energy dependent).

The overall photon detection efficiency for state-of-the-art IACTs is only 10-15% when averaged over the spectral range between 290 and 600 nm and is one of the aspects where new technological developments are being concentrated. In order to fully contain the shower image, a wide field camera of at least 3–5° field-of-view (FOV) is required. To observe significant sections of the sky, even wider FOV cameras are needed. Unfortunately this is very difficult to achieve. The use of Schmitt optics or a secondary mirror are both very costly, require much more precise optical elements, and would result in additional loss of photons. The optical image quality at the rim area of the 3–5° FOV camera should not degrade too much. This requires a rather large *f*-number in case of single mirror optics. As cameras can be quite heavy, a strong and heavy camera support is needed. The structure has to be anchored to the mirror support dish, resulting often in a large local mirror deformation. The light sensors must have ultra-fast response, on the order of a nanosecond, to reduce the large NSB as well as to provide some information about the timing structure differences in showers originating from different types of incident particles. In optical telescopes, sensors can integrate source photons over rather long times but the spatial resolution must be high. State-of-the-art IACTs have pixels ranging between 0.05–0.2° diameters; therefore a PSF of half a pixel size is sufficient. This allows for a considerably cheaper mirror construction. For large diameter mirrors a coarse isochronous mirror profile, with a time spread smaller than 1 ns, is required.

In addition, one of the main physics goals of MAGIC, besides the observation of steady or slowly flaring GR sources, has greatly influenced the design of the telescope: the observation of the prompt-emission from GR bursts (GRBs) and their prompt afterglow, which happens several seconds after the bursts. This requires the telescope to be repositioned as fast as possible to the sky position provided by an alerting satellite. To achieve a positioning time to any position on the sky map within much less than a minute requires both a very light-weight but steady telescope and a fast acting drive system. MAGIC uses therefore a low weight mirror as well as a low weight mirror support frame made from CFRP tubes.

Mounting and Drive System

The reflector frame satisfies three demands: it is very large, it is light-weight, and it can be repositioned extremely rapidly. The space frame is made from CFRP tubes and has a weight of only 5.5 tons. The CFRP construction is about three times stiffer and has less than a third of the weight of an equivalent steel construction. The structure has an alt-azimuth design and is mounted onto a circular rail 19 meters in diameter (Figure 4). The telescope can be moved from -80 to 105 degrees in declination and 450 degrees in azimuth. The camera is positioned 17 meters from the reflector and is carried by a single aluminum tubular arc. Pre-stressed steel cables keep the tubular arc in its nominal position. The weight of the camera is around half a ton. The reflector dish, under its own weight, slightly deforms locally from the ideal profile. The small deviations are corrected via a re-orientation of each mirror facet of the reflector through an active mirror control (AMC) system. Two motors control the telescope motion in azimuth and one motor the zenithal motion, with a maximum power consumption of around 7 kW per motor. The angular positions are controlled by absolute shaft-encoders of 14-bit precision/360 deg. In addition, a star-guider camera, mounted at the centre of the reflector, monitors the positioning of the telescope by triangulation with stars in the telescope FOV (Biland et al. 2007). The lightweight structure allows for very fast repositioning of the telescope to any position on the sky within only 20 seconds.

Camera and Readout

The camera of the MAGIC I telescope has a 3.6 deg FOV and comprises 576 PMTs in total, with 396 1 inch PMTs from Electron Tubes Ltd. of type 9116 covering the inner section (up to 1.2 deg radius) and 180 PMTs of type 9117 and 1½ inches for the outer part of the camera. The hemispherical cathode of the PMTs was coated with a diffuse lacquer

Figure 4: Rear view of the MAGIC telescope.

doped with P-Terphenyl, shifting the short wavelengths UV component of the Cherenkov light into the spectral range of maximum sensitivity to increase the overall quantum efficiency (QE). In order to minimize losses due to dead space between the densely packed PMTs, hex-to-round light concentrators were used (also called Winston cones), which increased the detection efficiency by deflecting photons such that their trajectory passes the semitransparent photocathode twice. The mean QE of the entire system (mirror, camera window, cone, effective PMT QE) between 300 and 600 nm is around 15%. The 6 dynode PMTs are operated at a low gain of around 40,000 in order to allow observations at modest moon shine (Albert et al. 2007). By observing under moderate moon conditions, MAGIC increased the duty cycle by 40%, extending the overall exposure time from 1100 to around 1500 hours/year. The fast PMT signal is converted into optical signals by means of large dynamic vertical cavity surface emitting lasers operating in analog mode and routed via 160 m long optical fibers to the counting house 80 meters away. At the counting house, the optical signals are converted back into electronic signals, split and routed to the trigger branch and to the fast digitization and recording system. The trigger is twofold: in level-0 the PMT analog signals are converted into logical signals once they exceed a 5-8 overlapping photoelectrons threshold. The level-1 trigger is dedicated to a time-coincidence topological reconstruction: when a number of pixels (adjustable by software) are activated at the same time according to a next-neighbor logic (2NN-5NN) a trigger signal is generated. The typical trigger rate is about 150-250 Hz interleaved by frequent calibration events. Upon a trigger, the data received are digitized by ultra-fast FADCs at 2 GHz sampling rate. Since 2005, this system has replaced the original digitizing system of only 300 MHz sampling rate and allows a much better timing measurement and a considerable reduction of the night sky light background. As a result, the overall telescope sensitivity has increased of a factor of two (Albert at al. 2008a).

Figure 5: The camera of MAGIC 1 with open lids.

The camera of MAGIC II has adopted quite a few new solutions with respect to MAGIC I that strongly boost its properties. The number of pixels has increased to 1039. They all have the same size of 1 inch.

Pixels are grouped into 169 clusters that constitute an operating unit. The trigger area has increased and consequently the sensitivity of the camera. The PMTs were replaced by Hamamatsu R10408 6-stage with hemispheric photocathode, operated at 40,000 gain, with photon quantum efficiency (QE) of 32% at 350 nm. In the future, we plan to substitute all photomultipliers with avalanche photon detectors (APDs), with expected QE larger than 50% and increased single photon resolution.

The Mirrors of the MAGIC Telescopes

Although the frame is almost identical, MAGIC II is a different system with respect to MAGIC I, where all the subsystems have seen an advanced technological improvement. Also the technology of the mirror facets has improved. MAGIC I and II both have a parabolic layout profile of 17 m aperture and 17 m focal length, and are tessellated with spherical mirror facets. The choice of the parabolic profile is motivated by the necessity of maintaining the time structure of the Cherenkov signal from atmospheric showers. The typical duration of the Cherenkov flash is about a few nanoseconds. When the telescope aperture exceeds about 10 m, the time delay between photons hitting the center of the mirror and its rim exceeds the Cherenkov flash duration in case of spherical reflector. This prevents a correct primary reconstruction and heavily undermines the background discrimination. The parabolic profile is isochronous, but presents larger aberrations than the spherical ones, in particular at off-axis incidence.

In the following, we will restrict the discussion to the mirror facets of the MAGIC II telescope. MAGIC II hosts two types of mirrors: a diamond-machined all-aluminum sandwich, and a cold-slumped glass-aluminum sandwich.

MAGIC II Aluminum Mirrors

Each mirror is a square of 0.985 meter per side and has a spherical profile whose radius of curvature is optimized for the position in the telescope to best approximate the paraboloid. The mirrors are an all-aluminum, lightweight sandwich assembly of two aluminum skins and a hexcell honeycomb structure in between (Figure 6). The mirrors are assembled with aeronautic glue 3M AF-173-2K and cured in an autoclave at 140° C and 3-5 bar pressure for several hours. The concave surface of the mirror is later milled with a rotating diamond-head machine at LT-Ultra in Germany. The milling machine is based on the fly-cutter technique, characterized by floating rotating parts over a thin air layer to reduce friction. The diamond-milling provides an average surface roughness of 4 nm and a mean reflectivity between 300 and 600 nm (where most of the Cherenkov light is concentrated) of 85%. The surface is further coated by a thin layer of quartz (with some admixture of carbon) for protection against corrosion and acid rain. Each mirror focuses light within 0.5 mrad.

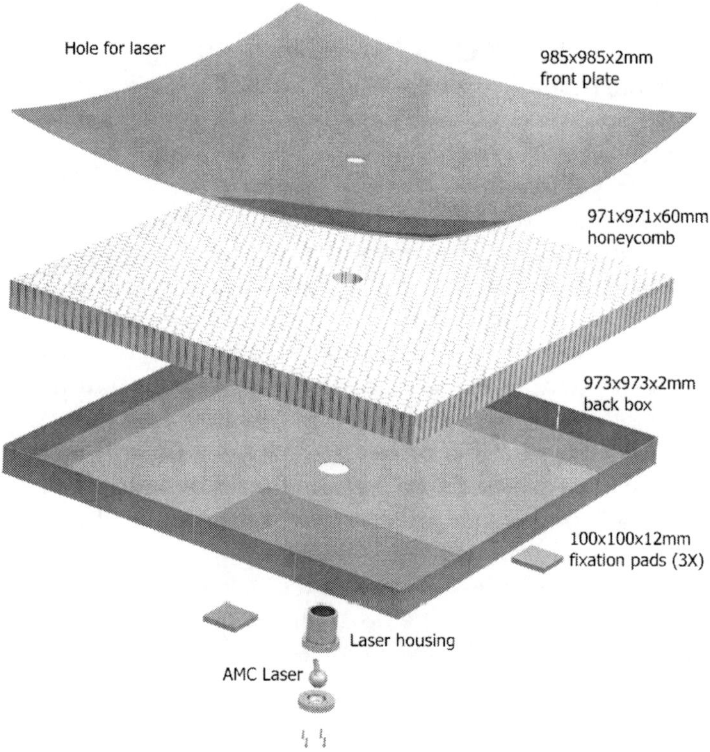

Figure 6: Sketch of an aluminum mirror of MAGIC II.

This technique was developed at the Max Planck Institute in Munich, Germany and at Istituto Nazionale di Fisica Nucleare (INFN) in Padua, Italy in 2000-2003. The aluminum alloys were chosen to guarantee high rigidity, easy machinability, and reduced degradation from atmospheric agents. It is in fact necessary to recall that these mirrors have to survive in the open air for several years, exposed at night-day temperature variations of up to 30° C, survive occasional ice deposits, snow, and strong rains, resist acid rains and *calima* (an ultra thin wind of sand coming from the Sahara desert), without major degradation.

For the back panel, a square bottom aluminum box, 973x973x61.9 mm and 2mm thickness was used. The aluminum alloy is 5754 H1111,[3] having Mg as the principal admixture and characterized by excellent resistance to corrosion, easy foldability, and resistance to operation at high temperatures as needed during the mirror assembly. The strength is almost comparable to steel. The thickness of the aluminum sheet is 2 mm. Three small fissures are equally spaced on each lateral side of the box plus one on each corner. Each fissure is of 56 mm height and 3 mm width and allows the bottom box to bend and match the curved shape of a spherical mold of 35 m radius during the sandwich assembly. The honeycomb is a hexcell structure, 971x971x 60 mm of type CRF 1/4 ACG .003 P 5.22 (respectively: foil treatment, cell size [inch], aluminum foil type, foil thickness P=perforated and density [lb/ft^3]). The honeycomb was chosen to maximize the stabilized compression, with a Young Modulus of 1000 MPa. It is micro-holed for ventilation, pressure equalization, and evacuation. The upper skin is composed of an aluminum alloy of type 6082–T6 that offers good weldability, brazeability, corrosion resistance, formability, very good machinability, and is also heat treatable. The principal admixtures are of Si and Mg. It has a dimension of 985x985mm and 3 mm thickness. It is produced as a flat sheet of material, and is slightly bent to match the concave mold. At the center, this reflects into a maximum deformation of around 3.6 mm from the flat shape.

At the center of the mirror, after the assembly of the sandwich, a hole is drilled to house a laser for the AMC. The laser points to the telescope camera and automatically orients the mirror to match the best focus during tracking (Biland et al. 2007). Three aluminum pads 100x100x12 mm are glued inside the mirror between the honeycomb and the bottom box. On each pad, the actuators of the AMC motors are screwed. The 3M Scotch-Weld Structural Adhesive Film AF 163-2K is used to glue the sandwich. It is a thermosetting modified epoxy structural adhesive film designed for honeycomb sandwich constructions. It has a high fracture toughness and peel strength, excellent resistance to high moisture environments, short cure time, and it is capable of low-pressure bonding and vacuum curing. It is deposited between the top plate and the honeycomb and between the honeycomb and the bottom box. Once the mirror is assembled, an epoxy glue is used to seal from

---

[3] We use the International Alloy Designation System standards for the definition of the aluminum alloy: the wrought aluminum is named with a four digit number to identify the alloying elements, followed by a dash and a letter identifying the type of heat treatment and a 1 to 4 digit number identifying the specific temper.

the outside the sandwich at the perimeter and over the lateral fissures. We used 3M Scotch-Weld Epoxy Adhesives DP190 (gray). It is a flexible weld with high shear and peel strength.

The sandwich is inserted into an autoclave where a cycle of high pressure and temperature cures the adhesive: the temperature is raised from 20° C to 125° C in 35 min and at the same time the pressure is raised to 5 bar. Temperature and pressure are maintained constantly for 90 min and then released back to room conditions in another 35 min. Once assembled, the sandwich is also called a *raw-blank*. The raw-blanks have an average radius of curvature of around 35 m.

To reach the exact value of curvature for the position of the mirror in the telescope, which ranges from around 34 m to around 36 m, the surface of the mirror is milled. This is done with a diamond-machine, which operates with a fast rotating diamond head that polishes the mirror surface. The surface acquires a roughness of a few nanometers and a reflectivity of around 90% at 400 nm (and an average over the Cherenkov range of 85%). The milling technique is called fly-cutting because the moving parts of the machine rotate on a thin air layer suppressing oscillations and offering the required stability. The mirror is fixed onto a mold mounted onto a granite table that floats on an air layer and slowly rotates back and forth on a horizontal plane at a velocity of around 1 m/h. The head of the diamond machine rotates extremely rapidly (@ 20,000 turns/h) around an axis that is inclined with respect to the mirror's optical axis. The combination of the two movements determines the desired mirror radius of curvature. The diamonds have a spherical shape with a radius of curvature of the head of around 100 mm. At the passage of the diamond over the surface, some material is removed. The milling of the surface proceeds in several cycles until the desired radius of curvature is reached. Usually it takes about one working day to complete one mirror.

At the Fraunhofer Institute (IFAM) in Bremen, Germany, the surface of the mirror is later coated through a vacuum deposition of quartz with an admixture of carbon. The technique is called plasma enhanced vapor chemical deposition (PEVCD). The surface to be coated must be cleaned in order to guarantee the chemical reaction between the quartz and aluminum. This is done with a chemical treatment. Subsequently the mirror is mounted inside a large autoclave. In PEVCD, first the mirror surface is cleaned with a hydrogen gas heated to a plasma state by alternating electrodes in a medium vacuum ambient condition (< 10 mbar). After the surface has been cleaned, the quartz is vaporized and deposited onto the surface. The deposition of the quartz layer introduces interferences between the incidence and reflected light, which depend on the wavelength. The choice of the width was chosen to minimize the interference in the region around 300 nm where most of the Cherenkov light from atmospheric showers is concentrated. For vertical incidence, the optimal layer thickness is 100 nm.

One of the main goals of this technique is to reduce aging. While normal aluminized glass plate usually has reflectivity losses of around 10% a year, aluminum mirrors show degradation of only 1% a year. On aluminized surfaces, an eventual surface defect lets the oxidation of the aluminum creep along the surface (it cannot enter the glass) and the damage increases rapidly. On the other hand, a defect in solid aluminum can creep vertically and the damaged area is smaller (Figure 7). This fact is particularly relevant for IACTs that have a very large reflective surface that cannot be aluminized as often as optical telescopes. The overall reflector has a point spread function (PSF) for a point-like source at infinity of 10 mm at the focal plane, which is to be compared with the diameter of the camera pixels of 33 mm. We recall that the Cherenkov photons are affected by intrinsic blurring due to the Coulomb scattering of shower particles in the atmosphere. Therefore the optical quality demands are much less demanding than for optical telescopes. The interested reader can find more information in Doro (2009).

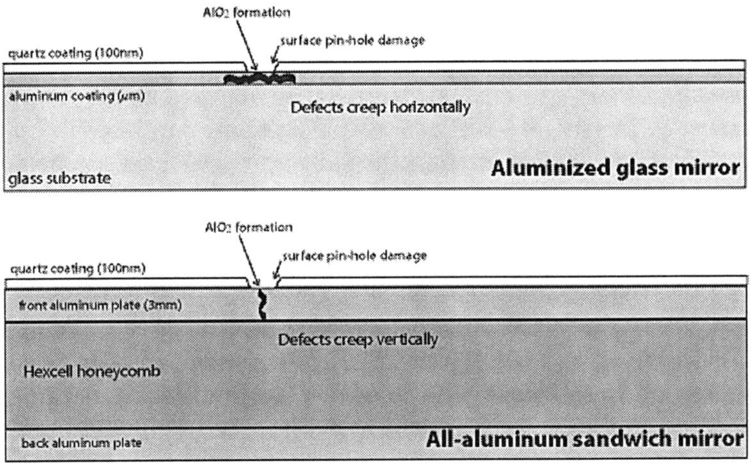

Figure 7: Schematic of the creeping of the defects on aluminum surfaces.

MAGIC II Glass Mirrors

The cold glass slumping is a new method derived from a similar technique proposed for the manufacture of X-ray optics of the XEUS mission that happens to fit the requirements of Cherenkov telescopes. It has been developed by MediaLario Technologies in collaboration with INAF-Brera Astronomical Observatory. A thin glass sheet, about 2 mm in thickness, is elastically deformed so as to retain the shape imparted by a master with a convex profile. The master has to be worked with the same optical precision needed for the mirrors, because every defect on its surface will be reproduced on the glass. The master is made of aluminum and its surface is diamond milled in the same way as for MAGIC aluminum mirrors. With large curvature radii, the glass sheets can be pressed against the master using vacuum suction. In order to provide the necessary rigidity, a honeycomb structure is glued on the deformed sheet under the vacuum forces. Lastly, a second glass sheet is glued on the top of the honeycomb structure, in order to form a sandwich. The gluing process is realized by heating the aluminum mold to around 80° C. After this stage the panel is aluminized and coated with quartz to protect the reflecting surface. By careful control of the spring-back effect during the gluing process, one can obtain several mirrors with different radii of curvature from a single master. We used this feature in MAGIC II production to obtain 12 different curvatures within the angular resolution specifications with only two masters.

Selected Scientific Results with MAGIC

The physics program developed by the MAGIC telescope includes astrophysics surveys of galactic and extragalactic targets like supernova remnants, binary systems, neutron stars, active galactic nuclei, and gamma ray bursts. MAGIC can also investigate fundamental physics questions such as dark matter, the transparency of the universe, and photon transport properties. Hereafter we highlight two particularly important recent results.

In 2008, MAGIC detected very high energy gamma rays from a giant flare of the distant quasi-stellar radio source (also called radio quasar) 3C279 (Albert et al. 2008b), located at a distance of more than five billion light-years (a redshift of 0.536). No quasar had been previously observed at these frequencies at such a large distance. Because high energy gamma rays are expected to be absorbed by the UV and near-IR photons of the extragalactic background light (EBL) through pair production, this observation has revealed a much more transparent universe than expected, as shown in Figure 8. The results support, at higher redshift, the conclusion drawn from earlier measurements that the observations of the Hubble Space Telescope and Spitzer correctly estimate most of the light sources in the universe. The derived limits are consistent with the EBL evolution corresponding to a maximum star-formation rate at redshift $z \geq 1$.

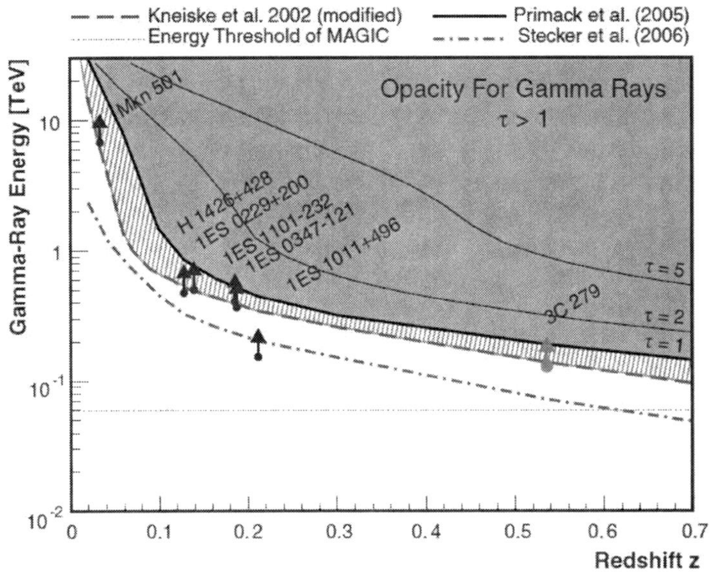

Figure 8: The gamma-ray horizon. The redshift region over which the gamma-ray horizon can be constrained by observations has been extended up to z = 0.536.

A second important result was obtained the same year. At the end of the fusion processes of stars with masses larger than 5 to 9 M, dramatic events of star collapse occur. After having burnt the lighter elements, the massive star shows a degenerate iron core. There is no nuclear process that can release energy and the iron is destroyed by photo–disintegration into neutrons and helium: $\gamma + {}^{56}Fe \rightarrow {}^{4}He + {}^{4}n$. This process also takes away thermal energy and weakens the internal pressure support. The star starts to collapse. Free electrons combine with protons in inverse beta–decays: $e^- + p \rightarrow n + \nu_e$ until all the matter from the star is converted into neutrons. The collapse continues until the core density reaches the nuclear density. No further compression is possible, and all the in-falling material is bounced back outwards. Depending on the primary star mass, the neutron protostar becomes a spinning neutron star (pulsar) or collapses into a black hole.

The other half of the mass is ejected outward and starts to form a propagating nebula called a supernova remnant (SNR). Some 99% of the initial gravitational energy is released as neutrinos, 1% as kinetic energy of the remnant particles, and only 0.01% of the energy goes into radiation. One fundamental question about pulsars concerns the mechanism of their pulsed electromagnetic emission. The gamma ray spectrum at the highest energies carries important information on this question. By developing a new electronic trigger, MAGIC lowered the energy threshold down to 25 GeV. In this configuration, pulsed gamma rays emission from the Crab pulsar at energies larger than 25 GeV were detected (Aliu et al. 2008), revealing a relatively high cutoff energy in the spectrum.

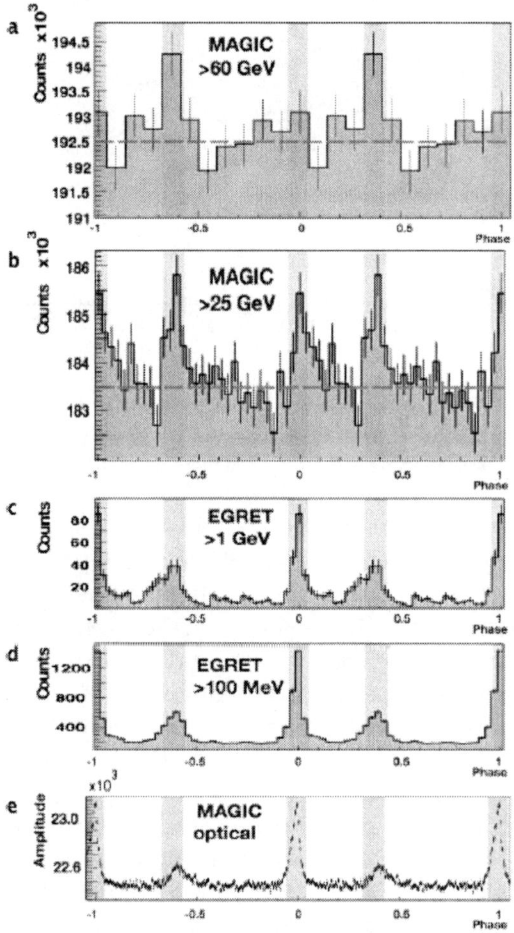

Figure 9: The pulsed emission observed by MAGIC and in different energy bands.

From a theoretical point of view, the spectral cutoff is explained as a combination of the maximum energies that electrons can reach because of the balance between acceleration and radiation losses, and the absorption of the emitted gamma-rays in the magnetosphere. Absorption is controlled by two mechanisms: magnetic e+-e− pair production in the extremely strong field close to the pulsar surface and photon-photon e+-e− pair production in dense photon fields. If, for a young pulsar like the Crab with a magnetic field of the order of 10^{12} to 10^{13} G, emission occurs close to the stellar surface (polar-cap models), then magnetic pair-production attenuation provides a characteristic super-exponential cutoff at few GeV. If emission occurs farther out in the magnetosphere, at several stellar radii or close to the light cylinder (slot-gap and outer-gap models), then absorption mainly arising from photon-photon collisions sets in at higher energies and produces a shallower exponential cutoff. MAGIC results favor the outer gap scenario of acceleration of cosmic rays, and exclude the polar-cap scenario as a possible explanation of our measurement. The emission therefore occurs far out in the magnetosphere. The high cutoff energy also challenges the slot-gap scenario.

Future Prospects

The IACT technique is now well established. The so-called generation IV of IACTs will likely start observation around 2015 (CTA in Europe and AGIS in USA). These new instruments will have a factor 10 increase in sensitivity by using a large array of IACTs of both large and small reflector size to allow observations on the high energy and the low energy domain of very high-energy gamma rays (hundred of TeV and few GeV respectively). Chances for additional improvements are promising as new technologies are emerging such as new high quantum efficiency photon detectors, low power GHz digitizers for each pixel, higher level triggers, and better mirror optics.

Acknowledgments

The author acknowledges the MAGIC collaboration and in particular A. De Angelis, E. Lorenz, and M. Mariotti for useful discussions.

Author's Note

Michele Doro can be reached at University of Padova & INFN Padova, via Marzolo 8, 35131 Padova (Italy). Tel, +30.049.827.7211, michele.doro@pd.infn.it.

References

Abraham J. and the Pierre Auger Collaboration. 2007. Correlation of the highest energy cosmic rays with nearby extragalactic objects. *Science, 318,* 938–943.

Albert J. and the MAGIC Collaboration. 2007. Very high gamma-ray observations during moonlight and twilight with the MAGIC telescope. *Astroparticle Phys.* submitted. e-Print: astro-ph/0702475.

Albert, J and the MAGIC Collaboration. 2008a. FADC signal reconstruction for the MAGIC telescope. *Instru. and Meth. in Phys. Res. A 594 (3),* 407-419.

Albert, J. and the MAGIC Collaboration. 2008b. Very high energy gamma rays from a distant quasar: how transparent is the universe? *Science 320* (5884), 1752-1754.

Aliu E. and the MAGIC Collaboration. 2008. Observation of pulsed γ−rays above 25 GeV from the Crab pulsar with MAGIC. *Science 322* (1221).

Aliu, E. and the MAGIC Collaboration. 2009. Improving the performance of the single-dish Cherenkov telescope MAGIC through the use of signal timing. *Astropart. Phys. 30,* 293-305.

Barrio, J.A., Blanchot, G., Börst, H.G., Blanch, O., Bosman, M., Bradbury, S. M., Cavalli-Sforza, M., Chilingarian, A., Contreras, J. L., Cortina, J., Dosil, M., Feigl, E., Ferenc, D., Fernandez, E., Fernandez, J., Fonseca, V., Gebauer, H. J., González, J. C., Haag, E., Holl, I., Hrupec, D., Ibarra, A., Karle, A., Kornmayer, H., Krawcynski, H., Llompart, X., Lorenz, E., Magnussen, N., Mariotti, M., Martinez, M., Merck, M., Meyer, H., Mirzoyan, R., Moralejo, A., Möller, H., Müller, N., Odeh, T., Ostankov, A., Padilla, L., Petry, D., Plaga, R., Prosch, C., Raubenheimer, C., Rauterberg, G., Sawallisch, P., Schmidt, T., Turini, N., and Wacker, A. 1998. The MAGIC Telescope, design study for the construction of a 17 m Cerenkov telescope for gamma-astronomy above 10 GeV. Internet <http://wwwmagic.mppmu.mpg.de> accessed 10 January 2010.

Biland, A., Garczarczyk, M., Anderhub, H., Danielyan, V., Hakobyan, D., Lorenz, E., and Mirzoyan, R., on behalf of the MAGIC Collaboration. 2007. The active mirror control of the MAGIC telescope. *Procs. of the 30th International Cosmic Ray Conference* (ICRC-2007), Merida, Mexico.

Buckley, J., Byrum, K., Dingus, B., Falcone, A., Kaaret, P., Krawzcynski, H., Pohl, M., Vassiliev, V., and Williams, D.A. 2007. The status and future of ground-based TeV Gamma-ray astronomy. A White Paper commissioned by the Division of Astrophysics of the *American Physical Society.*

Doro, M. 2009. *Novel Reflective Elements and Indirect Dark Matter Searches for MAGIC II and Future IACTs (PhD Thesis).* University of Padova, Italy.

Gaisser, T. K., 1990. *Cosmic Rays and Particle Physics.* Cambridge, England: Cambridge University Press.

Hess, V. 1912. Uber Beobachtungen der durchdringenden Strahlung bei sieben Freiballonfahrten. *Physik. Zeitschr., 13,* 1084.

Hillas, A. M. 1985. Why is Cygnus X-e (with "related sources") a highlight of cosmic-ray astrophysics? In *Proc. of the 19th ICRC,* La Jolla, CA. Internet <http://adsabs.harvard.edu/full/1985ICRC....9..407H> accessed 12 January 2010.

Lombardi, G., Zitelli, V., and Ortolani, S. 2008. El Roque de Los Muchachos site characteristics. III. Analysis of atmospheric dust and aerosol extinction. arXiv:0802.3947 [astro-ph].

Paneque, D., Gebauer, H. J., Lorenz, E., and Mirzoyan, R. 2004. A method to enhance the sensitivity of photomultipliers for air Cherenkov telescopes by applying a lacquer that scatters light. *Nucl. Instrum. Methods A. 518,* 619-621.

Weekes, T., Cawley, M., and Fegan, D. 1989. Observation of TeV gamma-rays from the Crab nebula using the atmospheric Cerenkov imaging technique. *Astroph. J. 342,* 379.

Alt-Az Telescopes

The plywood prototype of the radial flux direct drive telescope motor. The materials for the motor itself only cost about $300. The high resolution, on axis encoder was more expensive, however. Dave Rowe designed the motor and built the wooden prototype.

The final version of the azimuth direct drive for the Cal Poly 18 telescope. Dan Gray at Sidereal Technology welded up the six-legged steel support, each with an adjustable leveling screw.

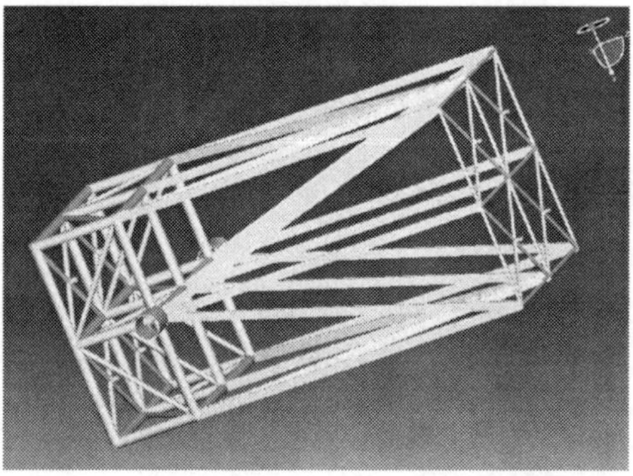

Josh Schmidt, a mechanical engineering student at California Polytechnic State University, designed this "four shooter" optical tube assembly (OTA) using finite element analysis (FEA). He varied tubing diameters, wall thicknesses, and geometric layout to minimize prime focus deflections as the telescope changed its altitude.

At an Alt-Az Initiative workshop in June 2007, Russ Genet (left) stands on a 24 inch diameter Inca bearing while Dan Gray spins him around. The workshop was held at California Polytechnic State University in San Luis Obispo. This very bearing was later used in the wooden prototype of the 1 meter portable photometric telescope.

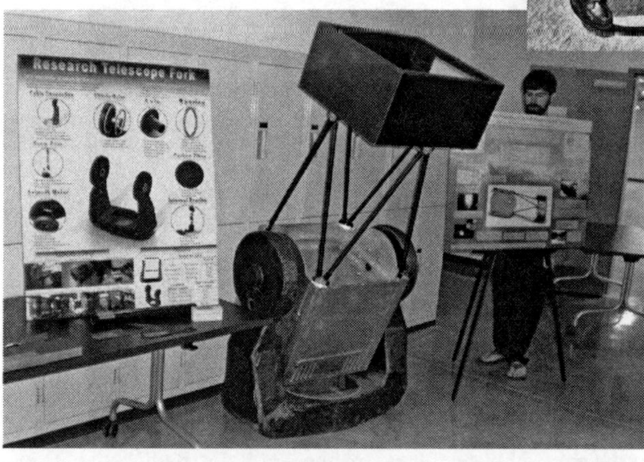

The Cal Poly 18 telescope's optical tube assembly and fork on display at a senior projects fair at California Polytechnic State University. Seven Cal Poly mechanical engineering students worked on the design under the supervision of John Ridgely, a professor of Mechanical Engineering at the university, and Russ Genet, a Research Scholar in Residence at Cal Poly.

Dan Gray (second from left) invented the "string" telescope. Three truss tubes in compression push up against the ring at the top of the telescope, while six "strings" in stress pull down at angles. The result is a surprisingly stiff, lightweight structure.

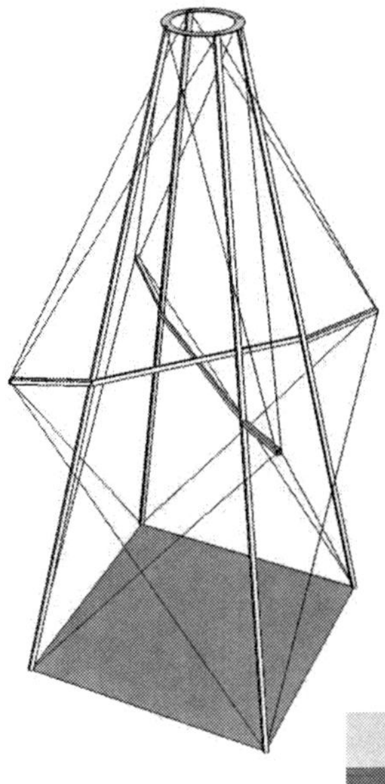

Don Peckham used Google Sketch Up, a 3D drafting program, to show a two-bay quadrapod truss with guy cables. On a per pound basis, the guy cables provide added stiffness thanks to their relatively large angles with respect to the optical axis

Howard Banich at Pine Mountain Observatory and the portable 28 inch telescope he designed and built. Howard has practical advice for those who wish to design large portable telescopes that will actually be used rather than gathering dust because they are too much of a pain to transport and set up.

FRuss Genet, Jo Johnson, Dave Rowe, and Kevin Harris stand in front of a mid-size alt-az telescope being built at EOS Technologies telescope plant in Tucson, Arizona.

Kevin, Dave, & Russ inspect a mirror support system for a large mirror. The central column provides radial support, while the arms provide axial support when the rest of the wiffletree structure is added.

Roger Angel, Russ Genet, and Dave Rowe discuss mirrors for mid-sized alt-az telescopes. Roger described a very interesting solar power project he was working on that uses very thin glass mirrors made from plate glass.

Russ Genet, Brian Mason, and Jo Johnson sitting in the Cassegrain instrument cage on the 4 meter telescope at Kitt Peak National Observatory. The instrument at Cass focus is a speckle interferometer camera which obtains images at very high speed. The speckles on thousands of fast frames are analyzed to determine the separation and position angles of close visual double stars.

Al Chambers at Pine Mountain Observatory (PMO) and a Celestron NexStar 5 purchased by Russ Genet at RTMC for a song and traded to Al for a baritone ukulele. The original control system had long been lost, but Al connected up a Sidereal Technology control system. It worked great. A number of observers with telescopes outside of the big dome for the PMO 32 inch telescope noted that the dome was moving around with great frequency. What was going on? Inside, they found Al and his 5 inch telescope in the center of the observatory floor. Moral of the story, with apologies to Teddy Roosevelt: "Observe small, but rotate a big dome."

Russ Genet stands beside an early version of the 1 meter photometric telescope. Actually this is just a cardboard cutout of Russ standing beside a 1/8th scale model of the telescope. The photo was realistic enough so that several folks thought Russ had built this monstrosity. Fortunately the final telescope is more compact.

Chris Estrada tests the fit of the mirror box cover for the wooden prototype of the 1 meter photometric telescope. The mirror box was built from Baltic birch plywood at the wood hobby shop at Edwards Air Force Base. Reed Estrada, who also worked on the telescope, is a test pilot at Edwards.

Part III

LIGHTWEIGHT MIRROR DEVELOPMENTS

Chapter 15

An Introduction to Lightweight Mirror Developments

Russell Genet[1], Andrew Aurigema[2], Steve Badger[3], Mel Bartels[4], K. Lisa Brodhacker[5], Rodolfo Canestrari[6], Peter Chen[7], Mike Connelley[8], David Davis[9], Mauro Ghigo[6], Greg Jones[10], Tong Liu[11], Eric Mendez[1], Giovanni Pareschi[6], Terry Richardson[12], David Rowe[13], Josh Schmidt[1], Kiran Shah[14], and Efrain Villasenor[1]

[1]California Polytechnic State University, [2]OTF Designs, [3]Pittsburgh Corning, [4]BB Astrosystems, [5]Lander University, [6]Brera Observatory, Italy, NASA Goddard Space Flight Center, [8]NASA Ames Research Center, [9]Toledo Scope Werks, [10]Eclipse Technologies, [11]Hubble Optics, [12]College of Charleston, [13]Sierra Monolithics, [14]Chroma Systems

Introduction

The Alt-Az Initiative (www.AltAzInitiative.org), launched in June 2007 at workshops in Portland and San Luis Obispo, is a catalyst for the development of lightweight, low cost, modest aperture (2 meters and less) alt-az research telescopes. To date, the Initiative has convened two full conferences and twelve workshops, launched or completed twelve technical initiatives, and released an edited book, *Small Telescopes and Astronomical Research* (Genet, Johnson, and Wallen 2010).

The Initiative's first alt-az technical demonstration telescope, the Cal Poly 18, recently saw first light. It has no gears, belts, or friction wheels; instead it incorporates direct drive motors and high resolution encoders into the telescope's structure. The structure, designed by California Polytechnic State University students using finite element analysis, has an unusually high natural frequency. The use of direct drives combined with a stiff structure counters wind gusts.

In May 2008, the Alt-Az Initiative issued a challenge to develop lightweight, low cost, 1.5 meter, transportable telescopes that could, as a minimum, produce useful scientific results in at least one area of research. We reasoned that the cost, aperture, and transportability requirements were difficult enough to require the innovative application of technology, yet not so difficult as to discourage a spectrum of low-budget efforts to meet this challenge. Key is the development of lightweight, low cost primary mirrors. This chapter, with an eye on the 1.5 meter challenge, introduces a number of approaches undertaken by Initiative members to reduce mirror weight and cost.

We recognize that some of the primary mirror approaches we are investigating will only be applicable to less demanding, dedicated, special-purpose research telescopes (i.e. light buckets), and in all likelihood will not meet the stringent demands of high optical quality, general purpose telescopes. Telescopes that must perform well across a wide range of instruments/scientific programs—including some that are optically demanding—may inherently be more expensive than specialized telescopes dedicated to less optically demanding programs. There is, for example, a world of difference between 1.5 meter telescopes intended for high quality, wide-field imaging that requires sub-arc second image quality, and telescopes dedicated to on-axis applications such as near-IR aperture photometry with a 1 mm diameter photodiode or low resolution spectroscopy. Initiative members have an interest in both high and low optical quality telescopes, and both are being pursued.

Multiple Mirrors

The weight and cost of mirrors is roughly proportional to the third power of their aperture, while their light gathering area is, obviously, only proportional to their second power. Thus, other things being equal (including equivalent light collecting area), a telescope with multiple smaller mirrors will have mirrors that, in total, are lighter in weight and lower in cost than a telescope with a single large mirror—a fact not overlooked by the Keck, Thirty Meter, Giant Magellan, and other large telescope designers. Of course multiple mirrors create serious complications, and that is why they have rarely been used in 1 to 2 meter class telescopes. Initiative members are considering two multiple mirror approaches—one with multiple independent optical systems on the same mount (requiring multiple instruments or fiber feeds), and the other with multiple mirrors bringing light to focus on a single instrument or fiber feed.

Multiple Foci

Josh Schmidt, a mechanical engineering student at California Polytechnic State University, used finite element analysis to design the optical tube assembly for a "four shooter" 1.5 meter telescope. Each *f*/3.5 mirror was 0.75 meters in aperture—well within the state-of-the-art for affordable, lightweight mirrors.

Two instrumental modes were considered. In one mode, the primary mirrors were fixed in position with cameras placed at each of the four prime foci. Such a system could be used for simultaneous four color photometry or, alternatively, single color photometry with co-added images.

In the other mode, either the four primary mirrors or four small steering mirrors were tip/tilted to compensate for differential flexure between the four optical systems as the telescope changed altitude. In this mode, each optical system needed to have its own autoguider. A 1.5 meter equivalent "four shooter" could, for instance, feed an off-telescope spectrograph with four fibers aligned along the spectrograph's slit for optical efficiency (see Barry 1995 for a pioneering example).

One objective of Schmidt's (unpublished) analysis was to determine if the optical paths for multiple fiber feeds could be kept sufficiently aligned with fixed primary mirrors in a rigid structure to avoid the need for tip/tilt adjustments. Schmidt concluded this would be difficult, although his analysis did not entirely rule it out.

Single Focus

More traditionally, many multiple mirror telescopes bring light to a single focus to avoid the difficulties inherent with multiple instruments. To achieve the highest possible resolution (approaching the theoretical limit for their aperture), large multiple-mirror telescopes at prime mountaintop sites require their multiple primary mirrors to be co-aligned with a precision that maintains optical coherence, a demanding and expensive undertaking.

At less than pristine sites and, especially for less optically demanding dedicated applications such as on-axis aperture photometry or low resolution spectroscopy, there is no reason to maintain optical coherence—the resolution of a single mirror is more than sufficient for the task at hand. Similarly, while many general-purpose telescopes strive for wide fields-of-view, specialized telescopes may only require a very narrow, essentially just on-axis field-of-view.

High quality, large aperture, multiple mirror telescopes utilizing matching sets of aspheric primary mirrors, such as off-axis paraboloids, are difficult and expensive to manufacture. Multiple spherical primary mirrors, on the other hand, are relatively easy to produce, although the individual mirrors still need to be tip-tilted with considerable precision. Furthermore, the sizeable spherical aberration produced by a large effective aperture spherical primary with a fast effective primary f/ratio needs to be corrected—a difficult task for large fields-of-view. A spherical primary with aspheric secondary optical configuration, such as used in Pressman Carmichael systems, can produce reasonably wide fields-of-view, although the convex aspheric secondary is somewhat difficult to produce. However, if only a very narrow, prime focus field-of-view is required, then a two-element, all-spherical surfaces, refractive corrector can provide sufficient correction at an affordable cost. Initiative member Dave Rowe has designed such a corrector for a 1.5 meter, *f*/3.5 telescope.

Meniscus Mirrors and Active Optics

Small telescope opticians have extended the routine production of 2 inch thick Pyrex glass to 32 inch mirrors as fast as *f*/3.5 and, most recently, *f*/3.0. Such Pyrex mirrors have been pushed experimentally to an aperture of 48 inches. Although eminently suitable for multiple mirror 1.5 meter telescopes (such as one with four 30 inch mirrors), these Pyrex mirrors may not be suitable for single-mirror 1.5 meter telescopes because Pyrex sheet widths are normally limited to 48 inches and, of greatest import, a 60 inch (1.5 meter) two inch thick Pyrex disk weighs about 600 pounds—not exactly lightweight.

Initiative members are working on an alternative that may overcome these difficulties by switching from Pyrex to ordinary soda-lime float glass (soda-lime plate glass) and limiting mirror thickness to ¾ of an inch or, at most, 1 inch. Plate glass is low in cost, readily obtainable, and can be purchased in widths up to 96 inches. Of course a 60 inch mirror only 1 inch thick is very floppy and may require active control of some sort to maintain its shape.

Slumped Float Glass Mirrors

Initiative member David Davis, using low cost and easily constructed kilns he designs and builds himself, has worked out a process to slump common plate glass "table tops." A 60 inch diameter, ¾ inch thick glass disk from Glass Tops Direct costs $374 plus $135 shipping, and weighs about 200 pounds.

Initiative member Mel Bartels suggests that meniscus plate glass (soda lime float glass) mirrors represent a new approach for large, fast telescopes. The synthesis of thermally controlled mirrors (with fans), inexpensive plate glass, and computer controlled kilns for slumping is producing promising results.

Meniscus mirrors, other things being equal, are significantly stronger than flat back mirrors. Consider a piece of paper—it is very floppy. Contrast it with a cone shaped coffee filter, which is very strong. A 12 inch standard thickness Pyrex mirror is often used with a 9 point back support. A 13 inch diameter 1 inch thick meniscus mirror needs but a 3 point back support as determined empirically with a star test at a 1 mm exit pupil. Meniscus mirrors can be slumped to very fast speeds, maintaining constant thickness from center to edge.

Compared to Pyrex, plate glass mirrors are an order of magnitude less costly. Thin plate glass meniscus mirrors can be cooled rapidly with a fan, reaching equilibrium in minutes. However, plate glass mirrors have a pejorative reputation which may hamper their widespread adoption. They are difficult to figure due to their sensitivity to small temperature changes. These mirrors require aggressive measures to keep them at ambient temperature to properly test them. On the other hand, plate glass does grind and polish quicker than Pyrex, and some commentators state that it takes a better polish with less scattered light.

Active Primary Mirrors
As apertures increase, the downside of large aperture, lightweight, thin meniscus mirrors is that at some point they require active support. While such support is, obviously, a complication, the actual cost may be modest, and actively supported thin meniscus plate glass mirrors may be a cost-effective approach to achieving large apertures in lightweight, transportable telescopes.

Initiative member Mike Connelley has developed an "active" mirror cell for an 8 inch, *f*/4.5 plate glass mirror that is ½ inch thick. His goal was to determine whether he could correct astigmatism and perhaps a few higher order aberrations (such as trefoil) by hand. To this end he built a very simple warping cell. He reasoned that if he could warp the mirror to compensate for the astigmatism in his eyes, then he might be able to compensate for the warping of a much larger mirror due to changes induced by gravity loading.

His mirror is glued onto 8 aluminum pads with silicone adhesive. Each of 8 screws is threaded through a hole near the tip of each of 8 wooden fingers. The head of each screw is captured in the aluminum pads so that each screw is free to turn inside the pad but can't move around (side to side or in and out). Each pad is made of two plates. The lower plate (cell side) has a large countersunk hole for the machine bolt head. The upper plate (mirror side) glues to the top to trap the bolt head. Turning the screw changes the separation between a wooden finger and the mirror, deflecting the finger. The spring force made by deflecting a finger applies a force to the mirror via the screw connecting them.

Connelley's experiment was just a quick, low cost demonstration on a small telescope. Implementing active control "for real" on a big mirror would be more complex. For example, the warping plate probably wouldn't be just a chunk of plywood with N fingers sticking out; there might need to be an inner ring of actuators. In Connelley's experiment, the screws carried the full weight of the mirror even when the scope was pointed horizontally (i.e. when the mirror was vertical), so that the force on the screws was perpendicular to the screws. Although this wasn't unreasonable for an 8 inch mirror, it probably would not work for much larger mirrors. A sling or other radial support system would also be required. Finally, it might be inappropriate for a large research telescope to utilize hand adjustments. Such adjustments could be made via motors turning screws, or by forces imposed on the mirror via "voice coils," i.e. linear motors.

We are only concerned with making very low frequency (much less than once per second) adjustments of primary mirrors to correct for the lower order Zernike terms, especially astigmatism as altitude changes. The requisite adjustments could be determined by observing "off line" a number of bright stars distributed in altitude using an automated coefficient determination algorithm that minimizes the spread of the stellar image. During operation, the current altitude of the telescope would be noted, and the required settings interpolated from a lookup table. This approach is used by the eight meter Subaru telescope with recalibration of their table only required about once a year.

Two undergraduate electrical engineering students at California Polytechnic State University, Eric Mendez and Efrain Villasenor, are currently developing low cost electronic controls for voice coil meniscus mirror adjustments that feature computer (PC) control of D/A converters that translate interpolated digital values from a look up table to analog signals. These analog signals are amplified by power operational amplifiers that drive the active mirror's voice coils.

Active Auxiliary Mirrors

Initiative members Greg Jones and Bruce Holenstein are both working independently on an entirely different approach to active optical correction: a small, adjustable secondary semi-passive bimorph mirror. Instead of using actuators to apply forces directly to the primary mirror itself, a much smaller mirror near the instrumental payload is distorted via applied voltages in a manner that cancels out the gravity-induced distortions in a lightweight primary mirror. These corrections can be applied via a lookup table versus altitude as discussed above.

Jones' bimorph semi-passive mirrors consist of two thin disks: one an active piezoelectric material coated with a thin layer of conductive metal on each side, and the other a thin front surface mirror. The two disks are bonded together with a low shrinkage adhesive. The metal coating on the back of the piezoelectric disk is divided into sections. When a high voltage (100-350 volts) is applied to these sections, the piezoelectric material in each section contracts or expands an amount dependent on the applied voltage and its polarity. The combination of these expansions and contractions warp the thin mirror disc in a modal manner. A computer controls these voltages through digital-to-analog converters and high voltage operational amplifiers.

Glass Sandwich Mirrors

A sandwich mirror can be thought of as a thin meniscus mirror that has been sliced in half to form two very thin disks. When these two disks are separated by lightweight glass spacer material, the resulting mirror can be as stiff as a conventional "full thickness" mirror, yet as lightweight as a thin meniscus mirror. The potential advantage of sandwich over meniscus mirrors is that while their structure is more complex, they should not require active support. Initiative members have been pursuing two types of glass sandwich mirrors: one uses small-diameter, solid glass disks as the spacers, while the other uses a lightweight glass foam spacer.

Fused Sandwich Mirrors

Ritchey made the first "fused" mirror, an egg-crate structure—just one of his many innovations. The Hubble Space Telescope features what is, perhaps, the most famous fused mirror. Hextek mirrors are made from hollow glass tubes stacked between two thin glass plates. The entire assembly is fused together in a kiln while injecting gas under pressure into the tubes which, when molten, fuse together into hexagonal shapes—hence the company's name. While fused mirrors are lightweight, they are not generally low cost due to their intricate nature.

Initiative member Tong Liu has developed an unusually simple, low cost, lightweight sandwich mirror blank which is comprised of an optical front plate, a backing plate, intermediate reinforcing plate(s) if needed, and many small, identical disk-support spacers between the plates. The plates and spacers are all made from readily available and low cost soda lime float glass. The entire assembly is slumped and fused together in a kiln. By using the same low cost material for all the plates and spacers, a mirror blank of essentially any size up to the width limit of the basic material (96 inches) can be constructed. A key feature of Liu's approach is the use of thin plate glass material throughout the mirror in order to obtain favorable thermal performance.

Currently the size of Liu's sandwich mirrors is limited to 20 inches by his kiln and polishing machines. However, he is developing mirrors up to 40 inches in diameter, and a 60 inch sandwich mirror is not out of the question.

Foamglas Sandwich Mirrors

Initiative members have been investigating the use of foam glass as lightweight spacer material between the top and bottom glass plates of a primary mirror. Foamglas, a cellular glass product manufactured by Pittsburgh Corning, made its debut in 1942. Cellular glass has a unique combination of properties which includes being chemically stable, thermally insulating, compression-resistant, and nonflammable. One of the first uses of Foamglas was the flotation of anti-submarine nets, buoys, and minesweepers during WWII. Since that time, cellular glass has been widely utilized in many industries, primarily as an insulation material.

Steve Badger, Director of Development at Pittsburgh Corning, points out that Foamglas is made from sand, recycled glass, and other abundantly available materials which are mixed to produce a specific composition of molten glass. The glass is extruded into a hollow tube, ground into a fine powder, and mixed with a gas-generating agent. The powder is then put into molds which pass through a cellulating oven where the material is heated to produce a cellular glass block with millions of hermetically sealed glass cells. The Foamglas insulation blocks are then passed through an annealing furnace to allow carefully controlled cooling of the blocks. Finally, the blocks are cut to size, tested, and packed.

Soda lime Foamglas comes in three densities, and there is also a single density, low coefficient of temperature expansion (CTE) borosilicate version of Foamglas. The soda lime version of Foamglas is quite inexpensive, lightweight, readily available, and easily milled.

Initiative member Andrew Aurigema has been machining concave surfaces in the tops of Foamglas blocks, while David Davis has been slumping and fusing glass plates to Foamglas in his kilns. Both have been investigating the use of adhesives instead of kiln fusion to fasten glass plates to Foamglas cores. Attendees at an Alt-Az Initiative conference in Hawaii were intrigued when Davis pointed out that Foamglas sandwich mirrors float on water. He then proceeded to toss a Foamglas mirror blank on the floor without damage.

The brittle nature of the Foamglas cell walls allows easy shaping with simple power equipment. Final shaping can be done with simple flexible surfaces or the slumped glass shell itself. Aurigema found that an easy way to shape the surface was to build a pendulum cutter—a Dremel tool suspended on a wire rope or thin metal rod whose length matched the desired radius of curvature. He used ultra low CTE epoxy mixed with glass micro balloons as the bonding agent. The resulting "frosting like substance" spread easily and filled voids. This mixture is also suitable for sealing the outer side walls of a Foamglas core.

Replica Mirrors

We have, above, considered several ongoing developments that could significantly reduce the weight and cost of 1 to 2 meter mirrors. We now turn to methods that could, potentially, not only reduce their weight, but drastically reduce their cost through the elimination, during production, of grinding and figuring mirrors (albeit the convex master molds used in the replication process would still need, one time, to be ground and figured).

Nanocomposite Replica Mirrors

Initiative member Peter Chen has been developing very lightweight carbon fiber replica mirrors for several years and is achieving success with this approach. Previous work at NASA's Goddard Space Flight Center had demonstrated that diffraction limited optical telescope mirrors were achievable using a combination of polymer resins and graphite fiber laminates. However, tests showed that these 'composite mirrors' became severely distorted when subjected to moderate cooling (from 20°C to 0°C). The cause of this undesirable behavior was traced to an inherent incompatibility between the two basic components of a composite mirror.

Over the past four years, major steps have been taken to resolve this problem. A new class of 'nanocomposite' replica mirrors has been developed using a radically different structure incorporating cryogenic polymers, carbon nanotubes, and carbon nanofibers. Tests have shown that nanocomposite replica mirrors can withstand cooling by liquid nitrogen to -150°C without damage or distortion. Work is currently focused on the development of meter class mirrors, mirrors with built-in active figure control, and ultra-smooth mirrors for space ultra-violet astronomy.

Polyurethane Replica Mirrors

Initiative member Kiran Shah is experimenting with polyurethane replica mirrors. These mirrors have a solid polyurethane front surface supported by a dense polyurethane foam puck which can be either cast or machined. The concave top of this puck is coated with a solid, thin layer of polyurethane resin that is pressed against a convex master that has been coated with a release agent. Successive thin layers of solid polyurethane are added until the built up solid surface develops a reasonable inverse replica of the master.

This work is highly experimental, and Shah does not expect high optical quality. However, the cost to make these mirrors in volume could be extraordinarily low and, thanks to their very light weight, a number of spherical replica mirrors could potentially be assembled into an unusually low cost and lightweight, large aperture, multiple mirror telescope with an instrument and refractive spherical corrector at prime focus. The instrument might consist of a near infrared aperture photometer or low resolution spectrograph.

Glass Replica Mirrors

A team at the Brera Observatory in Italy, consisting of Rodolfo Canestrari—a doctoral student—and his advisors Mauro Ghigo and Giovanni Pareschi, is working on Foamglas sandwich mirrors that do not require optical finishing. In his approach, Canestrari is gluing to a Foamglas core two thin borosilicate glass sheets, each about 1.5 mm thick. The front sheet is pre-shaped to the desired optical figure. This step can be accomplished following either one of two different processes, both based on the replica concept.

In the first process, a precise copy is made from a high optical quality ceramic convex mold using a hot slumping process (Ghigo 2007) and then glued to a Foamglas substrate (Canestrari 2008). Although the mold itself can be quite expensive, this cost can be amortized through the mass production of many replica mirrors. In principle, the optical quality reachable on the replicated thin glass meniscus mirrors is high enough that no further figuring is required.

In the second process, a glass copy of a diamond milled aluminum convex master (Pareschi 2008) is formed at room temperature by adhering an initially flat thin glass sheet to the master using vacuum suction. A Foamglas substrate is then glued to the back side of this now slightly concave glass sheet. One side of the Foamglas core is pre-shaped with a computer controlled mill-like tool to match the curvature of the front glass sheet. The gluing is accomplished with very low shrinkage epoxy and thermal curing. Because the mirror quality is only that of float glass, and this glass must be flexed, this process is only suitable for less demanding optical quality mirrors (an arc minute or slightly better) with a long radius of curvature (tens of meters), such as those used in Cherenkov radiation telescopes.

With either approach, stiff and lightweight mirrors can be produced with mass-to-area ratios of only 15-20 kg/m^2.

Polymer Spin-Cast Mirrors

A liquid revolving at a constant speed naturally assumes a parabolic shape. Spin-casting is capable of pre-forming large monolithic glass mirror blanks (Hill 1989), and has been used by Roger Angel and others to produce large-diameter, lightweight, open-back, monolithic cellular glass mirrors, such as the 8 meter mirrors on the Large Binocular Telescope. Peter Wangsness has successfully applied this technology to 1 to 2 meter class mirrors. Although saving much material and grinding, such spin cast mirrors still require normal optical finishing.

A diffraction-limited surface can be created on spinning liquid mercury mirrors (Borra et al. 1989). If, instead of mercury, one spun a liquid polymer (epoxy), once it sets the spin table could be turned off and the parabolic shape should be retained. For optical quality, the ideal materials (chemical system) for polymer spin-casting should have low viscosity and surface tension, a high glass transition temperature, good strength and hardness, low shrinkage, and a slow reaction rate.

Archibald (1957) produced spin-cast epoxy solar collectors almost a meter across and concluded astronomical mirrors were possible. Ninomiya (1979) investigated this method with various materials and identified several problems. His epoxy mirror was superior to those made from other materials. Alvarez et al. (1993) produced 1.8 meter off-axis mirrors for millimeter wavelengths while Richardson and Griffiths (2001) fabricated mirrors for use at 2.5 microns. Chen et al. (2008) demonstrated a composite spin-cast mirror containing simulated Moon dust as a prototype for lunar solar collectors and telescopes.

A group in South Carolina headed by W. A. Scrivens (with Initiative members K. Lisa Brodhacker and Terry Richardson) has progressed toward spin-cast epoxy mirrors for visible wavelengths using a resin system of diglycidyl ether of bisphenol A (DGEBA) epoxy with 4,4-methylen-bis-3-chloro-2,6-diethylaniline (MCDEA) as the curing agent and an industrial dilutent, cyclohexane dimethanol diglycidyl ether, known as ED757. They use two spinning mechanisms with air-bearing supported platters with diameters of 0.7 and 2.0 meters. Though some problems remain, this group has obtained a sub-arc minute resolution image of the Moon with a masked-off epoxy mirror.

Conclusion

We expect that several but not all of the developments discussed above will lead to relatively low cost, lightweight primary mirrors for 1 to 2 meter class transportable telescopes. The achievable optical quality may differ greatly from one approach to another, so while some types of mirrors may be suitable for high quality optical imaging, others may only be suitable for the least optically demanding dedicated research applications such as aperture photometry or low resolution spectroscopy. Whatever the outcome of these developments, we hope that they will help pave the way toward larger aperture, lightweight, low cost, transportable telescopes.

Acknowledgements

We appreciate helpful reviews and suggestions by Richard Berry, Gary Cole, Robert Nelson, John Ridgely, and Vera Wallen.

References

Alvarez, D. L., Dragovan M., and Novak, G. 1993. Large off-axis epoxy paraboloids for millimetric telescopes and optical light collectors. Rev. Sci. Instrum. 64, 261.

Archibald, P. B. 1957. A method for manufacturing parabolic mirrors. The Journal of Solar Energy Science and Engineering 1, 102.

Barry, D. J. 1995. Design of and Studies with a Novel One Meter Multi-Element Spectroscopic Telescope. Doctoral dissertation, Georgia State University.

Borra, E. F., Content, R, Drinkwater, M. J., and Szapiel, S. 1989. A diffraction-limited *f*/2 1.5-meter diameter liquid mirror. ApJ 346, L41.

Canestrari, R., Ghigo, M., Pareschi, G., et al. 2008. Investigation of a novel slumping technique for the manufacturing of stiff and lightweight optical mirrors. SPIE Proc 7018.

Chen P. C., Van Steenberg M. E., and Oliversen R. J. 2008. Moon dust telescopes, solar concentrators, and structures. Bull. AAS 40.

Genet, R., Johnson, J., and Wallen, V., eds. 2010. The Alt-Az Initiative: Telescopes, Mirrors, and Instrumental Developments. Santa Margarita, CA: Collins Foundation Press.

Ghigo, M., Canestrari, R., Basso, S., and Spiga, D. 2007, Development of lightweight optical segments for adaptive optics. SPIE Proc 6691.

Hill, J. M., Angel, J. R., Lutz, R. D., Olbert, B. H., and Strittmatter, P. A. 1989. Casting the first 8.4-meter borosilicate honeycomb mirror for the Large Binocular Telescope. SPIE Proc. 3352, 172.

Ninomiya, Y. 1979. Parabolic mirror made by the rotation method: its fabrication and defects. Appl. Opt. 18, 1835.

Pareschi, G., et al. 2008. Glass mirrors by cold slumping to cover 100 m2 of the MAGIC II Cherenkov telescope reflecting surface. SPIE Proc 7018.

Richardson, R. L. and Griffiths, P. R. 2001. Generation of front-surface low-mass epoxy-composite mirrors by spin-casting. Appl. Opt. 40, 252.

Chapter 16

Figures of Merit for Light Bucket Mirrors

Bruce D. Holenstein[1], Richard J. Mitchell[1], and Robert H. Koch[2]

[1]Gravic, Inc., Malvern, PA;
[2]Department of Physics and Astronomy, University of Pennsylvania, Philadelphia

Introduction

Aberrations of the surface of a mirror have multiple expressions. "Light bucket" mirrors challenge traditional Gaussian- and diffraction-oriented aberration theory due to large amplitude, caustic ray-crossing aberrations. Typical light bucket mirrors may be made of metal, epoxy, foam glass, tessellated segments, and other techniques (Genet 2009) and have many waves of aberration. What is needed is a method to characterize the suitability of light bucket mirrors for an intended purpose, whether that be diaphragm-limiting photometry, lunar occultations, intensity interferometry, Cherenkov radiation detection, or other non-astronomical uses such as solar power collection. (Note that we utilize the familiar alternate term "diaphragm" to signify the typically circular isolator located in the focal plane before a photomultiplier cell or photodiode detector. The word "aperture" has too many astronomical uses.)

A mirror surface that will satisfy an astronomical observing program will collect photons from the stellar or other program object and deliver them to the detector, yet exclude enough foreground/background photons so as to yield an acceptable signal-to-noise ratio (SNR). Herein several figures of merit are developed for light bucket mirrors dedicated to astronomical use.

We intend this chapter to be the theoretical, analytical background for our second chapter that follows in a later section of this book (Holenstein et al. 2010), and also a general contribution which applies to all highly aberrated mirrors.

Circle of Confusion

With high-quality mirrors, a standard procedure is to calculate a peak-to-valley (PV) criterion or Strehl ratio (SR) for a mirror surface. A PV value for a mirror is simply the difference between the highest and the lowest aberrated wavefront points on an optical surface; PV values of ¼ wave or better yield acceptable images for essentially all cases. A far better measure of optical surface quality is the SR, which computes the peak intensity of the light within the Airy disk of a point source compared to the theoretical maximum delivered by an ideal optical system. An approximation to the SR is given by the Maréchal formula:

$$SR \approx 1 - (2\pi\sigma)^2 \qquad (1)$$

where the root-mean-square (rms) wavefront aberration σ, is measured in waves. This form of the formula is good for SRs down to only about 0.5. Adding a few additional terms to the SR expansion does not help compute meaningful ratios when aberration is extremely large.

With light bucke telescopes collecting photons from point sources rather than from extended sources, imaging is relatively unimportant, so the foregoing mirror surface parameters are not as useful as they should be. Instead, the characterization of the point spread function (PSF) due to transverse aberrations that matters in this case is the blur-spot diameter of the program object at the focal plane, also identified as the Circle of Least Confusion or just the Circle of Confusion (CoC).

All optical systems have a certain angular resolution. The human eye, for example, has a resolution of about 1′ for a person with 20/20 vision. The depth of field of an optical system is known from the distance that the head may move backward and forward and yet have the blur spot remain smaller than the resolution of the eye. Typically, photoelectric detectors look through focal-plane diaphragms (with or without intervening imaging) of the order of a few millimeters diameter and are used on telescopes with focal lengths of the order of meters. For example, a telescope with a focal length of 1 m with aberrations that create a blur spot of a milliradian (about 3′) in diameter will exclude few photons that fall outside a 1 mm focal plane diaphragm. Inversely, given the focal length and mirror diameter, the problem is to characterize the maximum surface aberrations that will produce a CoC that yields an acceptable SNR when imaged on the detector.

Aberration Characterization

All mirrors are plagued by both systemic and local sources of aberration. Systemic aberrations often result from design flaws in the figuring or support of the mirror. An improperly supported mirror may, for example, have a correctible astigmatism that extends over the entire mirror surface and an improperly figured mirror may have a spherical aberration or astigmatism that can be corrected with secondary optics.

In the case of light bucket mirrors having multiple waves of aberration, coherence of the wavefront is not maintained at the focal plane. Also, reflected rays may cross, making the application of traditional Gaussian optics and Seidel aberration theory problematical (Born and

Wolf 1999, 236-244). Since the aberrations of light bucket mirrors are extensive and have varying expressions across the mirror surface, a productive way to evaluate such mirrors is with a statistical approach. This approach divides the mirror into zones, analyzes representative samples of the zones, and then collates the samples into an overall expectation of the mirror figure of merit.

In our current program, we sample the quality of zones across an unobstructed mirror surface using a right-angle Bath Interferometer, and an interferogram analysis software program called Fringe XP (Rowe 2003) is used to calculate the Zernike coefficients of the interferograms made with the device. In practice, almost any type of interferometer and fringe analysis software may be used as long as the output is a reasonable representation of the mirror figure. The Zernike coefficients are the wavefront errors (in waves) applied to the Zernike circle polynomials, the first few of which are copied into Table 1 (Wilson 2007, 288-293; Schroeder 2000, 264-265).

Table 1: Zernike Polynomials

(n,m)	Polynomial		Type
(0,0)		A_0	Piston
(1,1)	$A_1\rho\cos\theta$	$A_2\rho\sin\theta$	X, Y Axes Tilt
(1,0)		$A_3\ (2\rho^2\text{-}1)$	Defocus (power)
(2,2)	$A_4\rho^2\cos2\theta$	$A_5\rho^2\sin2\theta$	0°, 45° Astigmatism
(2,1)	$A_6\ (3\rho^2\text{-}2)\ \rho\cos\theta$	$A_7\ (3\rho^2\text{-}2)\ \rho\sin\theta$	X, Y Axes Coma
(2,0)		$A_8\ (6\rho^4\text{-}6\rho^2 + 1)$	Principal Spherical

In the table, a parameter pair, (n,m), defines the form of a Zernike polynomial from the generating function (Wyant et al. 1992), A_i is the unnormalized Zernike coefficients (note that "Z_i" nomenclature is used in *FringeXP*), ρ is the fractional radius, and θ defines the azimuth angle about the face of the mirror. As is apparent, these low-order Zernike coefficients are useful because they correspond closely to the types of problems that mirror makers typically experience—astigmatism, coma, and spherical aberration. The *FringeXP* software also yields an rms mirror surface quality, σ, consonant with a desired conic section. Spherical surfaces are usually easier to fabricate, so for most of our explorations we have used a spherical conic section, $e = 0$.

For a spherical primary surface, the principal aberration is usually spherical, which produces a CoC between the paraxial and marginal rays with an angular size given by (Schroeder 2000, 57):

$$\theta = 1/(128\, f^3). \tag{2}$$

where f is the focal ratio. An $f/2$ spherical mirror produces spherical aberration of about a milliradian. Either a greater focal ratio or a static spherical corrector lens system may be used to minimize this systemic aberration if the conic section cannot be shifted toward a parabola.

Next, we consider the CoC caused by random mirror-height variations using geometrical optics. Light entering a telescope from infinity and striking a spherical mirror surface comes to a focus at a distance of $\approx 0.5\, R_o$, where R_0 is the nominal radius of curvature. However, for a real light bucket mirror, different areas of the mirror will focus light at different effective focal lengths due to the numerous aberrations afflicting it. In general, rays that strike the surface at an area experiencing +σ waves aberration above average will come to a focus at $F_1 \approx 0.5(R_o + \sigma)$ and likewise rays that strike the surface at an area experiencing -σ waves aberration will come to a focus at $F_2 \approx 0.5(R_o - \sigma)$. Thus, 68% of the photons striking the mirror will come to a focus within $\approx 0.5(R_o \pm \sigma)$, 95% within $\approx 0.5(R_o \pm 2\sigma)$, and 99.7% within $\approx 0.5(R_o \pm 3\sigma)$.

A rough approximation to the diameter of the CoC may be derived geometrically from the focal ratio and the multiple, n, of σ that corresponds to the desired encircled flux fraction:

$$d_{CoC,surface\ height}(n) \approx 0.5 n\sigma/f \tag{3}$$

Note that there are many other aberration sources besides the normally-distributed aberration caused by surface height errors. In particular, the local slope and micro-roughness of the reflecting surface may or may not be small-amplitude Rayleigh-distributed. Errors of this kind can result from various systemic or isolated defects in the mirror shape or surface substrate. Specifically, two important manifestations of large transverse aberrations result from local slope deviations from the ideal slope in the radial and azimuthal directions. A local slope flaw with a value of $\Delta\phi$ causes photons to reflect from the local area deviated by $2|\Delta\phi|$. If the local slope aberrations are Rayleigh-distributed with root-mean-square value $|\Delta\phi|_{rms}$, 39.3%, 86.4%, and 98.9% of the flux is contained in a CoC blur spot with an angular radius of $2|\Delta\phi|_{rms}$, $4|\Delta\phi|_{rms}$, and $6|\Delta\phi|_{rms}$, respectively, for 1, 2, and 3 standard deviations in the continuous Rayleigh distribution function. The diameter of the CoC at the focal plane due to local slope transverse aberrations is a direct function of focal length, F, and the multiple, n', of $|\Delta\phi|_{rms}$ that corresponds to the desired encircled flux fraction and is given by:

$$d_{CoC,local\,slope}(n') \approx 4n'F\ |\Delta\phi|_{rms}. \qquad (4)$$

In order to estimate $|\Delta\phi|_{rms}$, we must first consider the formal Zernike wavefront representation.

Zernike Wavefront Representation

Each author uses his individual nomenclature and normalization for the Zernike polynomial representation. Mahajan (2007) presents the following nomenclature and equations for the wavefront as a function of the non-rotationally-symmetrical Zernike polynomials, Z_j, and expansion coefficients, a_j:

$$W(\rho,\theta) = \textstyle\sum_j a_j Z_j(\rho,\theta), W(\rho,\theta) = \sum_j a_j Z_j(\rho,\theta), \qquad (5)$$

where the polynomials are written:

$$Z_j(\rho,\theta) = \sqrt{n+1}R_n^0(\rho), \qquad m = 0 \qquad (6)$$

$$Z_{even\,j}(\rho,\theta) = \sqrt{2(n+1)}R_n^m(\rho)\cos(m\theta) \qquad m \neq 0 \qquad (7)$$

$$Z_{odd\,j}(\rho,\theta) = \sqrt{2(n+1)}R_n^m(\rho)\sin(m\theta) \qquad m \neq 0 \qquad (8)$$

The indices n and m are positive integers and also include zero, $n - m \geq 0$, and ρ, θ are defined on the unit circle. The radial polynomials are given by:

$$R_n^m(\rho) = \sum_{s=0}^{\frac{n-m}{2}} \frac{(-1)^s(n-s)!}{s!\left(\frac{n+m}{2}-s\right)!\left(\frac{n-m}{2}-s\right)!}\rho^{n-2s}. \qquad (9)$$

The above representation of the Zernike polynomials is orthonormal as follows:

$$\int_0^{2\pi}\int_0^1 Z_j(\rho,\theta)\, Z_{j'}(\rho,\theta)\rho d\rho d\theta \bigg/ \int_0^{2\pi}\int_0^1 \rho d\rho d\theta = \delta_{jj'}. \qquad (10)$$

where $\delta_{jj'}$ is the Kronecker delta function. As a result, the representation for the aberration variance is very simple:

$$\sigma_W^2 = \langle W^2(\rho,\theta)\rangle - \langle W(\rho,\theta)\rangle^2 = \sum_{j=2} a_j^2 . \qquad (11)$$

The first fifteen Zernike orthonormal polynomials are represented in Table 2 along with the notation (Wyant et al. 1992) and the normalization factors needed for the *FringeXP* Zernike coefficients to match expansion coefficients used in equation (5).

Zernike Polynomial Gradient

The gradient vector of the wavefront representation presented in equation (5) is given by:

$$\nabla W(\rho,\theta) = \frac{\delta W}{\delta \rho}\, \boldsymbol{e}_\rho + \frac{1}{\rho}\frac{\delta W}{\delta \theta}\, \boldsymbol{e}_\theta \qquad (12)$$

where $\boldsymbol{e}_\rho$ and $\boldsymbol{e}_\theta$ are the unit vectors in the radial and azimuthal directions respectively. The mean-square wavefront gradient norm over the unit circle is:

$$\langle \|\nabla W\|^2 \rangle = \frac{1}{\pi}\int_0^{2\pi}\int_0^1 \|\nabla W\|^2 \rho\, d\rho\, d\theta. \qquad (13)$$

The physical rms transverse gradient of the surface aberrations is found from the following equation:

$$|\Delta\phi|_{rms} = \frac{\|\nabla W\|_{rms}}{D/2}, \qquad (14)$$

where D is the diameter of the mirror or zone that can be tested. Both the numerator and the denominator must be expressed in the same units, for example, waves. Substituting into equation (4), one finds that the local slope-induced CoC is then:

$$d_{CoC,local\ slope} \approx \frac{4n'F\ \|\nabla W\|_{rms}}{D/2} = 8n'f\ \|\nabla W\|_{rms}, \qquad (15)$$

where the rms gradient is given in units of distance over the unit circle, usually in waves.

Estimation of $\|\nabla W\|_{rms}$ is eased with the recognition by Southwell (1982) that the gradients of higher Zernike terms in equations (6), (7), and (8) may be expressed in terms of lower Zernike polynomial terms. The integration over the unit circle of the square of the norm in equation (13) then is simplified by the orthonormality relationship expressed in equation (10).

A direct calculation of the individual Zernike gradients using Cartesian coordinates is found in Table 8.3 of Hardy (1998, 277-279). Braat

(1987) provides a direct derivation for the variance of the transverse gradient norm in terms of Zernike coefficients. Braat's equation B3 is expressed as follows using terms and normalizations compatible with equations (5) through (9) in this chapter and Table 2:

Table 2: Orthonormal Zernike Polynomials

J	*n*	*m*	Type	Orthonormal Polynomial	*FringeXP* (n',m')	Coefficient	Normalization
1	0	0	Piston	1	(0,0)	Z_0	1
2	1	1	X Axis Tilt	$2\rho\ cos\theta$	(1,1)	Z_1	$1/2$
3	1	1	Y Axis Tilt	$2\rho\ sin\theta$	(1,1)	Z_2	$1/2$
4	2	0	Defocus (power)	$\sqrt{3}(2\rho^2 - 1)$	(1,0)	Z_3	$1/\sqrt{3}$
5	2	2	45° Astigmatism	$\sqrt{6}\rho^2\ sin2\theta$	(2,2)	Z_4	$1/\sqrt{6}$
6	2	2	0° Astigmatism	$\sqrt{6}\rho^2\ cos2\theta$	(2,2)	Z_5	$1/\sqrt{6}$
7	3	1	Y Coma	$\sqrt{8}(3\rho^2 - 2\rho)sin\theta$	(2,1)	Z_6	$1/\sqrt{8}$
8	3	1	X Coma	$\sqrt{8}(3\rho^2 - 2\rho)cos\theta$	(2,1)	Z_7	$1/\sqrt{8}$
9	3	3	30° Trefoil	$\sqrt{8}\rho^3\ sin3\theta$	(3,3)	Z_9	$1/\sqrt{8}$
10	3	3	0° Trefoil	$\sqrt{8}\rho^3\ cos3\theta$	(3,3)	Z_{10}	$1/\sqrt{8}$
11	4	0	Principal Spherical	$\sqrt{5}(6\rho^4 - 6\rho^2 + 1)$	(2,0)	Z_8	$1/\sqrt{5}$
12	4	2	0° Secondary Astigmatism	$\sqrt{10}(4\rho^4 - 3\rho^2)cos2\theta$	(3,2)	Z_{11}	$1/\sqrt{10}$
13	4	2	45° Secondary Astigmatism	$\sqrt{10}(4\rho^4 - 3\rho^2)sin2\theta$	(3,2)	Z_{12}	$1/\sqrt{10}$
14	4	4	0° Tetrafoil	$\sqrt{10}\rho^4 cos4\theta$	(4,4)	Z_{16}	$1/\sqrt{10}$
15	4	4	22.5° Tetrafoil	$\sqrt{10}\rho^4 sin4\theta$	(4,4)	Z_{17}	$1/\sqrt{10}$

$$\langle\|\nabla W\|^2\rangle = \sum_{l=1}^{\infty} 8l\left[\sum_{i=l}^{\infty}\sqrt{2i+1}a_{2i}^{0}\right]^2 + \\ + \sum_{m=1}^{\infty}\left\{m\left[\sum_{i=0}^{\infty}\sqrt{2(2i+m+1)}a_{2i+m}^{m}\right]^2 + \right. \\ \left. + \sum_{l=1}^{\infty}2(2l+m)\left[\sum_{i=l}^{\infty}\sqrt{2(2i+m+1)}a_{2i+m}^{m}\right]^2\right\}. \quad (16)$$

With normalizations appropriate for the *FringeXP* program, the rms wavefront gradient is estimated using the first eleven Zernike coefficients by:

$$\|\nabla W\|_{rms} \approx [Z_1^2 + 2Z_1Z_6 + Z_2^2 + 2Z_2Z_7 + \\ +8Z_3^2 + 16Z_3Z_8 + 2Z_4^2 + 2Z_5^2 + 7Z_6^2 + \\ +7Z_7^2 + 24Z_8^2 + 3Z_9^2 + 3Z_{10}^2]^{\frac{1}{2}} . \quad (17)$$

To reduce the effects of local slope errors on the CoC, a fast design may be chosen if it may be assumed that the spherical aberration identified in equation (2) can be correspondingly reduced or eliminated by secondary optics. In any case, the diameters of the CoC given in equations (3) and (4) must be interpreted as lower limits to actual experience in the lab or dome or field.

One can specify the approximate wavefront aberration limit due to surface height flaws as follows:

$$\sigma_{\text{limit}} \approx 2f d_{Aperture}/n . \quad (18)$$

So in order to concentrate 99.7% of a source's photon flux within the blur circle falling on the detector for a detector diaphragm of 1.0 mm and focal ratio of 2.0, an astonishing 1.3 mm rms of smooth aberration is permissible. Such a value corresponds to 2,600 waves at 500 nm. For the local slope aberrations case, the rms wavefront gradient norm limit is the following:

$$\|\nabla W\|_{rms,limit} \approx \frac{d_{Aperture}}{8n'f} . \quad (19)$$

In this case, for the same detector diaphragm, wavelength, and focal ratio, the rms gradient norm limit is about 42 waves in order to concentrate 98.9% of a source's photon flux within the blur circle falling on the detector. This latter design criterion is generally harder to meet than the σlimit, especially for mirrors with surface ripples, except in the case of high-quality, tessellated mirrors which may be mounted at different heights.

Until the actual PSF of the optical system is measured, a design factor of 5 to 10 should be factored into the calculated upper limit of the aberrations because of the unknown surface contributions to them. In the example given, the design limit should be 300 to 600 waves rms surface height aberrations, with a rms gradient norm of just 4 to 8 waves for a light bucket mirror. For star fields at high galactic latitudes and for programs concentrating on bright objects, the optical system just characterized will be sufficient to avoid accidental errors due to overlapping focal plane images. Even with panoramic detectors and with the same constraints, this conclusion remains unchanged.

Figures of Merit

For a given optical configuration and detector diaphragm, how will mirror aberration affect the SNR of program object measurements? It is necessary to consider the program object photons that are counted compared to the counts of the shot noises from the star and sky foreground/background, detector noise, and scintillation (Schroeder 2000, 433; Howell 2006, 73-77):

$$SNR = \frac{N_{Star+Sky} - N_{Sky}}{\sqrt{N_{Star+Sky} + N_{Sky} + N_{Detector} + S^2}}, \qquad (20)$$

where the sky measure is assumed to be made with the same effective detector diaphragm as the program object, and S models the scintillation noise.

With the approximation provided in equation (3) for surface height aberrations, the signal from the program object alone depends on the detector diaphragm and the mirror surface quality:

$$N_{Star} = F_{Star}\ \mathrm{erf}\left(\frac{\sqrt{2}f\ d_{Aperture}}{\sigma}\right) \qquad (21)$$

where the constant, $F_{Star,}$ represents the flux collected by the telescope from the program object. The error function, *erf*, applies because equation (3) is based on a normally-distributed random variable with one degree-of-freedom.

The sky foreground/background noise flux incident on the detector diaphragm is proportional to the squared diameter of the diaphragm:

$$N_{Sky} = \pi K_{Sky} \left(\frac{d_{Aperture}}{2F}\right)^2 \qquad (22)$$

where the constant, K_{Sky}, is expressed in terms of flux per steradian.

Figure 1 shows the normalized SNR as a function of surface height aberration for different size detector diaphragms used with a 1.6 m mirror having a focal length of 3.0 m at 500 nm wavelength and with a foreground/background that is bright compared to the star and detector noises.

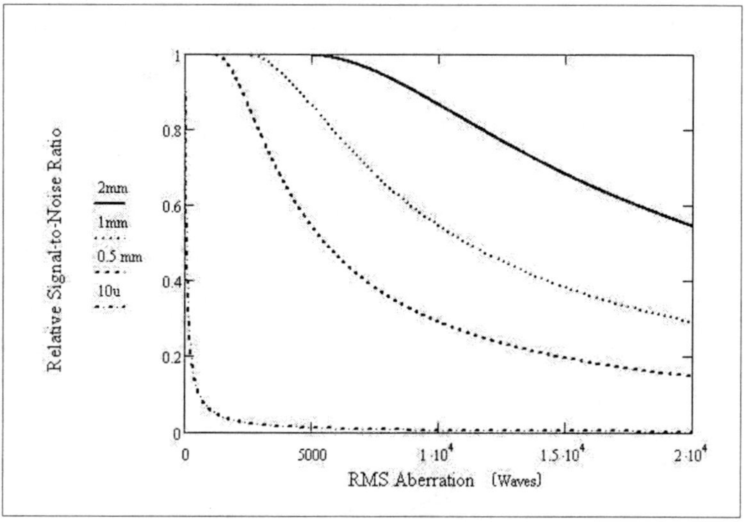

Figure 1: SNR figures of merit for sample cases of different sized diaphragms as a function of random surface height aberration where a bright point source is imaged by a 1.6 m *f/1.9* mirror. The plots are normalized to the case of the mirror with no aberration. A similar plot may be created for the rms gradient norm aberration.

One sees from this illustration that large mirror aberrations significantly reduce the SNR compared to a perfect mirror. For example, a detector with 10 μm pixels is totally unsuitable even when the aberration is modest unless binning, say 100 x 100 pixels, defines the area of the CCD detector used for counts of the program object and sky background. What is particularly interesting from the figure is that there is a knee in the fall-off of SNR for relatively large amounts of mirror aberration with larger detector diaphragms. For the specific example in the figure, 2500 waves rms aberration with a 1 mm detector diaphragm yields little reduction in the SNR and 5,000 waves aberration reduces the SNR by a factor of only about

15%. For bright program objects and long integration times yielding a high intrinsic SNR, one may decide that even larger mirror aberrations are permissible.

The Full Width Half Maximum for a normally distributed variable is given by $FWHM = 2\sqrt{2ln2}\sigma \approx 2.35\sigma$. A Gaussian PSF is two dimensional and has a Rayleigh cumulative distribution function given by $cdf = 1 - e^{-r^2/2\sigma^2}$. Therefore, with this type of PSF form, a radius of 0.5(FWHM) (i.e., one HWHM) encircles 50%, a radius of 1.0(FWHM) encircles 93.7%, and a radius of 1.5(FWHM) encircles 99.8% of the flux. The last value is considerably less than the 3.0(FWHM) conservative value Howell (2006, 116-117) gives for 100% of the flux. No matter what the factor, for the sample case given in Howell, how does the SNR relate to diaphragm size for highly aberrated mirrors? Figures 2 and 3 are graphs of the SNR given by equation (20) for stars of different brightness for the cases of 2500 waves rms surface height and 10 waves rms wavefront gradient norm aberrations at 500 nm. The SNR is plotted as a function of detector diaphragm diameter for four different apparent brightness stars (8th, 12th, 16th, and 20th magnitudes) in relation to a 21st magnitude per square arc second background imaged by a 1.6 m *f*/1.9 mirror, a 10 second integration, and 15% overall collection efficiency. Scintillation is modeled using equation (1) of Young (1967) at an elevation of 1000 m, and an air mass of 1.5.

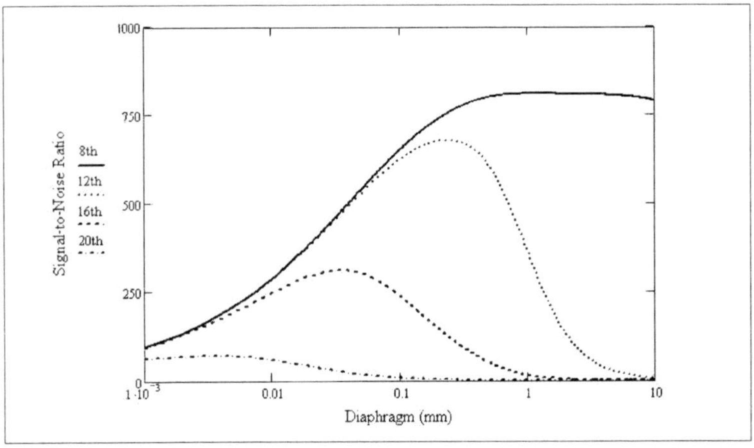

Figure 2: SNR as a function of detector diaphragm diameter for four different apparent brightness stars (8th, 12th, 16th, and 20th magnitudes) in relation to a 21st magnitude per square arc second background imaged by a 1.6 m *f*/1.9 mirror with 2500 waves rms surface height aberration. Scintillation is estimated for an elevation of 1000 m and an air mass of 1.5, with other parameters as described in the text.

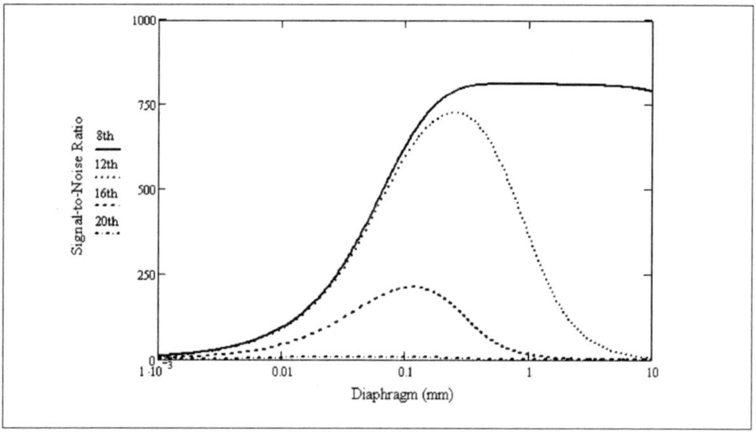

Figure 3: SNR as a function of detector diaphragm diameter for four different apparent brightness stars (8th, 12th, 16th, and 20th magnitudes) in relation to a 21st magnitude per square arc second background imaged by a 1.6 m *f*/1.9 mirror with 10 waves rms gradient norm aberration. Scintillation is estimated for an elevation of 1000 m and an air mass of 1.5, with other parameters as described in the text.

For both forms of aberration shown in Figures 2 and 3, the SNR grows as the diaphragm increases but eventually turns over as the shot noise contribution from the sky overtakes the signal flux. The plots for the fainter objects illustrate that light bucket mirrors are not desirable for use with low contrast objects. The SNR of near 750 for a 1 mm diaphragm with a bright object drops by about 50% for an object 4 magnitudes fainter. Note that the peak SNR moves to the left in the direction of smaller diaphragms for fainter objects, much as a damped harmonic oscillator experiences a frequency increase. For variable star radiometry a compromise between detector diaphragm sizes for the star, comparison, and check stars must be struck if the stars differ in brightness by considerable amounts.

Combining Results

In order to use equations (3) and (4), actual light bucket mirrors may need to be analyzed in zonal areas and then the results combined mathematically. Uncorrelated waveform aberrations add in quadrature. The overall rms waveform aberration may be calculated from the contributions of various interferograms taken over subareas, A_S, of the mirror surface, adjusted to the interferogram taken at the center:

$$\sigma^2_{Overall} = \frac{\pi D^2}{4} \frac{\sum_{All\ subareas} \sigma^2_{Subarea} A_S}{\sum_{All\ subareas} A_S} \tag{23}$$

An alternate approach using the blur spots of zonal regions may be added as follows:

$$d^2_{Coc,Overall} = \frac{\pi\, D^2}{4} \frac{\sum_{All\ subareas} d^2_{Coc,Subarea}\, A_S}{\sum_{All\ subareas} A_S} \qquad (24)$$

Use of equation (24) requires that the focal ratios of the individual zones, and not the entire mirror, be used in equations (3) and (4).

Note that actual results of using equations (23) and (24) will vary depending on the degree to which the aberrations of different zonal subareas are in fact independent and add in quadrature.

An overall strategy for light bucket mirrors is to use the focal ratio which minimizes equation (24) for the main highly aberrated primary mirror, and then to use high quality focal reducer and spherical corrector optics to improve the character of the PSF. For instance, using a *f*/2 spherical mirror instead of a *f*/4 spherical mirror will reduce the CoC blur spot caused by local slope aberration issues by a factor of two, but it will also increase the spherical aberration by a factor of 2^3, or 8. Figure 4 shows the effect of a combined CoC blur spot from spherical, surface height, and local slope aberrations as a function of focal ratio.

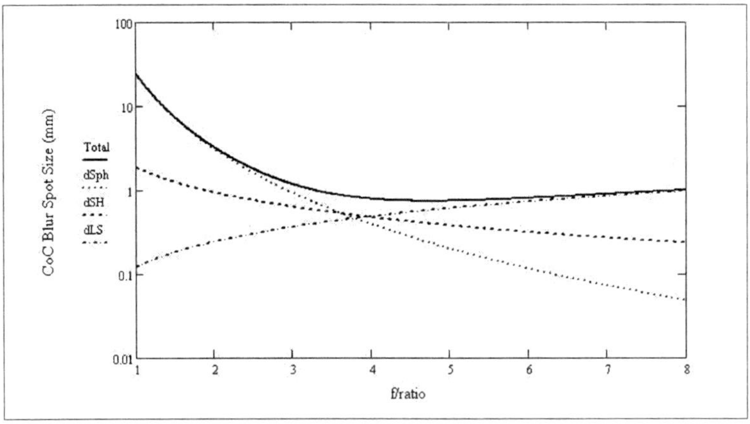

Figure 4: CoC blur spot size as a function of *f*/ratio for a 1.6 m spherical mirror with 2500 rms waves of surface height and 10 rms waves of gradient norm local slope aberrations. The symbols dSph, dSH, and dLS are the CoC blur spots from spherical, surface height, and local slope aberrations, respectively, and the total aberration includes all three sources added in quadrature. The surface height and the local slope CoC blur spots are three standard deviation sizes.

Figure 4 shows that systemic spherical aberration dominates the CoC blur spot size without secondary correcting optics for this example.

The crossover between surface height and local slope aberrations moves to the left as surface height aberrations diminish.

Secondary optics which correct for systemic spherical aberration may be much easier to fabricate than secondary optics for complicated surface height and local slope aberrations if the resultant CoC blur spot is still larger than the desired tolerance limit.

Also, note that high frequency surface defects, such as microripples causing diffuse reflection, do not have a ready solution. One must select the best substrate possible and polish out these problems if possible. In other words, light bucket mirrors may have considerable low-spatial frequency problems of manufacture or support which are fixable by mechanical adjustments of the primary mirror and/or secondary static-, active-, or adaptive-optics of one form or another. High–spatial frequency problems, on the other hand, have no ready solution which is economical in comparison to the cost of the light bucket mirror itself.

Telescope Tracking Errors

Telescope tracking errors make a program object drift across or even out of the detector diaphragm. Software that processes a CCD image may be able to compensate by moving the readout window on succeeding images. In contrast, diaphragm photometry with a single photodiode or photomultiplier detector will suffer unless re-centering of the program object occurs frequently. This is inherently a mechanical/electrical problem and not an optical one.

For light bucket mirrors the PSF extends far beyond the Airy disk. Figure 5 illustrates how flux is lost on one side yet gained on the other side of a drifting circular-shaped diaphragm.

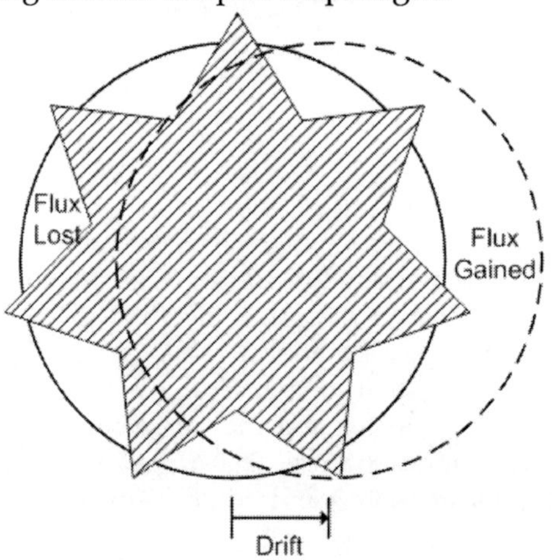

Figure 5: A circular diaphragm gains some program object flux as it drifts due to tracking errors. However, more flux is lost than gained due to the drift so there is always a net decrease in program object flux.

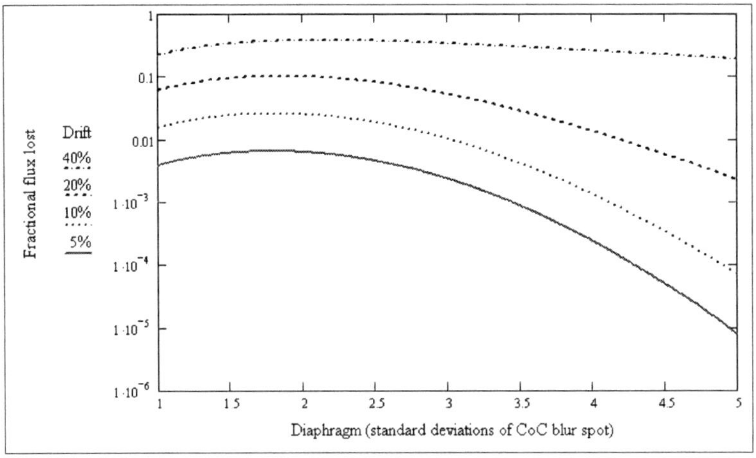

Figure 6: Fraction of flux lost in a circular diaphragm due to centering-drift errors at the focal plane as a function of diaphragm size in standard deviations of CoC aberration from a Gaussian PSF. The drift is expressed as a percentage of the diameter of the detector diaphragm.

Figure 6 shows the flux deficit that results from centering errors of a circular diaphragm as a function of diaphragm size and drift of the program object from the center of the diaphragm.

The figure shows that the fraction of flux lost initially increases with diaphragm size due to drift in an absolute sense also increasing (i.e. as a percentage of diaphragm diameter). The fraction of flux lost turns over rapidly as the diaphragm increases in size beyond two standard deviations of the CoC blur spot. For a mirror with a focal ratio of 2.0, and 15 waves rms local slope aberration at 500 nm, equation (15) gives the one standard deviation CoC as 0.12 mm. So, a diaphragm of 4 standard deviations is 0.5 mm and still yields less than 0.1% (i.e. one milli-magnitude) flux loss when the drift is about ten percent during the detector integration. However, if the drift in this example is allowed to extend to 20%, about 1.0% of flux is lost.

Conclusions

• Light bucket mirrors challenge traditional aberration analysis used for diffraction-limited general-purpose telescope mirrors. A statistical approach is needed to characterize performance properly.

• For a detector diaphragm of 1.0 mm and a telescope focal ratio of 2.0, a maximum of about 1.0 mm rms of smooth surface height aberration and 42 waves rms of local slope aberration is permissible for about 99% of a point source's flux to be collected.

• Large mirror aberrations significantly reduce the SNR of collected flux. The peak in SNR occurs at a detector diaphragm diameter considerably less than the size for collecting 99% of the source flux. Light bucket mirrors are not suitable for imaging applications nor are they desirable for use with low contrast objects.

• CoC blur spots from surface height flaws and local slope figuring issues may be quantified by combining interferometer results from multiple areas.

• Telescope tracking requirements are relaxed due to the large detector diaphragms used with light bucket mirrors.

Acknowledgements

The authors wish to thank editor Russ Genet and some members of the Alt-Az Initiative for suggestions and feedback on the contents of this text.

References

Born, M. and Wolf, E. 1999. *Principles of Optics, 3rd ed.* Cambridge: Cambridge U. Press.

Braat, J. 1987. Polynomial expansion of severely aberrated wave fronts. *J. Opt. Soc. Am., A4*, 643-650.

Genet, R. 2009. Lightweight mirror developments. Presented at the Society for Astronomical Sciences 28th Annual Symposium on Telescope Science.

Hardy, J. W. 1998. *Adaptive Optics for Astronomical Telescopes.* New York: Oxford University Press.

Holenstein, B. D., Mitchell, R. J., Holenstein, D. R., Iott, K. A., and Koch, R. H. 2010. Experiments with pneumatically-formed metalized polyester mirrors. In *The Alt-Az Initiative: Telescope, Mirror, and Instrument Developments*, eds. Genet, R., Johnson, J., and Wallen, V. Santa Margarita, CA: Collins Foundation Press.

Howell, S. B. 2006. *Handbook of CCD Astronomy, 2nd ed.* Cambridge: Cambridge U. Press.

Mahajan, V. 2007. In *Optical Shop Testing, 3rd edition*, ed. D. Malacara, 516-518. Hoboken WI: Wiley.

Rowe, D. 2003. *FringeXP: A Fringe Analysis Program, Ver. 2.* Internet <http://www.ceravolo.com/fringeXP.html> accessed 27 January 2010. Note that *OpenFringe* is an open source variant that is available at Internet <http://sourceforge.net/projects/openfringe/>.

Schroeder, D. J. 2000. *Astronomical Optics.* New York: Academic Press.

Southwell, W. H. 1982. What's wrong with cross coupling in modal wave-front estimation? *Proc. SPIE, 365*, 97-104.

Wilson, R. N. 2007. *Reflecting Telescope Optics I: Basic Design Theory and Its Historical Development, 2nd ed.* New York: Springer-Verlag.

Wyant, J. and Creath, K. 1992. In *Applied Optics and Optical Engineering, Vol. XI*, ed. R. Shannon and J. Wyant, 1-39. New York: Academic Press.

Young, A. T. 1967. Photometric error analysis. VI. Confirmation of Reiger's theory of scintillation. *AJ*, *72*, 747-753.

Chapter 17

Meniscus Mirrors

Mel Bartels

B B Astrosystems

Introduction

Meniscus mirrors are mirrors with a constant thickness. Since the face is concave, the back must be convex to maintain the constant thickness. Because of the curved back, meniscus mirrors cannot be held as normal during production or inside the telescope tube.

This chapter is the result of grinding and polishing a 13 inch [33cm] *f*/3.0 meniscus mirror. My goals were to study specific techniques for handling meniscus mirrors and to test the visual performance of an *f*/3.0 mirror. The mirror is 1 inch [2.5 cm] thick plate glass. As part of his ongoing experiments with mirror casting, Richard Schwartz took an old Coulter 13 mirror, slumping it over a mold in his oven.

Figure 1: The 13 inch meniscus mirror ready for aluminizing, showing the extreme depth in the center.

Annealing, Strain, and the Polarized Light Test

Strain or lack of annealing can result in the glass not holding its figure during polishing and later while in use in the telescope. This particular mirror exhibited no strain using the polarized light test. It is important to conduct a polarized light test to look for strain in mirrors that have been slumped. The slumping process must include re-annealing the glass.

The polarized light test is done using a large piece of polarizing material held in front of the mirror. A second piece of polarizing material is placed behind the mirror. The polarization of the blue sky is an acceptable substitute for the second piece. To conduct the test, rotate the front polarizing material while looking through it at the mirror. The mirror will appear to darken and lighten. Strains appear as black crosses in the center of the mirror, or curved black arcs near the edge. Extreme strain can cause rainbows to appear. If the glass looks clear, then it passes the polarization test and is deemed sufficiently annealed.

Why Meniscus – The Advantages and Disadvantages

Advantage: lightweight

Meniscus mirrors are lighter weight mirrors. Lighter mirrors mean lighter telescope tubes and mountings, which scale geometrically in weight. Lighter structures are not only easier to design but are lower in cost. Lighter mirrors also improve the dynamic performance of the telescope structure. Lighter mirrors reduce the moment of inertia, shortening the telescope's time constant. The telescope is literally stiffer. A stiffer telescope reacts quicker to guiding commands. A stiffer telescope also can quickly shed energy dumped into the structure by wind gusts, meaning faster dampening of vibrations. The result is a telescope that can operate in higher wind conditions. All this is because the mirror is lighter.

Advantage: strength

Meniscus mirrors are significantly stronger than their sibling mirrors. Pick up a piece of flat paper. See how floppy it is—it won't stay flat in your hand. Pick up a coffee filter that is coned shaped. It's very strong and hard to deform. Meniscus mirrors are similarly strong. The 13 inch diameter, one inch thick mirror has the strength of a much thicker piece of glass, about three inches thick. This was determined experimentally by examining the deformations via optical testing by placing forces on various parts of the mirror. This results in a much simpler and lighter weight mirror mounting, since the meniscus mirror can be supported on fewer points. For the 13 inch, a three point support is sufficient, with no deformation visible in the star test image at high magnifications of a 1mm exit pupil.

Advantage: fast focal ratios

Meniscus mirrors can be slumped to very fast focal ratios. Standard flat mirrors run into trouble as the sagitta, or depth of the curve, increases. With these mirrors, the center simply becomes too thin, limiting the focal ratio. Thicker flat glass can be considered, but the increased weight becomes a significant deterrent.

Advantage: cost
Finally, meniscus mirrors can be much less expensive. Because they can be slumped from flat pieces, thinner glass that is cheaper can be used. The difference is quite significant. A ¾ inch thick piece of plate glass costs $100-$200, with slumping doubling the cost. Compare this to a Pyrex sheet with enough thickness to account for the sagitta. The Pyrex costs an order of magnitude more.

Advantage: thin meniscus plate glass cools quickly
Plate glass today amongst some amateurs has a pejorative reputation. Pyrex itself is no angel when compared to zero expansion glass or to quartz glass. Consequently, many amateurs use fans to bring the Pyrex mirror into thermal equilibrium with the ambient air. One reason for making this 13 inch mirror was to assess plate glass under the same condition. I found that employing a fan moving a large volume of air cooled the mirror to ambient air temperature so as to give stable high power star images in about 15 minutes. The thinner meniscus glass speeds catching up to ambient temperature. A typical temperature difference between mirror and air was about 20 degrees Fahrenheit, with the air temperature dropping about 2 degrees Fahrenheit per hour.

Disadvantage: temperature problems during figuring
Plate glass does have its challenges during figuring and testing. Unless active measures are taken to bring the glass to room temperature, the figure will appear over-corrected when cooling to room temperature and under-corrected when warming up to room temperature. The mirror and pitch lap need to be at the same temperature or the mirror surface may not polish smoothly, even resulting in Dog Biscuit or Orange Peel. Consequently, active measures and time are needed while figuring plate glass mirrors. Fans to bring the mirror to thermal equilibrium, working in a polishing and figuring room that is of constant temperature, dunking the mirror in water baths at room temperature, and patience during testing are necessary to accurately figure the mirror and maintain a smooth surface. On the plus side, plate glass polishes quicker than other glass and anecdotally can take a more brilliant polish than Pyrex.

Focal Ratios through History

With meniscus mirrors comes the ability to work very fast focal ratios. Faster focal ratios promise lighter and more portable telescopes, or, equivalently, larger apertures for the same size package as smaller, longer focal ratio telescopes. Faster focal ratios have been a steady progression through history mainly due to eyepiece technology. The beginning of the twentieth century's *f*/10 became mid-century's *f*/7 and end-of-century's *f*/5. Today, wide angle eyepieces and coma correctors are optimized for *f*/4.5 and are often used on *f*/4.0. How fast can we go?

The fastest publicized scope for many years was a 12 inch [30 cm] *f*/3.75 featured in *Sky and Telescope* circa 1980. Lately several French telescopes have pushed faster, i.e. to *f*/3.25 and David Vernet's one meter *f*/3.0 (ADIA). Steve Swayze showed an 18 inch [46 cm] *f*/3.5 at the 2008 Oregon Star Party. The goal is that the image in a fast scope be indistinguishable from that in a slower scope, say *f*/5. From observer reports, Swayze's scope accomplishes this, as do the French telescopes.

Coma correctors: a necessary accessory

Naturally, a coma corrector like the TeleVue Paracorr or Keller's Astro System Austria design is mandatory (Astro System Austria 2008). Without a coma corrector, the 13 inch at *f*/3.0 gives perfect star images at the center of the field of view. As the star is moved towards the edge, coma appears. Half way to the edge, a star literally disappears in a smudge of light. With the coma corrector, it is an entirely different story. The star stays a dot all the way to the edge of super wide angle eyepieces. As scopes become faster, manufacturers will likely follow suit with coma correctors that work at ever faster focal ratios. French observers report that some eyepieces do not perform well with coma correctors at *f*/3.0.

The Finish Line

Our goal is a mirror practically indistinguishable from perfect. A perfect telescope focuses star light to a point, such that the light forms a diffraction pattern characteristic of a circular aperture. There's a central Airy disk surrounded by ever dimming rings. A lower quality optic brightens the rings, making fine detail harder to discern. Our mirror should present the same Airy disk with approximately the same brightness in the rings. Geometric based methods that calculate the path of the reflected light rays across the mirror face are popular and have a long history. These tests typically measure the longitudinal aberration, or the discrepancy between where the light ray travels compared to where it ought to be. However, without taking into account diffraction, the size of the Airy disk and the brightness of the surrounding rings cannot be determined with certainty, and indeed, in some cases, are non-intuitively correlated with the geometric ray tracing. In other words, the light does not go where the geometric ray traces say it goes, thanks to the diffraction of wave optics (Burrows 2009).

Even when the Airy disk and surrounding ring brightness are calculated, the problem remains of relating the test results to judgments at the eyepiece. How does a mirror with a Strehl ratio (a measure of relative brightness of the Airy disks and, by extension, of the relative brightness of the surrounding rings) of 0.90 compare to a mirror of 0.85—is the 0.85 mirror acceptable? This is a problem of mapping objective measurements to subjective expectations.

Star testing

Star testing provides a direct test of mirror performance. The tester judges the star image qualitatively. This reduces the problem of mapping objective measurements to subjective expectations to an easier-to-solve problem of deducing the mirror surface profile from the star test.

Figure 2: Star test setup, aimed at Polaris.

Star tests can be done at night on stars. For northern hemisphere residents, the North Star Polaris is a convenient stationary target. The star test is subject to poor seeing (scintillation caused primarily by temperature differences in the upper atmosphere) and difficulties in bringing the mirror to ambient temperature. Star testing indoors during the daytime in controlled conditions means surer results.

My setup is a finely figured 20 inch [51 cm] telescope placed horizotally with a pinhole at focus in place of an eyepiece. I make several pinholes of varying size in a small piece of aluminum foil and place it inside the focuser. The larger pinholes give a brighter star test image and are used for coarse testing, while the smaller pinholes which give a much dimmer star image can be used for precision testing. The mirror to test resides in a simple frame that includes a diagonal mirror and a high power eyepiece sufficient to achieve a 1 to 2 mm exit pupil. This test frame is placed horizontally and aimed directly at the reference telescope.

Figure 3: Pinholes in aluminum foil taped to bottom of focuser adapter.

This is very convenient and fast: I can make several figuring spells, testing each one, despite the plate glass temperature vagaries, in a single evening. As it turns out, this is important at f/3.0, because figuring is a ticklish process. I also use the Ronchi test as a qualitative test for surface smoothness and zonal smoothness.

Completing the Mirror: The Last Planner Approach

To finish a project doing only the work necessary at each stage, it is useful to imagine the finished product and determine what needs to be done. This then sets up an earlier stage that undergoes a similar treatment. Eventually we reach the beginning of the project, with a clear path of dependencies or what needs to be done in order to accomplish the next stage. The milestones in reverse for completing a mirror are: figuring, polishing, fine grinding, and curve generation. Each stage depends on the previous stage being completed correctly.

Figuring

In order to make an indistinguishable-from-perfect star image, the mirror surface must be accurate to a small fraction of the wavelength of visible light. The stage of adjusting the mirror surface by preferential polishing is called figuring, a phrase so called because of the mathematical figuring that goes into calculating the mirror surface from Foucault test measurements. To begin this phase, the mirror surface should be smooth and accurate to a wavelength of light.

Polishing

To achieve this preparatory to figuring stage, the mirror is polished to a shape that is smooth and spherical. The rate of glass removal during polishing is exceedingly small. It could take fifty years of non-stop polishing to polish a flat piece of glass to within a wavelength of light of the desired mirror profile. Using silicon carbide grit, the curve can be achieved in hours, albeit with heavy damage to the mirror face by the grit particles. This stage is called rough grinding. Meniscus mirrors avoid this stage completely and start with fine grinding.

Polishing with a pliable material like pitch results in a smoothly polished surface, accurate to a wavelength of light or better, that is ready to begin figuring. The act of polishing is both a mechanical and a chemical process (Bartels 2001).

Oversized laps and turned edges

Plate glass mirrors are notorious for turned edges during polishing. Flash polishing after each stage in fine grinding shows an even edge, so the turned edge must occur during polishing. I've found that an oversized pitch lap controls turned edges. Here, a 14.5 inch [37 cm] pitch lap was used, an oversize ratio of 11:10. Oversized ratios can be up to 6:5, the so called magic oversize ratio that automatically maintains a spherical shape at desired radius of curvature.

Figures 4: The pitch lap (left) compared to the oversized pitch lap (right).

Fine grinding

A series of ever smaller grits are employed in order to repair the damage caused by rough grinding, ending with aluminum oxide which leaves much smaller pits and few fractures compared to the silicon carbide. This stage is called fine grinding. I like to use three grit sizes: 220 silicon carbide, 500 silicon carbide, and 9 micron aluminum oxide. Grit of a particular size comes with a wide distribution of particle sizes. Typical are 20% of particles that are twice the stated size. Comparing particle sizes of 400 grit with 500 grit, the size ratio looks to be 4:5. But when looking at the 20% particle distribution, it is a nearly identical

9:10 ratio. Consequently it's wasteful to run through a long series of grit sizes, as commonly practiced: 220, 300, 400, 500, 600, 25 micron, 12 micron, 9 micron, and 5 micron aluminum oxides. The third and final grit that I use is 9 micron aluminum oxide. Ending with 9 micron instead of 5 or 3 micron reduces the chance of sticking on large blanks and controls scratching. Comparing 9 micron to 5 micron looks to be a nearly two times reduction in glass pit depth, but looking at the 20% particle distribution, the reduction is only one-third.

Many people are troubled by the prospect of fine grinding a meniscus mirror. I support the mirror with an old blanket, wiggling the mirror until a depression forms. I rotate the mirror 30-45 degrees every fifteen minutes. This prevents astigmatism from occurring via print through the mirror back. I use a full sized plaster and ceramic tiled tool on top of the mirror. Mirror on top can also be used, but the grit seems to fall down between the tiles requiring more grit and wets to complete.

Parabolizing

Because of the luxury that the radius of curvature can change, there are always three mirror profiles that can be aimed at. One is at best radius of curvature, one is just inside, and the other is just outside the radius of curvature. Every mirror surface profile can be remapped to one of these three profiles.

Parabolizing a sphere can be accomplished by deepening the center (reference profile is just outside radius of curvature), lowering the edge (reference is inside radius of curvature), or a combination of both. The latter is the traditional method. I examined all three methods, parabolizing the mirror a dozen times.

The process of parabolizing is one of demonstrating control over the mirror surface profile. That is, as parabolization progresses, each zone or area of the mirror should be partially parabolized evenly. If halfway through parabolizing, then each zone of the mirror should be 50% parabolized.

I found that deepening the center (a popular method with larger mirrors) resulted in the edge lagging far behind, and when the edge was parabolized, the mid zones were excessively deepened. Parabolizing by deepening the edge was very slow. The best combination was a slight deepening of the center to begin, followed by exclusively deepening the edge from then on. I was able to fully parabolize the 13 inch in about two hours using this technique.

Figuring

Once full parabolization is obtained, the goal is to maintain overall parabolization while adjusting individual zones. I discovered that straight through the center half long strokes on top of an oversized pitch lap maintained parabolization. Other approaches, like the traditional 1/3W wide strokes and using a parabolizing pitch lap, gradually destroyed the edge, either deepening or shallowing it.

Figure 5: Extreme parabolizing by deepening the mirror center.

I examined how best to adjust individual zones using area laps and accentuated action strokes. Area laps are laps that have pitch concentrated or removed in a zonal or circular pattern. For instance, it is common to find the 70% zone too deep. Scraping away pitch in the 70% zone results in an area lap that exerts little polishing action on the zone. This proved effective and could be controlled by limiting polishing time.

Another technique I investigated was accentuated action strokes. Let us say that the 80% zone was too shallow and needed deepening. By using short circular strokes that placed the 80% zone at the edge of the pitch lap, I was able to smoothly deepen zones. This technique proved highly effective with strong action in seconds to minutes.

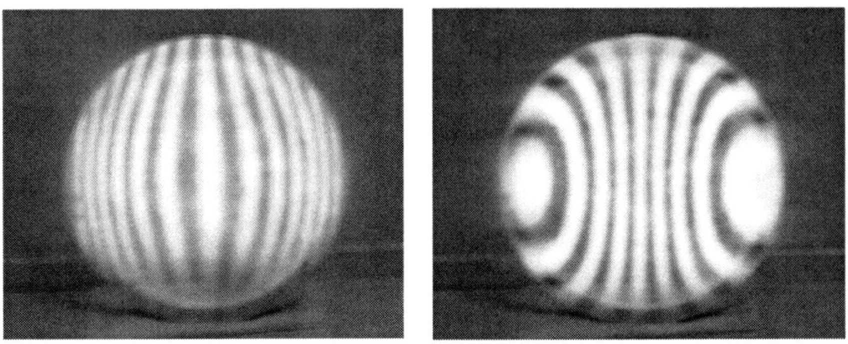

Figure 6: Ronchigram inside radius of curvature (left). Ronchigram outside radius of curvature (right).

Conclusion

Meniscus mirrors possess many advantages. Meniscus plate glass mirrors of f/3.0 can be figured to a high standard and in a controlled fashion. They produce stable high definition star images. Coupled with a coma corrector, the incredibly fast f/3.0 yields good images to the edge of the eyepiece's field of view.

Author's Note

You can contact Mel Bartels at mbartels@bbastrodesigns.com.

References

ADIA and Vernet, David. 2009. 1m Dobson. Internet <http://www.astrosurf.com/altaz/1000_e.htm> accessed Feb 2009.

Astro System Austria. 2008. Internet <http://www.astrosysteme.at/eng/correctors.html> accessed Feb 2009.

Burrows, Jim. 2009. Internet <http://home.earthlink.net/~burrjaw/atm/t_verse.lwp/t_verse.htm> accessed Feb 2009; Internet <http://home.earthlink.net/~burrjaw/atm/odyframe.htm> accessed Feb 2009.

Bartels, Mel. 2001. The chemical mechanical process of polishing. Internet <http://www.bbastrodesigns.com/chemical_mechanical_polishing.html> accessed Feb 2009.

Chapter 18

Active Optics for Thin Meniscus Telescope Mirrors

Michael Connelley

University of Hawaii

Introduction

Telescope mirror support systems as a whole can be divided into three types: passive, astatic, and active. Small mirrors of the sizes most familiar to amateur astronomers are made thick enough so that they will not warp under the influence of their own weight. The 6:1 diameter to thickness ratio rule reflects this philosophy. Recently, amateur astronomers have used finite element analysis tools, such as PLOP, to design and build more complex, albeit still passive floatation mirror cells to support increasingly larger and thinner mirrors. Large professional mirrors also made use of passive mirror cells until about 30 years ago. The advantage of a simple mirror cell is offset by the need for a thick, heavy mirror that cools slowly.

Astatic mirror cells typically use weights hanging from levers to more precisely support the weight of a mirror as the telescope moves across the sky than a completely passive mirror cell would. This is a purely mechanical system designed to apply a force to the mirror proportional to the sine of the elevation of the telescope.

An active mirror cell is used when a mirror is so thin, and thus flexible, that it cannot be satisfactorily support by a passive mirror cell. Thin mirrors are used to minimize the weight of the telescope, allowing the telescope to be stiffer, less expensive, and deliver superior image quality (see chapter by Mel Bartels in this book). The warping of the mirror is usually caused by changes in the gravity loading of the telescope as the telescope is pointed to different parts of the sky, by "potato chipping" or "folding" of the mirror as the telescope points to lower elevations, or by mechanical flexure of the mirror support. Active correction is typically on a time scale of seconds to minutes.

Active correction can be implemented in a number of ways. The mirror can be adjusted manually by the user. The mirror can be controlled automatically by a computer "open loop", where the forces applied by the actuators are based on a look-up table. All large professional telescopes with active optics work in this mode. The mirror can also be controlled by a computer in "closed loop", where the wavefront is continuously examined and the mirror is adjusted to optimize the delivered wavefront. Adaptive optics (which should not be confused with active optics) seeks to correct for atmospheric turbulence (seeing) and works at kilohertz frequencies.

In the 1970s, several infrared telescopes were being built and the detector technology at the time did not allow for high resolution imaging. One of them, the 3.8 m United Kingdom Infrared Telescope (UKIRT), was built as an infrared "flux collector." It was designed to use a thin (28.7 cm at the edge, 19.2 cm in the center), somewhat flexible mirror since the flux collector was built on a tight budget and the detectors couldn't make use of good image quality anyway. The mirror is supported on a bed of 80 actuators. The image quality is inspected monthly which, due to a number of recent upgrades, is really very good. Since then, many large observatory telescopes have used thin meniscus mirrors with active control. For example, the Subaru Telescope is an 8.3 meter telescope with a meniscus primary mirror just 20 cm thick (40:1 diameter to thickness ratio). The mirror is fully supported and controlled by 264 actuators. The observatory has developed a look-up table that describes how much force each actuator should apply to the mirror as a function of telescope elevation. Although the active optics system is run open-loop (they do not observe a star to control the mirror), the forces applied by each actuator is controlled by a closed loop. The 8.2 meter meniscus mirrors of the Gemini telescopes are also actively controlled, but the majority of the weight of the mirror is supported by air pressure that is carefully regulated. Of the telescopes larger than 5 meters, seven have a slumped monolithic meniscus mirror under active control: the Subaru 8.3 meter telescope; the Gemini North and South 8.2 meter telescopes; and the four VLT 8.2 meter telescopes. One other has a classical thick mirror (the 6 meter BTA in Russia), four have segmented mirrors (the two Keck telescopes, the Hobby-Ebery Telescope, the South Africa Large Telescope, and the Gran Telescopio Canarias), and four have honeycomb mirrors (MMT, the Large Binocular Telescope, and the two Magellan Telescopes).

Until recently, amateur mirrors have been clearly in the domain where passive mirror support is fine. Even as mirrors up to ~1 meter have been made with 2 inch (~5cm) thick glass, passive floatation cells provide adequate support. However, the amateur telescope, and in particular the large truss Dobsonian, has hit a scaling limit. Corning didn't make 2 inch thick Pyrex glass larger than 48 inch sheets, limiting the size of a blank that could be cut to about 1 meter. Now they don't make 2 inch thick Pyrex at all. It is difficult to figure or support mirrors as large as 1 meter and as thin as 2 inches (5 cm) without inducing astigmatism.

Until recently, such mirrors tended to have focal ratios near f/5, which makes for very long telescopes, requiring a very tall ladder (and a stout heart) to look into the eyepiece. A few opticians do make large mirrors as fast as f/4, and in some cases slightly faster, allowing the telescope to be much shorter. Nevertheless, these mirrors, and the resulting telescopes, are heavy (~600 lbs. to ~800 lbs. for a conventional Obsession style telescope), bulky, and difficult to handle. Future large telescopes will likely have faster primary mirrors, and may not be Newtonians in order to bring the eyepiece closer to the ground (Kriege & Barry 1997).

Mirrors 2 inches thick also take a long time to cool down. While thinner mirrors cool faster, data by Greer (2004) show that as thickness decreases below 1 inch, thermal advantages are not as significant. Therefore, cooling time alone is not sufficient motivation, considering the significant problems involved with making and using such thin mirrors. However, the potential savings in mirror weight, and thus the weight of the entire telescope, drive us towards exploring thinner mirrors.

The Experiment

The Ultimate Goal

My vision is of a meter class portable telescope primarily for visual use. It will have a thin (<1 inch) fast (~*f*/3) slumped meniscus primary mirror that is actively supported. We choose a slumped meniscus mirror since they can be lighter than honeycomb mirrors, can be made without concern about print through, and the mirror blanks are much less expensive. A segmented mirror (in the style of Keck) is not considered, since each segment would be an off-axis asphere. A design with a segmented spherical primary could be chosen, but they tend to require a number of correcting elements, although multiple spherical mirrors might be fine for "light bucket science." The focal lengths of the segments must match to high precision, and co-phasing the segments requires a very complex computer control system. Although co-phasing of the mirror segments is not necessary if the user only wants the light gathering power of a larger telescope and is willing to have the angular resolution of the telescope limited by the diffraction limit of the individual segments, the segments of the mirrors must nevertheless be carefully aligned so that the images from each segment overlap at the focal plane.

The telescope would be a corrected Dall-Kirkham optical design with a Nasmyth focus. A Dall-Kirkham is chosen since its spherical convex secondary mirror is easier to fabricate than a classical Cassegrain or RC's hyperboloidal secondary, and because the primary mirror is less aspheric than the primary of a classical Cassegrain or RC, again easing fabrication. The resulting off-axis coma might be addressed with a commercial coma corrector, such as a Paracorr, if the telescope is designed carefully. The Nasmyth focus was chosen to avoid perforating the primary mirror and to have the eyepiece at a convenient height and location. The telescope would be on an alt-az mount that is driven on both axes for sidereal tracking. Since the telescope is meant to be portable, easy to set up in the field, and easy to use, I prefer to build an active optics system that is as simple as possible. This means no cameras, computers, complex electronics, or power supplies. The error in the wavefront would be seen by eye and corrected by hand.

The Immediate Goal

We expect that mirror flexure of a large meniscus mirror will be dominated by low order aberrations (primarily astigmatism and trefoil) as the mirror "potato chips" or "folds over" when the scope is moved to lower elevation. There may also be residual astigmatism and other low order aberrations left from figuring. The immediate goal is to demonstrate that low order aberrations such as astigmatism and trefoil can be correctly interpreted by eye and corrected by hand. The correction must be stable enough that the correction can be made on a star, then the telescope moved to a nearby target without adversely affecting image quality. Since the test will be on a much smaller scale than a 1 meter telescope, these goals should be easily met, as all things become more difficult with larger telescopes. The experiment should also be quick and inexpensive to implement, again because all things become more difficult with larger telescopes. I simulated correcting for the astigmatism of a large meniscus mirror by actively correcting the 8 inch mirror of my telescope to compensate for the astigmatism in my own eyes.

The Design

In order to do this experiment quickly and cheaply, I modified a Coulter Optical Odyssey 8 telescope. This is an ideal telescope for this experiment since it wouldn't have been a terrible loss if the experiment failed, and the telescope has a plate glass mirror 14 mm (9/16") thick that should be relatively flexible. Also, this telescope never had very good image quality, so this experiment could actually make an improvement.

Since I was primarily interested in correcting astigmatism, I chose to use 8 screws evenly spaced around the perimeter of the mirror to apply the correcting force. These 8 screws allow astigmatism to be corrected along 4 unique position angles, which I expected to be sufficient. The key component of the active mirror cell is the flex plate, which is made of 1/2" thick plywood and has 8 fingers (Figure 1). Each screw threads through a threaded insert near the tip of each finger. Each screw has a machine head, which is trapped between two metal plates such that the screw is free to rotate but otherwise cannot move. This is the one place in the mirror cell that requires some precision to minimize slop. The mirror itself is glued to these metal plates with silicone adhesive (Figure 2). This mirror cell has no edge support for the mirror; the glue bond supports the weight of the mirror when the mirror is vertical. The 8″ mirror is small and relatively thick enough to not need a floatation cell—the weight of the mirror is only supported by the 8 screws.

Figure 1: The back end of the mirror cell. The 8 screws protrude through oversized holes to allow the mirror assembly to tilt without the screws hitting the side of a hole. The three pairs of collimation bolts are seen toward the center of the back plate. The wing nuts are secured to the screw with a hex nut and provide a better grip on the screw.

Turning the screws changes the separation between the mirror and the wood finger, thus flexing the finger which in turn applies a spring force to the mirror. The width and thickness of the wood fingers were calculated such that 1.5 turns of a 20 turns/inch screw would result in a 1/4 wave deflection in the mirror surface (1/2 wave on the wavefront). Turning opposing screws in the same direction while turning the orthogonal pair of screw in the opposite direction warps the mirror to correct for astigmatism. An important detail is that the screws act by applying a force to the edge of the mirror, and the user adjusts that force by turning the screw. An active cell that worked by trying to move the edge of the mirror by a certain distance would work very poorly, since the distances involved (a fraction of the wavelength of light) are so small.

Figure 2: A view of the active mirror cell before the mirror was glued onto the 8 aluminum pads with silicone adhesive. The aluminum pads, which trap the machine heads of the screws, are free to spin on the screw heads. Each screw passes through a threaded insert near the tip of each wood finger of the flex plate. The round wood disk behind the flex plate is the collimation plate.

Figure 3: A view of the mirror cell with the mirror glued on. The aluminum pads can just be seen under the edge of the mirror.

Although the mirror could be collimated by appropriately turning the screws, for the sake of simplicity I kept the collimation adjustment separate from the active mirror correction (Figure 3). To do this, I glued a separated collimation plate below the flex plate with a 1/8" spacer disk between them to allow the fingers to flex without hitting the collimation plate. Threaded inserts in the collimation plate allow the usual three pair of push-pull collimation bolts to align the mirror. I drilled large holes in the solid wood back plate of the Odyssey 8 to be able to turn the 8 screws from the outside. These holes were oversized to allow the whole mirror assembly to tilt without the 5" long screws hitting the side of a hole.

Lessons Learned from On-Sky Testing

After reassembling the telescope, I was pleased to find that the telescope still balanced, focused, and collimated well. Pointing the telescope at Polaris with a 35x eyepiece (to give a 5.8 mm exit pupil so that the observed image quality was limited by my eye's aberrations) and removing my glasses, I was able to cancel the astigmatism in my eye in less than a minute after a few active mirror adjustment iterations. I looked at Polaris inside of focus so that the image isn't reversed, put my hand in front of the telescope so that I could determine the orientation of the pupil to the image I was looking at, and adjusted the 8 screws accordingly. For example, the astigmatic image of a star inside of focus looks like an oval. If I put my hand into the bottom of the incoming beam, and I see the shadow of my hand on the long axis of the oval, then I know to push on the top and bottom of the mirror and pull on the left and right sides of the mirror. With the above active cell design, this is done by loosening the top and bottom screws and tightening the left and right screws, leaving the other 4 screws untouched.

Being reasonably pleased with myself, I decided to increase the magnification to 203x to look at the Trapezium. I was greeted with nearly the worst image quality of any telescope that I've ever seen! Back on Polaris, I could see at high power that the mirror figure was truly dreadful, with severe higher order distortion of the mirror. I was able to correct for these aberrations with a bit more tweaking of the screws using the same procedure I had used to correct for astigmatism. After establishing the orientation of the pupil, I pulled on the part of the mirror where the edge intrafocal image was too close to the center, and pushed on the parts where the edge of the intrafocal image was too far from the center. After this second round of corrections, Saturn showed the best image that I've had with that telescope. Even though Saturn was 90 degrees away from Polaris, the image quality remained excellent.

What has this experience taught us? Achieving the main goal of the exeriment on the first attempt so quickly demonstrates that astigmatism can be easily corrected by hand for a small mirror. The correction of the astigmatism at low power was relatively crude, resulting in the poor image quality at high power. If the optical aberrations are in the primary mirror rather than in the observer's eye, then it would be present at all magnifications, and most apparent at high power. As such, correcting for the mirror's aberrations should be done at high power, or whatever power the observer will be using on the next target(s). We learned that the correction to the mirror was stable enough to adjust the mirror while observing Polaris, then move to Saturn (at roughly the same elevation, but ~90 degrees away) while maintaining excellent image quality. Having 8 adjustment screws means that they can induce higher order aberrations, which they can also correct. I was able to correct for the astigmatism in my eyes at low power and improve the image quality of the telescope at high power. Finally, I also learned that it would be very nice to be able to adjust the mirror cell while looking into the eyepiece.

Toward Larger Mirrors

Everything gets more difficult as aperture increases. Larger mirrors are more flexible. To put this into perspective, the flexure of a beam under a given force increases as the cube of the length of the span. And larger mirrors are heavier, increasing that force to flex the mirror even more. As the diameter to thickness ratio decreases, it becomes increasingly easy for mirror flexure to excite higher order aberrations. Countering this requires adding more actuators to the mirror cell. Furthermore, the static support of mirrors becomes more difficult as they get thinner.

Results from PLOP

As was pointed out by Mark Cowan, an active mirror cell has to support its mirror at least as well as a passive floatation cell would. As an example, if finite element analysis (i.e. PLOP) shows that a certain mirror would need a 27 point floatation cell, then an active cell for that mirror should also support the mirror on at least 27 points. We used PLOP to determine how many floatation points would be required to support a meniscus mirror. We approximated a meniscus mirror by entering into PLOP a very long focal length so that the mirror would have uniform thickness. Analysis by PLOP for a 20 inch (50 cm) mirror 0.4 inch (1 cm) thick shows that it can be supported by an 18 point cell with a 1/10th of a wave wavefront error. A 20 inch mirror can be a mere 5 mm thick and weigh only 5 lbs. and still be supported with a 1/13th of a wave wavefront error using a 54 point cell.

However, things turn bad very quickly as mirrors get larger. The wavefront error for a 1 meter mirror of 1 cm thickness is 1/3rd of a wave for a 54 point floatation cell. For comparison, a 20″ mirror 1 cm thick on a 54 point cell would have a wavefront error of only 1/50th of a wave! A 1 meter mirror would need to be 2 cm thick and weigh 80 lbs. in order for a 54 point cell to support it with less than a 1/10th of a wave wavefront error. Clearly, a meter class mirror needs to be supported at more points (which increases the complexity of the cell), or be thicker (which increases weight and cool down time), or it should not be supported on discrete points. The accompanying chapter by Mel Bartels in this book describes how meniscus mirrors are stronger than a flat glass disk of the same thickness and diameter (which is how I simulated a meniscus mirror in PLOP). Thus, the actual wavefront errors for a meniscus mirror will likely be less than as stated above. Detailed finite element analysis may be required to determine the wavefront error due to sagging for a meniscus mirror.

A Proposed Design

These discussions lead us towards a design of an active cell that can be used on a portable meter class telescope in the field. Instead of 8 screws, this design has 18 actuators in two rings; 12 around the outside and 6 around the inner ring. Although my 8 inch mirror was easily controlled with 8 screws, I suspect that meter class mirrors will require more degrees of correction to adequately control the mirror. It is desirable to have 12 actuators around the outer edge of the mirror to control astigmatism along 6 unique position angles and to control trefoil along 4 unique position angles. As such, the 8 fingered flex plate I made for my 8 inch experiment wouldn't work, since it cannot be easily adapted to a system with two rings of actuators. If each actuator is topped with a triangle with three support pads, we have 54 points of support. But, rather than a triangle, each actuator is topped with a Y shaped flex plate which converts the linear motion of the screw to a force to be applied to the mirror.

The thickness of the flex plate is chosen so that one turn of the screw yields a fraction of a wave of correction, and the width of the fingers are chosen so that they all give the same force when they flex by a given amount. In cases where a mirror is small or thick enough to be supported on 54 discrete points, the mirror may simply be glued to the tip of the fingers of each Y by silicone adhesive. In cases where 54 discrete points are insufficient, something such as small air bags or foam pads could be used to distribute the force more evenly across the back of the mirror. Instead of a series of triangles and pivot points, the screws thread through a single lightweight rigid plate, possibly honeycomb aluminum. The edge of the mirror is supported by a steel cable sling, in part because the mirror is too thin for the usual nylon strap.

There are several ways to apply a force to the mirror. A simple way is with a motor turning a screw that pushes against a spring. This mechanism converts a rotational motion to a linear motion, then finally to a force between the rigid plate and the mirror. Motors are inexpensive, readily available, and only require electric current when you want to move them. Voice coils are another alternative. They apply a force directly rather than converting a displacement to a force via a spring. They are simple (no moving parts) and light. However, they require a constant current to hold position, which creates heat. While an observatory-housed telescope has abundant access to electric current, portable telescopes powered by batteries must use their power more carefully, especially since such a large telescope will have other functions that also require power (e.g. drive motors, heaters, go-to computer). How and where the telescope will be used and kept are very important considerations in designing an active mirror control system.

An equally important part of the design is the user interface. The user shuld have a hand-held control box to make adjustments to the mirror while observing a star. At a minimum, the box should have a master power switch and a way to increase or decrease the force applied to each actuator. These switches or knobs should be arranged in the same way as the actuators on the mirror to make using the box more intuitive. In addition to this, it would beneficial to have a way to correct astigmatism with one switch. Rather than fiddling with 18 switches, the box can have a knob to set the position angle of the astigmatism to apply, another knob to set the amount of correction (rarely adjusted), then a switch to increment/decrement the amount of correction. The user would look at a star, determine the position angle of the astigmatism, set the position angle knob appropriately, and then while looking at the star use the increment/decrement switch to cancel the astigmatism. A similar arrangement could correct trefoil with one switch as well. There will likely be residual higher order aberrations, which could then be corrected with the switches for the individual actuators.

Automated Mirror Correction

Ultimately, we would like to have fully automated mirror correction. Such a system could be operated in either an open loop or closed loop fashion. For an open loop system, the user generates a look-up table of actuator forces or positions as a function of telescope elevation (assuming the telescope is alt-az mounted). The telescope itself would need to be very stiff so that the mirror flexure is repeatable each time the telescope is used. This would be particularly challenging for a telescope that is portable and is set up each time it is used. There would also need to be a way to ensure that the actuators are applying the proper force. For example, the active optics system on the Subaru Telescope is run in open loop, but each actuator is controlled in closed loop to make sure it is applying the proper force. A closed loop system would remove the requirement that the flexure of the mirror be repeatable. However, such a system would require a wavefront sensor and a control computer, both of which consume power and weight. A wavefront sensor should be mounted so that the mapping between the detector in the wavefront sensor and the mirror actuators remains unchanged. The wavefront sensor could be fed by a flip mirror or a beam splitter in the light path before the eyepiece. Ideally, with such a system the observer could point the telescope at a star, push a button, and the rest would be handled by the computer.

Future Challenges

The above discussions assume that an active mirror cell can compensate for low order aberrations (e.g. astigmatism and trefoil) in large meniscus mirrors with smooth figures (i.e. no high order aberrations such as zones, turned edge, or spherical aberration). My demonstration of this on an 8 inch telescope suggests that active correction may be successful on larger mirrors. If an active mirror cell can correct for low order aberrations caused by flexure, then it can also correct those same aberrations if they are left in the mirror after figuring. It becomes increasingly difficult to figure mirrors without low order aberrations as they get larger and thinner. Using an active mirror cell can increase the tolerable figuring error to low order aberrations, making such large and thin mirrors easier to fabricate, and thus less expensive. However, there is always value in making a mirror that has a good figure rather than counting on the cell to fix a mirror with a poor figure. It may be very difficult for an optician to tell if a mirror has unacceptable higher order aberrations that an active cell could not correct if it has significant lower order aberrations that the optician expects the active cell to correct. Nevertheless, if making large meniscus mirrors without low order aberrations proves to be too difficult, time consuming, or expensive, then using an active mirror cell to compensate for these errors may be the best solution.

Another challenge is figuring large meniscus mirrors that are optically perfect, or at least where the only imperfection is a small amount of astigmatism. This challenge includes making a mirror that is fast, is well corrected for spherical aberrations, and is free of zones or a turned edge. To this end, Mel Bartels has demonstrated figuring a 13 inch f/3 mirror that is 1 inch thick. This begs the question: what is the practical limit for the thickness of a large mirror? Many related parameters affect the solution to this question. If a mirror is to be corrected by hand, then the number of actuators should be small, such as in the design above. It is unreasonable to expect a person to interpret the image and correctly control a large number of actuators. If correcting the mirror by hand limits an active cell to, for example, 18 actuators, then a ~1 meter mirror that is ~1 cm to 1.5 cm thick would be about the limit due to unacceptable sagging of the mirror between the support points. If the active optics system is under computer control and has a much larger number of actuators, then the mirror could be much thinner. Then the limiting factor would be controlling higher order aberrations during the figuring process. However, all of this is theoretical speculation backed by a little math and an experiment with a cheap 8 inch telescope. Considering that the Subaru telescope, with 264 computer controlled actuators, has a diameter to thickness ratio of 40:1, it would be overly optimistic to expect amateur made telescopes with active optics to have an angular resolution better than about 0.5 arc seconds (seeing limited at a good site) with a mirror having a diameter to thickness ratio greater than 40:1 in the immediate future.

The static support of large meniscus mirrors is a significant challenge. Using non-discrete support points (e.g. air bags, foam pads) may help. If a mirror sags significantly on a 54 point cell, that error has a very high spatial frequency that will throw the light far away from the diffraction core. Since large mirrors will not be diffraction limited, the question is how badly can the mirror sag at high spatial frequencies before the mirror is no longer seeing-limited? Some telescopes, such as UKIRT and IRTF, were never designed to have good image quality, and thus the tolerance on how much the mirror can sag is much looser.

We also do not know how practical it will be to actively correct the figure of a large meniscus mirror by hand. Such a mirror's figure may change significantly with small changes in elevation or even temperature. It would be annoying at best, and completely impractical at worst, if the observers had to retune the mirror cell for every target that they wanted to look at. In this case, computer control of the mirror (discussed above) would be necessary. One possible reason that the correction to the 8 inch mirror was stable is that the 8 inch mirror is small and light enough not to warp under its own weight. This will certainly not be the case for large mirrors.

Summary

The Dobsonian revolution allowed telescopes larger than ~12 inches to easily be within reach for most amateur astronomers and telescope makers. The invention of the open truss Dobsonian allowed amateurs to build portable telescopes with mirrors up to 1 meter. However, at this point the Dobsonian revolution appears to have stalled. Conventional Dobsonians of this size or beyond are too long, heavy, and bulky to be practical. A significant technological change is required to push through the 1 meter barrier, or to even make 20 inch to 1 meter telescopes more portable and easier to use. That change is the development of lightweight optics. Active optics is one of several ways to implement a lightweight mirror, and has many pros and cons. Active optics allows the use of a lightweight meniscus mirror without any backing material. Controlling a meniscus mirror does not necessarily require a computer or a wavefront sensor, and the technology can be highly portable. Active optics controlling a meniscus mirror may present the lowest areal density of all lightweight optics options. However, the mirror figure must be closely monitored and corrected, requiring a skilled user if the correction is being done by hand. If collimating your 10 inch Dobsonian is frustrating, active optics is not for you. The practical use of active optics on large meniscus mirrors relies on several key points that have not been tested. For example, can thin, fast meniscus mirrors be fabricated in a cost effective manner? Is it practical to use an active mirror cell in an amateur telescope to compensate for flexure and low order figuring errors in a large mirror?

This brings us to the final question. Is it really worth it? Unfortunately, the answer to that question depends on the results of several experiments currently underway or which will happen in the near future. Foamglas core mirrors are currently being pursued by several groups (see accompanying chapters). If these efforts are successful, then amateur astronomers could enjoy the benefits of large, lightweight mirrors with a simple passive support system that requires no attention on the part of the user. Our next step is to build a 20 inch-class telescope with a meniscus mirror to answer some of the questions stated above. Stay tuned.

Acknowledgments

I am grateful to Mark Cowan and to Joshua "Beefcake" Walawender for their helpful and insightful comments.

References

Greer, B. 2004. *Sky and Telescope*. June, 122.

Kriege, D. and Barry, D. 1997. *The Dobsonian Telescope*. Willmann-Bell: Richmond, Va.

Chapter 19

Hogging With Heat

Slumping Meniscus Mirrors in a Homemade Kiln

David Davis

Toledo Scope Werks

Introduction

The heart of any telescope, professional or amateur, is the optics. The optics must be perfect for precise resolution and must be manufactured accordingly. In reflecting telescopes, these optical components are mirrors.

Because of the tolerances demanded by the optics, the material used as the mirror substrate must be stable and should not deform. The combined experience of amateurs and professionals alike has found that, if glass is used for the mirror substrate, a safe size for the substrate can be found in the ratio of one inch of thickness for every six inches of diameter (Howard 1959, Sidgwick 1971). This is a safe thickness ratio, but not the only ratio possible. Amateurs have pushed this limit for many years, and mirrors with a ratio of one inch of thickness for every twelve to fourteen inches of diameter are quite common. Even in the early years of the hobby, successful mirrors with a ratio of 1 inch for 27 inches diameter were made (Ingalls 1947b). The question is, how thin can we go? Would a mirror of one inch of thickness for every 36 inches of diameter be possible? Would even thinner substrates work? Do the thinner substrates offer advantages that cannot be found in thicker material?

These questions are not new ones. One has only to look in various periodicals associated with science and astronomy to find this discussion has been carried on in earnest for a long time. From the May 1947 issue of Scientific American, an article by Albert Ingalls states, "Averaging once a year someone asks this department why telescope mirrors have to be so thick …." (Ingalls 1947a).

There are several good reasons to explore the idea of using thinner mirrors. Thin mirrors cool faster and so come to equilibrium quicker. Thin glass is lighter in weight and this makes the telescope into which it is mounted easier to move, setup, and manipulate. The cost of thin glass is proportionately less.

One objection to using thin glass for this purpose is found in classic mirror forming techniques. The first step in making a telescope mirror involves grinding a curve of appropriate depth (Howard 1959; Ingalls 1932a). This is often referred to as hogging out the mirror. Hogging removes glass at the middle of the blank and makes the blank thinner overall. With very thin glass of large diameter, this step could, in theory, grind completely through the substrate. Therefore a new method of creating this curve must be used in order to utilize thin glass in telescope construction.

Such a method is heat-forming the curve. The blank is curved using heat and so it retains its thickness across the entire diameter (Lowell Observatory 2005). The curved shape also adds strength and shifts the center of gravity toward the back of the mirror, causing it to hug the mirror cell when it is placed into the telescope. And it is not some wild novice that first suggested doing this. In the past, the pundits of thick glass have suggested using this technique when thicker blanks were unavailable. In January 1932, in his monthly column in Scientific American, Ingalls suggests using Pyrex pie pans to manufacture a mirror, since thick blanks " ... cost so much" (Ingalls 1932b). Pie plates are but thin kiln-formed glass.

Another objection to the use of thin glass in mirror making involves an artifact called print through. As a piece of glass sits on a point, for example the suspension point in a mirror cell, the glass above that point appears to be flexed upwards because the glass around the suspension point sags downward due to gravity. This apparent "hump" is print through. Thicker glass has negligible print through as the glass does not sag appreciably around the point on which it sits.

This is a classic example of mixing two technologies and suffering the results. The cell, noted above, is designed for thick blanks. Thin substrates require a different mounting procedure. Many large telescopes employ thin substrate mirrors. They use an active cell, air pads, and radial mounting to support the thin glass, and print through is not an issue. There is however, a boundary past which print through will almost certainly occur in simple mounting cells, and this boundary appears to be very near 3/8ths of an inch of thickness. This has been established by the author using ultra-thin mirror blanks of common size and weight. Larger diameter blanks, due to the weight of the glass, could require greater thickness, or could be supported by copying the active mounts of the large telescopes offered as examples above.

Finally, consider the material costs. It is not uncommon to spend from five hundred to one thousand dollars for a mirror blank of the more common sizes, 12 to 24 inches in diameter, and prices of ten thousand dollars and up are not unknown for blanks 36 inches and larger in diameter. Using a 2001 catalog and price guide as my source, I found that a 32 inch blank from Newport Glass (1999) had a delivered cost of $5850.00. A glass table top ½ inch thick by 36 inches in diameter from Glass Tops Direct (2008), an on-line supplier, has a delivered price of $114.95. Even factoring in the cost of heat-shaping the curve, the savings are remarkable.

Pyrex Pie-Pan Mirrors

I began exploring thin glass mirrors with Pyrex pie pans, just as Ingalls suggested. A pie pan is nothing more than cheap, thin, heat-formed glass. They are very common and the typical price is $1.99 for a used one and $3.99 for a new one. They yield a 7 inch *f*/5 mirror. I have made several mirrors from this material and they produce good quality optical elements.

The surface used is the outside bottom curve on the pan. The sides are retained both for control of the grinding process as well as mounting the finished mirror. The glass is thin and some different techniques need to be considered for working the glass. Excessive pressure will deform the glass and introduce astigmatism and other errors. Grinding with the mirror on the top will help to prevent these errors from developing. Placing the tool on top flexes the surface of the glass and can cause astigmatism. Some uniformly distributed weight is required to make the work progress at a reasonable rate. The weight is also required for pressure during the polishing and figuring of the glass. The simplest solution proved to be filling the pan with water, choosing the amount and weight depending on pressure required. The tool must be full size. This supports the glass completely during the grinding process and keeps the curve true. I cast tools from rapid set grout and then glued on broken glass pieces for the grinding surface.

Small mirrors need small glass pieces so they come into good contact in a reasonable time. Otherwise, the pie pan will be ground too thin while bringing the tool into full contact with the mirror surface.

A machine will yield more consistent results but most of the mirrors I made from pie pans were done by hand. Because the sagittal curve is cast in, the job of the optician becomes one of optically smoothing the glass surface. When the pans are formed, the glass develops stretch marks as it is worked. These artifacts of the manufacturing process have to be ground out. The zones range in depths of one or two hundredths of an inch and appear randomly on the surface. However, there are two zones, diametrically opposed, that are the result of stamping the manufacturer's name on the bottom of the pan. These are always the deepest artifacts and the last to disappear. Final thickness of the pan bottom, after figuring, averages 4mm to 5mm.

A 16 Inch Mirror

After finishing my first two pie pans, I accepted a 16 inch kiln-formed mirror project that had not been completed. This was a piece of soda lime glass that had been kiln formed to an *f*/4 curve. Final figures revealed an *f*/3.55 curve. The slumping tool was included and I surfaced this with glass pieces and pressed it into service as the grinding tool. This project was meniscus-shaped and that, coupled with the thin glass, offered very little edge to control the grinding process. Starting thickness was 9mm, final thickness was approximately 8mm.

Since the shape and the edge of the glass made machine-working the mirror impractical, I ground and polished this project by hand. The glass was extremely flexible and I had to pace my strokes accordingly. To increase friction between my hands and the glass, I wore rubber-surfaced gloves. To spread the contact, I worked with fingers spread, keeping the weight off of the heel of my hands. Work progressed normally and I noticed the same artifacts found on the pie pans in evidence on the slumped piece. Grinding is a very wet process and I found no anomalies in the process; however, polishing and figuring allowed the heat from my hands to build up on the back of the mirror and this changed the figure. I discovered I could warm the back of the mirror with my hands and change the figure of the front of the mirror. Consequently, polishing and figuring sessions had to be shortened and I cooled the mirror between wets in a bath of room temperature water.

Kilns

Once bitten by the thin meniscus mirror bug, I began to search out or create other kiln-formed mirror projects. An immediate hurdle to overcome was the kiln-forming process itself. Kilns capable of larger diameters are expensive to purchase and operate. The owners of these kilns were justifiably concerned with letting me slump one of my projects in their equipment. The "warm glass" world is replete with horror stories of kilns destroyed by unknown users and hidden flaws in forms and mold that burst under heat and ruined a kiln. After spending most of a year looking for a kiln operator who would do my projects, I decided I would have to do the work myself. Due to the cost of the equipment needed to form the mirrors, this decision also included building the kiln.

Kiln-forming telescope mirrors requires the glass, a form to shape the glass, and a kiln with precise enough controls to heat and cool the glass without shock. As I looked for a kiln to do my projects, I began to acquire the glass. Initially, when I had hopes of renting time from an experienced kiln operator, I looked only for the glass for finished mirrors. At the top end of my project list, I was thinking of a one meter mirror;

at the bottom end of the list was a 24 inch diameter mirror. The most common source for thin glass in these sizes is glass table tops. I quickly found a 36 inch disk with a thickness of 1/2 inch. This was found at a thrift store for $5.00. Watching the local thrift stores and occasionally searching on-line sites, I purchased three 42 inch disks and one 48 inch disk, all second hand, all at pennies on the dollar. As it became apparent that I would have to do my own kiln work, I began to get a range of glass, everything from old pot lids to microwave turntables. These would be used to learn the craft of heat-forming glass.

A kiln is little more than an insulated box with a heat source. Ancient kilns were built on the chimney principle, with the fire box located at some distance from the kiln. Many of these were dug into a hillside to take advantage of the natural draft and the fire proof and insulating ability of the soil. Of the many heat sources available to fire a kiln, I chose electric coils. They were readily available, the price of electricity at my home is quite cheap, and a coil is a simple and clean piece of technology to use.

The controller for the kiln was not so easy to find. Initially I thought I would use pyrometric cones. I had been studying warm glass techniques on-line, and cones seemed to be an elegant and inexpensive solution to the control problem. However, cones are but indicators of heat benchmarks. They do not tell the operator where the kiln is now, only where it has been. Among the literature I studied were accounts of some of the early kiln-formed mirrors that had inexplicably broken while in the field. The suspect in these failures was thermal shock induced during the kiln-forming process.

Cones would not be precise enough to prevent this flaw, so I decided on an electronic controller. The controller would have to be a portable one, as I had not settled on a final design for the kiln box. Cost was another concern; these controllers can cost in the hundreds, even thousands of dollars. Ultimately I settled on a controller built from parts purchased over time over the Internet. This not only allowed me to design the configuration of the controller but gave me the opportunity to become familiar with the circuit and possible failure points. Total cost was $285.00, and of this amount the controller board was the most expensive at $149.00. There is no way to mitigate this cost, save getting a cheap used kiln with a controller attached or borrowing one. However, a club could spread the cost out and lend the controller to members for their projects.

The kiln box was the final issue. Standard soft kiln bricks can be found for two or three dollars each. I read about these finds but never enjoyed one of my own. Soft kiln bricks that I found had a typical price of $4.95 a brick. The bricks are two to three inches thick by four and one half inches wide by nine inches long. If I was going to form a 48 inch disk, I would need a minimum inner diameter of 56 inches to allow air flow. This meant a rough inner circumference of 175 inches. Given a nine inch brick, this meant about 20 bricks per ring. With each brick lying flat I could get around two inches of height per ring. For airflow I needed twelve inches of height overall. That means six rings or 120 bricks, or roughly $600.00 for the walls of the kiln. The top and the bottom would be an additional cost and suddenly the cost of the box was way above all of the others. I was not even sure I would do more than the few disks I had and then what would I do with the kiln?

One day while reading about kilns on the Internet I found a page on the subject of Raku, a type of pottery making. These artists were making kilns from a ring of wire covered with ceramic wool. The cost of the wool was around $100.00 for fifty square feet. Looking up the information on ceramic wool I found it would handle heat up to 2500° F and two layers had the insulting equivalence of a soft ceramic fire brick. What I realized was I could make a kiln of any size or shape using wire and the ceramic wool. It should last well, as I was using clean heat and never going higher than about half the rated temperature. I could resize the kiln at will to match my current project and store it in a box on the shelf when not in use. My box of ceramic wool cost me $130.00 delivered. The lid to the kiln was a wire square covered on both side with ceramic wool. For the floor of the kiln I used four inches of dry mineral sand with a few common fire bricks on top of the sand to support the slumping mold.

Slumping Molds

The slumping mold proved to be another problem looking for a solution. All of the molds I was familiar with were positive hard molds, shaped either by hand or machine to as precise a curve as possible. The material had to be able to handle the heat of the kiln cycle without warping, shrinking, or cracking. It had to be inexpensive and easy to form and work. Ceramics seemed to be a natural solution. Yet, I found the ceramic material hard to use and unforgiving. It had a high shrinkage, a tendency to warp as it dried, and if it were poured or layered too thickly it cracked. It shaped well and could be brought to a good curve.

I experimented with several ideas to form the ceramics and ultimately developed a process using sand. I would pound sand into a circular ring of metal. I then cut a "target" curve into a thin piece of scrap wood. This curve was the sagittal depth needed for the mirror focal length. I would screed the curve into the sand, forming a simple mold into which I poured multiple thin layers of ceramic slip. Each layer was allowed to dry before the next layer was added. Once I had a thick enough blank,

I would remove it from the form and shape it to the final figure. There was no need to fire the mold to use it. The slumping temperature was well below the firing temperature of the ceramic, and I used the molds I made unfired, save at slumping temperature. Consequently, they stayed soft and could be easily re-figured and easily broken.

One day while preparing to pour a mold, I suddenly thought about the kiln-forming process as a whole. Regardless of how smoothly I ground the mold, stretch marks caused by the slump still formed in the glass surface and had to be ground out. Precision was not important. I was hogging the curve of the mirror with heat. All I needed was a basic curve of the right depth and such a curve was right in front of me, the curve I had formed in the sand with my wood screed. Why not slump into the sand? It already formed the bottom of my kiln. I could pack it into a ring placed into the kiln and then screed the curve into it. The ring, if made just less than the circumference of the glass blank, could form the edge of the mold and would hold the weight of the blank. I immediately tried this solution and it worked perfectly. I have not made a hard mold since.

In summary, my kiln system is a simple ring of wire of appropriate diameter for the project. I cover it with a wall of ceramic wool, inside and out, by wrapping the wool around the outside of the ring and folding it over into the kiln. I tie my coils to the wall of the kiln with iron tie wire used to tie rebar. I poke two bronze rods through the ceramic wool and make my electrical connections via those rods. I lay some simple thin fire bricks in the bottom and place the iron ring for the mold on the bricks. I use construction strapping to form the rings. I fill in the kiln bottom to the top of the ring with dry mineral sand. I dampen the sand and press it down hard, then screed the target curve into the sand. I lay the glass on top of the ring and put the cover on the kiln. I put the temperature probe in the middle of the cover pointing down into the center of the kiln. I'm ready to fire.

Kiln Operation

As I noted earlier, a sudden fracturing of some kiln-formed mirrors has been reported. These were undoubtedly thermal stress fractures introduced during the firing sequence. Thermal stress zones are not necessarily visible cracks. They are zones where the crystal structure of the glass has been weakened by uneven heating or cooling. Later these zones can form a crack due to the uneven heating of the glass when the mirror is in use. This failure is easy to avoid: heat and cool slowly and gently.

Because there is moisture in the kiln, specifically in the sand mold, the first step in firing the kiln is to drive this moisture out of the kiln. I heat the kiln to 300°F (149°C) and hold the temperature there for 60 minutes. This drives the water out of the kiln. During this step I use a single layer of ceramic wool to close the kiln. Once the kiln has dried out, I place a second layer of wool on top of the kiln to help it retain heat at the higher temperatures.

When heating glass, care must be taken not to stress the glass too much. A rule of thumb is to allow 60 minutes per quarter inch of thickness. Another is no more than 1.5 degrees per minute. It is essentially the same rule. The idea is to allow the heat to penetrate the glass so that the inside of the glass sheet is as close to the temperature of the surface of the glass as is possible. A rise of several degrees per minute would leave the interior lagging well behind the surface and introduce thermal shock. The 60 minute rule is very conservative and in practice works well. For glass up to ½ inch of thickness, 100 to 150 degrees of heat per hour is a safe increase. In thicker glass, 100 degrees per quarter inch is probably as fast as the temperature should be increased. Art glass projects are increased much faster than this but the glass is thinner and the results are less critical.

In a similar manner, the loss of heat must also be controlled. Here the steps must be smaller and more precise, as the cooling at the top end of the cycle both relieves and entrains tension in the glass.

Glass in general makes the transition from a solid to liquid over a range of temperatures. At the bottom of this range, around 700°F (370°C), the glass becomes soft enough that it can no longer suffer thermal shock. This is a general benchmark and specific temperatures are used for specific types of glass. Once you reach this benchmark, you can increase the temperature at a greater rate and it will not damage the glass. However, it will not cause the glass to reach a thermal balance any faster. The glass can only absorb or lose heat so fast. The top end of this transition range is in the area of 1100F (594C). At this temperature, the glass has fully softened and begins the transition to a molten state. Once you have reached the top end of this temperature range, the glass will slump fully into a form, regulated only by time. The longer the glass is held at this temperature, the more completely it will settle into the mold.

This range, 700oF to 1100oF generally, is the anneal/strain range for the glass. As the glass cools from 1100oF to 700oF, crystal formation occurs and strain is introduced into the glass. Annealing allows the glass to cool at a consistent temperature inside and out and so mitigates this strain development. Once the glass crosses the bottom threshold, around 700oF, no further strain is introduced, but no further annealing is possible. Any strain left in the glass is trapped there, as are any thermally shocked areas.

In general, cooling should be very conservative to achieve the best anneal possible. A general rule of thumb is: cool at half your heating rate. My suggestion is, go slow, you cannot over-anneal. One last step should be considered. The glass should be held at the top of the anneal range for some period of time (soaked), to be sure the glass is at a uniform temperature inside and out and to be sure that the reassembly of the crystal structure of the rigid glass is complete. Generally, about 60 minutes for every quarter inch of thickness is needed.

I use a very simple program to kiln-form my glass. I heat the glass at 100° F per hour up to 1100 degrees. I hold it there for one hour to allow a full slumping of the glass and to be sure the glass temperature is uniform. I then cool at 50° F per hour, down to room temperature.

Once the glass has been slumped to the proper curve, it can be worked with the established mirror grinding techniques noted above.

Conclusion

I hope I have caused some excitement and possibly a desire to try one of these thin mirror projects. As the process develops and comes into the mainstream of the ATM community, very thin large aperture mirrors could become common in amateur telescope making.

References

Glass Tops Direct. 2008. Internet <http://www.glasstopsdirect.com/glass-round-36.php> accessed November 1, 2008.

Howard, N.E. 1959. *Standard Handbook for Telescope Making.* New York: Thomas Y. Crowell Company.

Ingalls, A. G. 1932a. *Amateur Telescope Making, Book 1.* New York: Scientific American Press.

Ingalls, A. G. 1932b. Making mirrors from Pyrex pie plates. *Scientific American,* January. Reprint.

Ingalls, A. G. 1947a. Mirror Thickness. *Scientific American,* May. Reprint.

Ingalls, A. G. 1947b. Making Thin Mirrors. *Scientific American,* June. Reprint.

Lowell Observatory. 2005. Lowell Observatory announces completion of Discovery Channel telescope primary mirror blank. *SpaceRef-Space News as it Happens,* October 20. Internet <http://www.spaceref.com/news/viewpr.html?pid=18084> accessed November 1, 2008.

Newport Glass Company. 1999. Web site, on-line catalog. Internet <http://www.newportglass.com/angwcat.htm> accessed November 1, 2008.

Sidgwick, J.B. 1971. *Amateur Astronomer's Handbook.* New York, Dover Publications Inc.

Chapter 20

Glass Foam Composite Mirror Developments

Andrew Aurigema, OTF Design
David Davis, Toledo Scope Works
Russell Genet, California Polytechnic State University

Introduction

The key to lightweight, low cost, transportable, "large" (1 to 2 meter) aperture telescopes is, of course, lightweight, low cost primary mirrors. There are many approaches that have been taken in the past and are currently being pursued in the development of lightweight mirrors. They include meniscus mirrors formed from slumped soda lime glass, sandwich mirrors made from layers of glass separated by glass spacers and fused together, carbon fiber composite mirrors, spin-cast epoxy mirrors, and many more. A particularly interesting development is foam glass mirrors that are comprised of a glass front plate backed by lightweight foam glass.

Foam glass mirrors are not actually a new idea. During the OPEC energy crisis several decades ago, the US Department of Energy generously funded many solar power developmental projects—one of them being modest work on foam glass mirrors. The late Peter Manly discussed foam glass mirrors in his 1995 book, *Unusual Telescopes*. Pete did not mean "unusual" in a derogatory sense, only inferring "interestingly different." As so often happens, however, it takes the right combination of people and events, after previous false starts, to make a technology come alive.

The current foam glass mirror development started when one of us (Andrew) developed an 18 inch diameter, very fast (about *f*/0.8), Plexiglas "mirror" called a "Starnomenon" for educational and public outreach. In use, a Starnomenon is placed on the ground or a low table, and observers stand around the device and look down (instead of up) to see the reflected night sky. The stars magically float about a foot above the mirror.

Andrew and Russ wondered if these vacuum formed, thin shell, acrylic mirrors could be scaled up to a diameter of 1 meter with sufficient accuracy to form a roughly focused, on-axis spot, even if one that only a photometrist (that would be Russ) could love. They agreed that if Andrew could produce a 1 meter Plexiglas mirror Russ could use for photometry (i.e. a light bucket mirror), Russ would purchase it for $500 and see what science it could accomplish. If this mirror worked, it would instantly fulfill Russ' dream for lots of cheap photons. Four of these lightweight 1 meter mirrors in a multiple mirror telescope would have an equivalent aperture of 2 meters.

Figure 1: A photograph looking down into Andrew Aurigema's Starnomenon. Without straining one's neck looking up, one can see the entire sky while comfortably looking down. A group gathered around the Staromenon can point to and discuss various stars and constellations which seem to float above the surface of the mirror.

Andrew painstakingly developed a 1 meter mold to be used in the acrylic forming process, where a hot piece of acrylic is placed on the mold and a vacuum is then applied to pull it tight against the mold while the plastic cools. The result, they reasoned, could be mass produced, lightweight, dirt cheap mirrors that only needed aluminizing and overcoating—no optical figuring would be required.

About a year later, Russ heard of another amateur astronomer, David Davis, who was slumping sizeable pieces of thin plate (soda lime) glass in a large homemade kiln in his garage. How thin, David wondered, could such mirrors be made? Could they be made from low cost plate glass table tops? Dave had a collection of glass table tops he had picked up for a song at thrift shops and used furniture stores.

Thin Meniscus Mirrors

Thin meniscus mirrors offer a new range of development opportunities for the amateur telescope community to explore. Some of the positive aspects of this technology include:

- These mirrors are relatively light in weight.
- Thin glass cools quickly, allowing rapid equalization with night temperatures.
- If made from soda lime, the glass for the mirrors is cheap and easily obtained.
- Thanks to advances in kiln forming, apertures of one meter and larger can be created and figured.

Figure 2: Andrew Aurigema working on the mold for his vacuum-formed acrylic (Plexiglas) 1 meter mirror. He is checking the surface with a plumb blob suspended from a tripod.

In spite of these positive reasons to use a thin meniscus mirror, supporting a thin meniscus mirror is the hurdle that limits the acceptance of truly thin and hence very lightweight mirrors.

Mirror cells which utilize simple mounting points or whiffletree configurations can distort a meniscus mirror and ruin its figure. Extremely thin meniscus blanks are nearly impossible to figure, as they flex under the pressure of polishing strokes. Two mirror histories from David's experience provide examples of these issues. One involved a

406 mm (16 inch) meniscus mirror with a finished thickness of 8 mm (5/16 inch). This mirror never performed to expectations due to the distortion caused by the various cells into which it had been mounted. Only when a cell with memory foam in lieu of a whiffletree was used—supporting the curve with broad areas of pressure—was the mirror performance worthy of comment. Even then, the adhesive used to hold the mirror to the cell pulled the middle of the mirror down as it cured and affected the figure.

Figure 3: The thin plastic 1 meter mirror did not work out. It was way too floppy. The Young's modulus of Plexiglas is not something to write home about.

Another meniscus project, which was broken before it was finished, involved another 406 mm blank with a thickness of only 3 mm. This project could not be figured due to the flexure of the blank. Each time a different support scheme was attempted, new zones would form as the blank was figured.

Supporting Meniscus Mirrors with a Backer

In November 2008, Russ introduced David to Andrew, and Andrew and David began to explore possible solutions to the problem of supporting very thin meniscus mirrors. Andrew had been doing independent research on possible backing plates for thin glass. His initial designs used various resins and fillers to create lightweight backing plates to add thickness to the glass. These designs would be created separately and then attached to the meniscus blank prior to optical work being performed. They agreed they should, as a team, go where the data and materials would lead them. Russ retained his role as a cheerleader and kibitzer.

If a glass blank has a thickness ratio of 25 mm for every 150/200 mm of diameter, i.e a ratio of 6:1 to 8:1, flexure is not an issue. There is sufficient thickness to resist "drooping" due to gravity. David had explored the idea of adding thickness, but only after the mirror had been ground and polished. Andrew pointed out that this would not work. Binding the glass to a thicker substrate after the optical work had been completed would only induce tension due to the adhesive used and cause distortion in the figure. This made sense based on David's experience with the 406 mm x 3 mm blank mentioned above. Andrew also suggested that the adhesive layer should be as thin and uniform over the area of contact as possible to avoid distortion.

Historically, composite mirrors of this type have not worked well. In 1928 George Ritchey made a composite blank using glass lapped to close tolerances glued together with Bakelite cement. This mirror held its figure for a number for years, but ultimately came apart at the glue joints. Kirkham followed suit and glued several disks of glass together using Baltic pitch to create a disk of sufficient thickness to grind and polish. It separated and lost its figure in 5 years. Although such mirrors could sometimes be figured and mounted, the stresses induced by varied coefficient of thermal expansion (CTE) values between the materials ultimately caused the adhesive joints to fail. One day the mirror would literally fall apart. We hoped that by using a flexible adhesive we could avoid this failure.

The first test was to be a lattice configuration made of epoxy and fillers. The mirror would be made from a 560 mm by 6 mm plate glass table top from David's glass collection. After kiln forming, it would yield a 535 mm *f*/3 meniscus. The backer plate would be epoxy and carbon fiber cast in a milled foam mold. The primary concerns were matching the curve of the kiln-formed glass and matching the CTE of the materials as closely as possible to minimize strain. David had been experimenting with foam insulation as a stiffener and sent an email to Andrew commenting on the light weight and stiffness of the spray foam and rigid foam backer David had built.

Enter Pittsburgh Corning Foamglas

As he worked on the design of this first backer assembly, Andrew remembered a glass foam product he had read about a number of years before. In his email to David, he said, "Thinking about foam and glass brought it to mind, so I Googled it." Andrew's email was titled, "Thinking Way Outside the Box." Foam glass is a common industrial insulation made by Pittsburgh Corning, and marketed under various trade names, the most common being *Foamglas.* Intrigued by the product's data sheet, Andrew acquired a test piece and ran some milling tests on it. The material milled easily and accurately with both a pendulum router and a CNC mill. Andrew forwarded some samples to David that he also milled into various shapes using a pendulum router as well as a lathe and common mirror grinding techniques. The material had a consistent grain and took detail well. After some discussions via email, they agreed to set the resin-backer project aside and test the limits of glass foam as a backer. This decision was based on several characteristics of the glass foam material:

- It was made from soda lime glass and thus shared a common CTE with the thin plates David was to provide.
- It was non toxic, non reactive, waterproof, and lightweight.
- It was unaffected by any of the adhesives we proposed to use.
- It was a common material available from suppliers throughout the United States at an average cost of $5.00 per square foot for four inch thick material.

Figure 4: Andrew's CNC mill with a block of Pittsburgh Corning Foamglas in place.

Figure 5: An *f*/3 convex surface Andrew milled into a block of Foamglas.

One concern was the apparent softness of Foamglas. Russ contacted Steve Badger, the Director of Research at Pittsburg Corning, who pointed out that there were other grades of the material available as well as another glass base used in its manufacture—borosilicate. Borosilicate can be advantageous because it has about one-third of the thermal expansion of soda lime glass. However, the borosilicate Foamglas is more expensive and difficult to obtain than soda lime Foamglas. Furthermore, borosilicate plate glass is many times more expensive than soda lime plate glass. Russ obtained an 18 inch diameter Borafloat, donated by United Lens, but it broke in transit and that was the end of our experiments with borosilicate foam and plate glass. Others, however, might want to pursue this avenue further.

Pittsburgh Corning Foamglas Fusing Experiments

Andrew forwarded a milled 6 inch *f*/1 test puck of Pittsburgh Corning Foamglas to David and asked if a glass disk could be heat fused to the front of the puck. The material was soft enough so that once David slumped a front plate he could lap it into close contact for the fusing run. The foam is glass and should react to a fusing cycle just as any other glass would. However, it is also a foam and could disintegrate once heated in the kiln. Russ asked Steve Badger at Pittsburgh Corning if Foamglas had been fused by them into larger pieces. Steve confirmed that this was the case.

Since glass foam should retain its structure in the kiln, David wondered if the milled puck could be used as the mold for slumping. If so, could the slumping and the fusing be combined into one process? This is not usually done because the glass softens and slumps first, and only later gets sticky enough to fuse. At the higher fusing temperature, the glass can run if held at too high a temperature for too long. In other words, the front plate could end up as a puddle of glass in the middle of the foam puck. Another unknown was how long the glass should be held at the fusing temperature. The glass foam, being an insulator, would not transmit heat throughout the puck as easily as the solid glass would transmit heat through its thickness.

David ran some simple heat transmission tests on the foam in his kitchen oven by heating the oven to 200°F and inserting three cubes of glass foam measuring two inches on a side. At 20 minute intervals he pulled one block out and tested the internal temperature with a probe thermometer marked so he would insert it only to the depth of the center of the block. After 20 minutes, the first block measured 68°F at its center. At 40 minutes the second block measured 100°F at its center. Finally, at 60 minutes the last block measured 110°F at the center. Given that the surface should be at 200°F by this time, this was good evidence of a significant lag in thermal transmission through the glass foam. Also of interest was the fact that in the short time that the block was in the air being measured, it cooled enough to be held comfortably with bare fingers. Heating the foam was going to be an issue.

On the first kiln run David conducted two tests. The first test, a slumping and fusing test, involved placing a disk of 2 mm thick glass on the test puck of glass foam. Could it be used as a slumping mold? Would the glass foam puck fuse to the slumped glass disk? The second test involved taking two small pieces of glass foam and laying them one on top of the other to see if any fusion occurred between the glass foam at the temperature cycle he was using.

The temperature cycle was a best guess, based on fusing theory and the material David was using. That is, fusion would only occur at the contact zone and only in the top 1 or 2 mm. To prevent the glass plate from melting down and running to the low point of the *f*/1 curve, as well as to prevent too much damage to the glass foam, the heat would be held at the fusing temperature for a minimal period of time. The 2 mm plate should be of equal temperature throughout in 5 minutes or less. Ten minutes should allow the fusing to occur; so top soak was set at 15 minutes. In theory then, one should be able to run the strain and anneal cycle based on the fused glass thickness, about 5 or 6 mm, and ignore the other 30 mm of glass foam.

David could not do it. Gut instinct told him it would not be a long enough soak or a gentle enough cool down if he only used that thickness to calculate cool down. Instead, he held the puck at 1096°F for two hours to equalize the heat as best he could and then cooled at 35°F per hour to room temperature. It worked out to be a 30 hour cycle, which seemed safe but long. To shorten the kiln time, the heating rate was

Figure 6: Fused thin glass plate and Pittsburgh Corning Foamglas. Reflections show a potato chip surface to the blank. Notice the reflection of the board on the left and the "puddle" reflection of the paper roll on the right. This was an excellent fusion, but the final curve made the puck unusable.

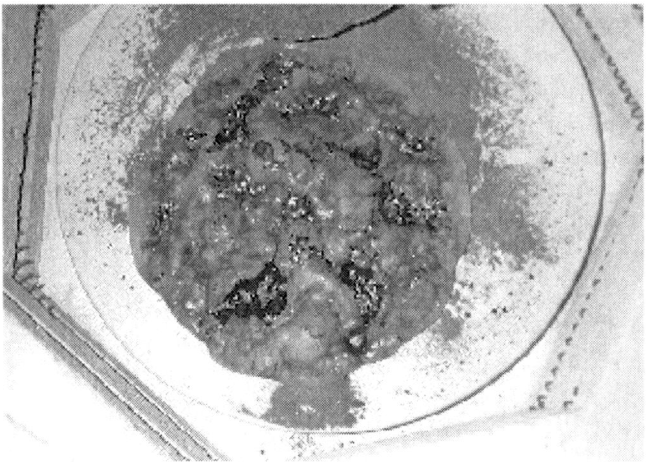

Figure 7: Too much heat too long resulted in a melted glass foam blank. A baby Horta is one possible outcome of a bad kiln test, but warped foamglas is more likely.

increased to 150°F per hour, reasoning the glass plate was the only thing in there that might be affected by thermal shock, and it could handle that rate easily due to its thinness. This cut the time down to 25 hours. The kiln shutoff was set at 600°F, knowing the kiln would only lose about 150°F per hour. The cycle details were as follows:

- Heat up at 150°F per hour to 1300°, hold for 15 minutes
- Cool as fast as possible to 1096°, hold for 2 hours
- Cool at 35°F per hour to 600°F
- Shut down

The cycle took 21 hours 35 minutes to complete, and the kiln was cool enough to open in 33 hours. The fusion of the plate to the glass foam was complete. The slump was perfect with no trapped air, voids, or bubbles visible under the glass. The puck had turned green in color (a common artifact according to Pittsburgh Corning) and showed no traces of distortion. David attempted to pry the glass plate off the glass foam, using his hands, and could not. He tried to pry it off using a screw driver and chipped a small piece off the edge of the glass plate as well as leaving an imprint of the screw driver blade in the glass foam. He tried to break the plate by grabbing the edge with a pair of vise grip pliers, but only chipped the edge of the plate. He tossed the puck at the wall, and it bounced off and landed on the floor. The glass foam was dented but the glass was unharmed. He dropped the puck onto a concrete floor. The glass foam and the glass plate bounced but were unharmed. It appeared we had an indestructible combination. Later abuses at the Galileo's Legacy conference in Hawaii finally chipped a significant piece from the glass plate, and David was finally able to pry the plate off the front of the foam puck. The foam surface under the glass was smooth and equal, showing no large divots or humps in the surface of the foam. The glass retained small pieces of the glass foam, showing that the fusion of glass to foam had been complete across the entire surface.

Figure 8: Successful glass and foam fusion run.

The two small pieces of glass foam used in the second test had fused at their thinnest point, and only there. The foam was distorted and green in color. David broke them apart and noted that the fracture depth was about 2 to 3 mm. This suggested to him that the heat required to fuse the pieces had only penetrated a millimeter or two into the foam, and only at the thinnest point.

Two more quick tests were run in the following days. A thicker front plate, measuring 6 mm thick, was fused onto a piece of glass foam that had a random curve milled and pressed into the surface. The glass was essentially optically flat, and gave a good reflected image. The test was to see if the thicker glass would complete a full slump into the complex surface curves, leaving no gaps. Would it fuse as well as the thinner glass? Did the heat cycle distort the optical surface of the glass? The glass settled into all the various curves and divots. It fused well. It did lose some of its optical qualities, in that the image lost some of its definition. This could have been caused by de-vitrification of the glass.

The second test involved six wedge shaped pucks of the glass foam that formed a larger circular puck. These were held in proximity with a paper tape. A piece of 2 mm plate was laid on top of the composite puck and the resulting group was fired with the same program as the previous two tests. The glass slumped and fused well. The six pucks separated from each other by an average of 1 mm during the firing cycle. Visual inspection showed that the foam had shrunk and distorted during the cycle. There was far more surface distortion on the glass than in previous tests.

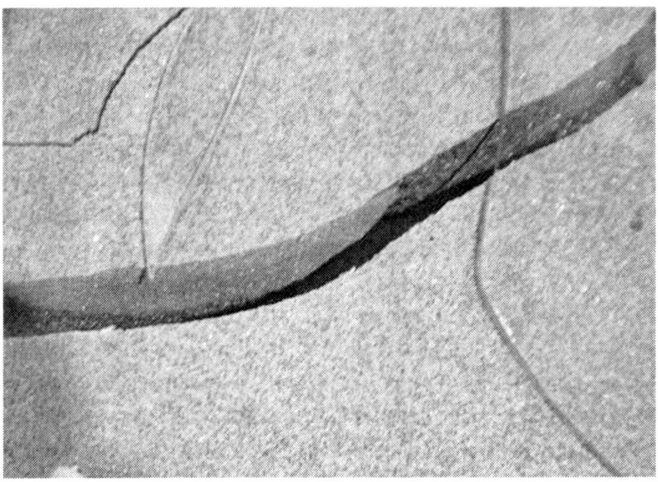

Figure 9: Visible here is the extreme heat failure of a foam disk with multiple fractures.

This proved to be a foreshadowing of the future of this research. Subsequent tests revealed many short-comings in fused glass foam and glass plate mirrors. The four primary problems are listed below:

- The foam structure is not consistent. For use as an insulator an inconsistent foam structure is not a problem, but using the foam as a component of a composite mirror was doomed by this factor.
- The heat range of the glass foam was not great enough to allow a good fusion in thicker blanks.
- The glass foam did not heat consistently in thicker blanks.
- Small pieces could be fused into larger pieces but the joint was poor, easily broken, and the foam was badly distorted by the heat.

Of these issues, probably the key limiting factor is the inconsistent structure of the foam. Various heat cycles and milling processes yielded some workable but temperamental solutions to the puck thickness and survivability problems. However the results were never the same between kiln cycles. The cells of the glass foam are irregular in size and shape and under a slight vacuum after manufacture. When the foam softens in the heat of a kiln cycle, the cells fail in irregular cascades. This left us with warped pucks, caved pucks, divots, cracks, shrinkage, and the occasional melt down. And yet when it worked, the foam provided an excellent supporting back to a thin meniscus front plate.

Use of Adhesives Instead of Fusing

Given our difficulties with fusing, we began to explore glued blanks. David created a dozen glued mirror blanks at his workshop and ground some of them as well as sending others out to selected amateurs to test. Andrew created several blanks at his workshop and ran environmental and stress tests on them, sending some to David for grinding. As noted earlier, glued blanks have failed consistently over the years. Reviewing those failures, David found that the adhesives had been very rigid, even somewhat brittle. His test pucks were assembled in groups of four with three different adhesives: polyester foam, polyester caulk, and silicone caulk. All three adhesives showed an excellent bond to the glass and they filled in the top of the foam puck, strengthening it and providing a uniform surface. All three had better flexibility at cure than the historic adhesives that had been used. Andrew chose various epoxy resins for his adhesives tests and a commercial two-part silicone adhesive.

Of the adhesives, two types failed almost immediately: the epoxies and the polyester foam. The epoxy resins shrunk as they cured and tore away from the foam. The top layer of foam cells were shattering during the curing cycle due to tension. The polyester foam bonded well, but was irregular in its expansion. This pulled on the glass and created zones that changed as the foam cured over time. Finally, under the stress of grinding and polishing, the bond would fail and the glass would fall off. Of the remaining adhesives, the consumer silicone was too flexible and allowed too much movement of the glass during figuring and in usage. This left two adhesives that showed good promise, the polyester caulk and the commercial two-part silicone. At this point in the research, a result unrelated to adhesives changed our focus for composite mirrors.

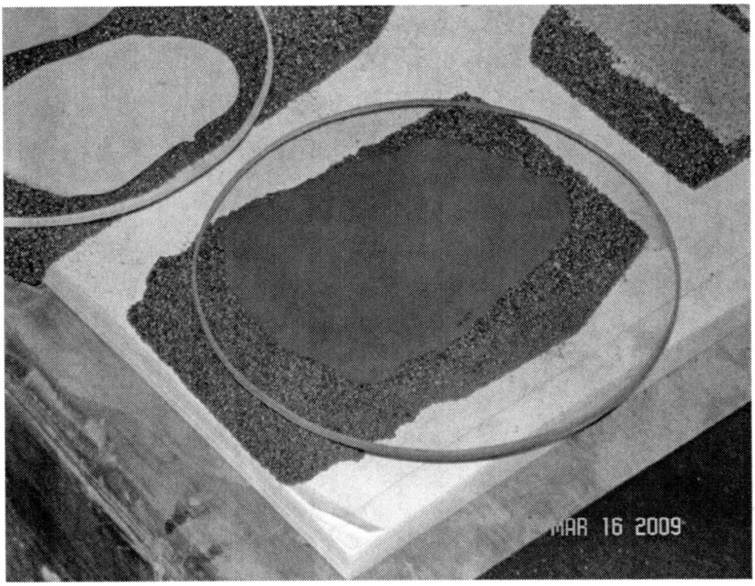

Figure 10: Glass plate and foam glass adhesive tests examined various adhesive candidates.

Enter Tessellations

Part of David's work was finding heat cycles to form the glass front plates for the blanks. To this end he was also experimenting with kiln size and design and the general heat forming processes. To build a 1.5 meter mirror, would slumping a 1.5 meter disk of glass and then gluing it to a foam backer with whatever adhesive worked best be successful? Up to this point David had worked with 42 and 48 inch disks. However when he received a 60 inch disk it became quickly apparent that this was a piece of glass that would require special handling techniques. The diameter made it too unwieldy to be handled by one person, and the weight of the blank crushed the original form he had shaped to slump it. While working with this piece of glass, which ultimately broke, David realized that this procedure was beyond the capabilities of average amateur mirror makers. Entire material handling systems would need to be built just to move and control the glass during forming and grinding, and the cost, though modest, would not be inexpensive.

Figure 11: 18 inch tessellated prototype. Small, thin, hex-shaped pieces of glass were adhered to a Foamglas substrate.

Andrew has had a long-term interest in joining mirror segments together to form a single mirror, something he had discussed with Russ for some time. Andrew had expressed an interest in tessellated mirrors since his first contact with David. Enamored by the Keck 10 meter tessellated mirror, Andrew pushed for development of this process. David initially resisted. The thought of grinding and figuring 100 mirrors to the same specifications was a nightmare for an amateur.

After the 60 inch disk had broken, and while David was licking his wounds, he had a flash of inspiration. Why not grind the 100 mirrors in a block in the same manner as commercial lens production, where lens blanks are mounted on a mandrel of known radius and ground with a common tool? This yields a batch of lenses with identical focal lengths. And what better block to mount the 100 mirrors on than the glass foam cell that would become their permanent home?

Several questions needed to be answered, but ever the empirical experimenter, David assembled an 18 inch test puck that very afternoon. The puck consisted of a block of 4 inch Foamglas, 18 inches square, with an *f*/3 curve ground into it. David gathered all the scrap glass he could find, and cut out 39 three inch hexagonal tessellations. These were glued to the foam with the only adhesive on hand, the dreaded consumer grade silicone adhesive Napa Ultra-Black Engine Gasket Maker.

After three days of curing, David began grinding the tessellated blank. Crude as it was, the surface was brought to a smooth curve in about three hours of grinding with a full size tool and 220-grade silicon carbide grit. Then it was 400-grit for an hour of wets. Finally David put the blank on the grinding machine and ran an hour of 25 micron aluminum oxide wets as a pre-polish. The blank was very smooth, and when wet reflected a solar disk about the size of a nickel. A ring lap began the polish. Total polishing time spent on this first attempt was 21 hours.

There is a moment in mirror making when the surface begins to act as a true mirror, bringing some object to focus. In a darkened room with a bright target, the focus jumps off the mirror and floats in space in front of the mirror. David remembers that moment with this first tessellated blank. The target was a fig tree some 50 feet away, and he could see the individual leaves floating in front of the mirror. He took a plano convex lens and, using it as a simple eyepiece, brought the leaf to focus. The veins in the leaf could easily be distinguished.

Ronchi tests showed a very rough figure. Each tessellation had an individual Ronchi pattern, but there was not a consistent Ronchi pattern on the whole surface of the mirror. The tessellations were not coming to a single focus—they were not coherent. David dropped back to 500-grit and ground the mirror on the machine using a 4 inch ring tool to smooth out the curve. After 12 hours of polishing, he had a Ronchi pattern over the whole surface, but several tessellations still appeared dark—perhaps they were lower and not as well polished. So David ground the surface with the ring tool again and returned to polishing. Fifteen hours later he had a consistent polish and a Ronchi pattern over the entire mirror surface. Although the figure was a bit deeper than a sphere and was not corrected yet, David quickly cobbled together a test scope and went hunting for stars. He was able to separate Mizar and Alcor in the handle of the dipper, but could not split Mizar into a double. Still, that was 12 arc minutes of resolution and a reasonable validation of the idea. David shared this early success with Andrew and Russ.

An Improved Tessellated Foam-Backed Mirror

We began to explore tessellated mirrors in earnest. The 18 inch mirror was crude, and the repeated grinding had brought the tessellations to a thickness of just 1/32 of an inch. The glass tessellations had been placed too closely together, and as the heat and cooling cycles of late summer took their toll, coherency in the mirror image was lost. The silicone had continued to cure and shrink, and this pulled some of the tessellations out of focus. Still, it was a great start, and we ran with it.

Figure 12: Tessellated test blank made by Andrew, showing the blank floating in a fish pond.

Andrew milled a precise puck from the glass foam and used an optical grade two-part epoxy/silicone adhesive to attach plate glass hexagonals that had been cut by water jet. These had been coarse ground on the back to prevent double reflection, and then mounted on the puck with a few thousandths of an inch separation between them. The adhesive layer was assumed to be about three thousandths of an inch at most. The radius of curvature of the foam puck was roughly 90 inches. Each little hexagon sat on a flat facet in the foam that was a chord of the sphere. Andrew sent the blank to David for optical work.

The blank took 33 hours to finish. Nothing remarkable in the grind, but perhaps that is the remarkable thing. It yielded a working mirror that resolved images down to 5 arc seconds or less. This was an incredible achievement. Using scrap glass and commercial foam insulation, we had made a working telescope mirror. And a very light weight one at that—it floated.

Figure 13: Early Ronchi test of a tessellated mirror.

A 1.5 Meter Foam Glass Mirror

Using what he learned from these mirrors, David set out to make a 60 inch blank that could be ground and polished. He glued 4 inch thick Pittsburgh Corning Foamglas blocks together, creating a square 72 inches by 72 inches. From this, a disk 63 inches in diameter was cut. A ring tool was used to mill an *f*/3 curve into the foam and 6 inch tessellations were glued to the surface using polyester caulk. A wooden disk 60 inches in diameter was glued to the back of the puck to protect it. Total cost of the blank was $477.50. The weight of the blank was 165 pounds.

The concept behind the blank is that the glass foam is equivalent to the mirror cell of a simple Newtonian telescope that one might buy retail. In those scopes, the primary mirror is frequently glued to the cell with some adhesive. Little concern is given to distortion since the mirror is thick enough to stand on its own and not be affected by the mismatch of CTEs. In the same way, the 6 inch tessellations are thick enough to avoid distortion by any action of the adhesive. Nearly 100 small mirrors had been mounted on a cell and the next step was to grind them to a common focal length.

The advantage of a "tessellated" mirror comprised of hexagonal glass segments is that the segments can be directly fastened to the curved surface of the underlying support structure, such as Foamglas, without being slumped. If the hexes are small enough, not much glass will be removed during grinding and, as a result, quite thin, straight pieces of glass will suffice. The result can be an unusually lightweight mirror.

Figure 14: Hexagonal segments of thin plate glass can be cut out via a water jet, a surprisingly inexpensive process.

Figure 15: Three of David Davis' experimental tessellated foam glass mirrors. David brought the mirror in the upper right to Pine Mountain Observatory during summer 2009 in a test telescope he threw together in a couple of hours. David and Russ observed a double star with it.

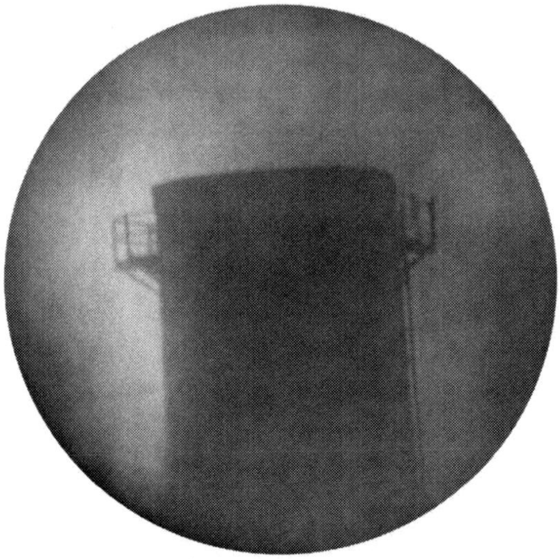

Figure 16: This smokestack was photographed by a test "telescope" with an uncoated tessellated foam glass primary mirror. The smokestack was some 0.658 miles from David's house per Google Earth. The pipes on the top of the smokestack are common 1.5 inch chain link fence posts. The shot is a-focal with a hand held camera. The spherical mirror's resolution is about 6 arc seconds. Although the mirror was not fully corrected, this photograph is still an impressive testament to the potential for foam glass mirrors.

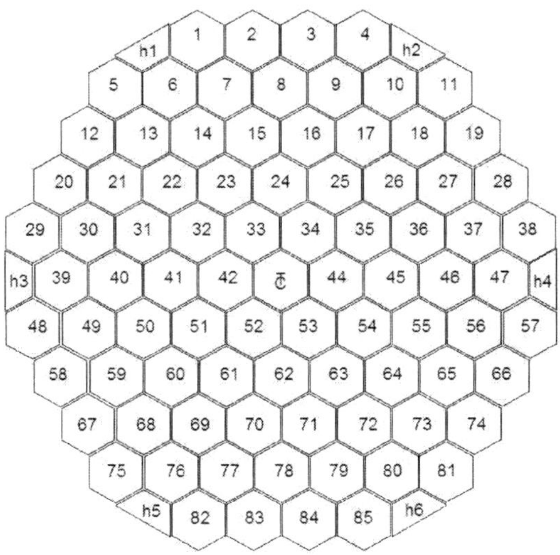

Figure 17: The layout of the 85 full hex segments and 6 half hex segments on David Davis' 1.5 meter Foamglas mirror.

David is well along on making a lightweight 1.5 meter mirror he calls "Tess" from pieces of Pittsburgh Corning Foamglas glued together and faced with a number of hex-shaped thin glass segments adhered to the foam glass in a tessellated pattern rather similar to Keck. Tess has 85 full hexagons and 6 half hexagons.

Figure 18: David's Foamglas substrate. You can see where blocks of Foamglas have been fastened together. You can also see the pronounced sagitta of David's *f*/3 mirror.

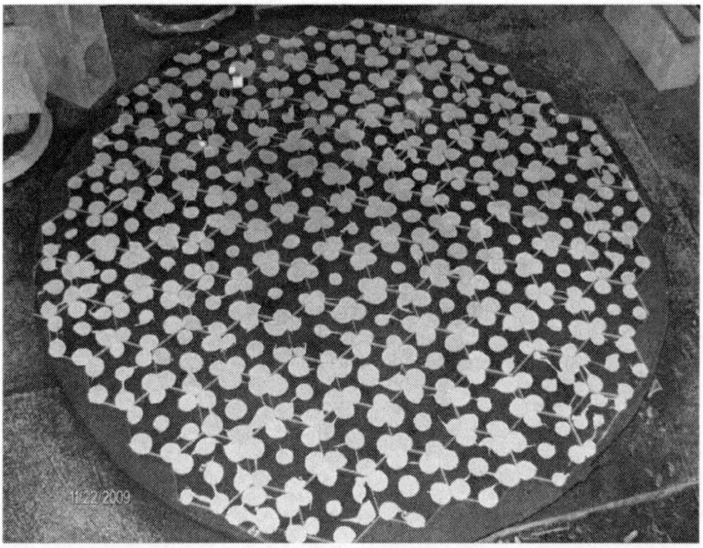

Figure 19: If you look closely, you can see that each clear glass hex segment has been adhered to the Foamglas substrate at seven support points with an adhesive that David selected, after some experimentation, to be sufficiently stiff yet not too brittle.

Figure 20: At the beginning of the grinding process you can see where some of the hexes, which are slightly higher than others, have been ground off somewhat more. Of course they will all be evenly ground and polished in the end.

Starstone Mirrors

Meanwhile, Andrew uncovered another type of foamed glass manufactured by Earthstone from recycled colored glass. Unlike the closed cell Pittsburgh Corning Foamglas, Earthstone is an open cell foam. Russ contacted the CEO of Earthstone, George Morandin, and found that they were about to start up a large kiln at a landfill outside of Albuquerque, New Mexico, where they would be able to make much larger pieces of Earthstone—up to 1.8 meters wide and 30 cm thick. A large contiguous piece of Earthstone would avoid the need for adhering smaller pieces together to form a mirror blank.

George pointed out, however, that in Europe, foam glass types of material were commonly adhered together with sodium silicate (also called water glass). The pieces of foam glass are covered with a "soupy" layer of sodium silicate and placed in an oven heated to about 300°F for about an hour. The pasty sodium silicate is transformed into a hard, glassy substance. Sodium silicate is often used to repair exhaust pipe leaks. The exhaust heat causes the transition to the glass-like foam.

Andrew experimented with mixtures of sodium silicate, micro-balloons, and ground up foam glass. Foam mirror cells can be "painted" with this sort of mixture, which hardens to a glassy surface when heated in an ordinary oven for about an hour. Such mixtures can also be used to fasten glass plates to foam glass substrates, although Andrew prefers direct kiln fusion as it allows slow annealing of the mirror.

Figure 21: The continuous batch process kiln used to make Earthstone foam glass. The finely ground up colored glass is mixed with a gasifying agent in the hoppers at the top of the stairs on the right which feeds the kiln underneath. The units stretching off to the left are temperature controlled stations that gradually reduce the Earthstone's temperature. Finished Earthstone comes out at the far left where it is sliced off. This plant is in Austria, but the one in New Mexico will be identical.

Earthstone has located a large new kiln at the main Albuquerque landfill not only to be close to their prime material source, New Mexico's used beer bottles, but to also tap into the waste methane the landfill produces. Normally such methane is simply burnt off—carbon dioxide being a much less potent greenhouse gas then methane. Earthstone, however, is utilizing the methane to fire their kilns. Recycling at its finest! A landfill of broken beer bottles and rotting garbage will produce large, pristine astronomical mirrors which astronomers will use to study distant galaxies—garbage in, galaxies out (GIGO).

Figure 22: A "Starstone" (patent pending) mirror made by Andrew Aurigema. The lightweight foam glass has been CNC-milled out to improve air circulation and lighten the mirror. The edge of the thin glass front can be seen on the left of the picture.

Figure 23: Side view of one of Andrew's foam glass mirrors. This one happens to be made of Bathstone, a material similar to Earthstone. The top glass plate is very thin as compared with the foam core, keeping the overall mirror weight down. The sides have been sealed with a compound that includes sodium silicate (water glass) that turns glassy hard when heated in an ordinary oven for about an hour.

Figure 24: Andrew fuses, in a kiln, the thin top glass plates of his mirrors to the foam glass CNC-milled supporting substrate. A test mirror was broken in half and turned upside down to show the complete fusion of the front plate to the foam support.

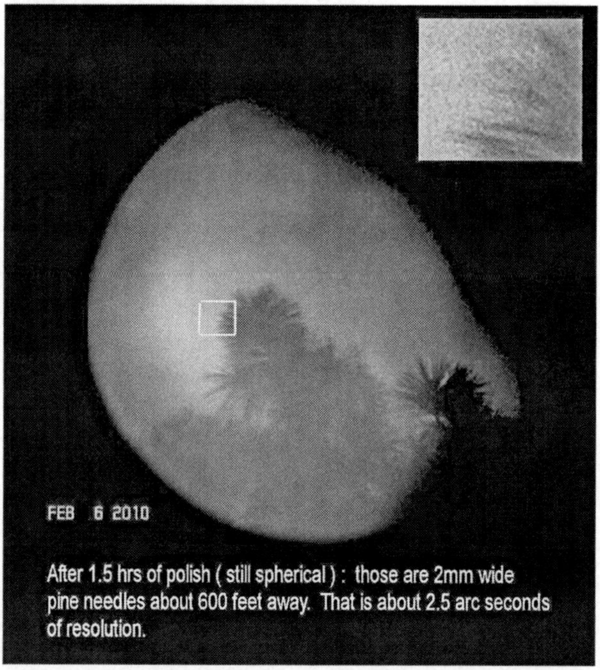

Figure 25: Early optical testing of a 9 inch *f*/4 "Starstone" mirror in a simple telescope during the Alt-Az Initiative's Florida Fringe Festival in Feb 2010. There were only 1.5 hours of polishing, and the mirror was still spherical. Those are 2 mm wide pine needles about 500 feet away. The resolution is about 2.5 arc seconds.

Andrew, who has a vacuum chamber for coating mirrors up to 60 inches in diameter in his garage (well actually his hanger), has recently added a large kiln and mirror making machine to his growing collection of telescope making toys. He is serious about producing large foam glass mirrors!

Figure 26: "Baby George" is Andrew's 68 inch diameter x 68 inch deep vacuum coating chamber, seen here with a 44 inch diameter acrylic mirror.

Figure 27: Raw foam material is CNC-milled to a very specific shape prior to pore sealing with a sodium silicate solution. This is Andrew's "baby" milling machine. A much larger one, capable of handling 1.5 meter foam mirror pucks, is under construction.

Figure 28: Andrew's home-built Mirro-O-Matic style mirror grinding machine. It is capable of automated grinding and polishing of 60 inch diameter mirrors. The arm and controls can be seen on the right. The "kiddy pool" is to catch the slurry that comes off a mirror during grinding. It doubles as a dog-washing station.

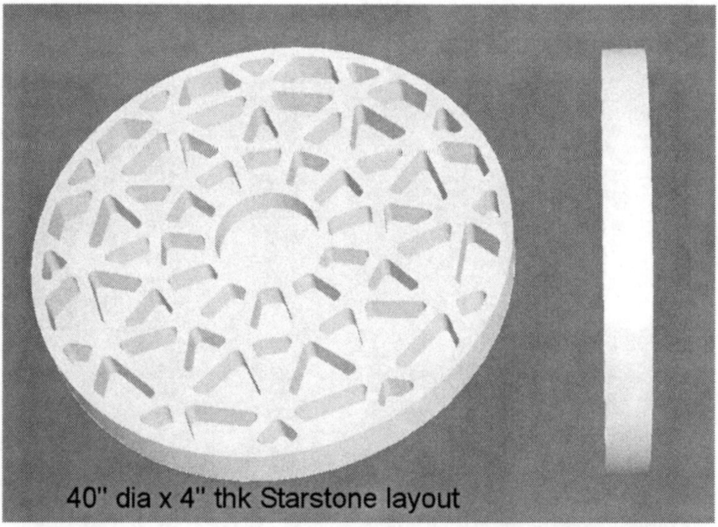

Figure 29: The layout above is one that Andrew is considering for his production 1 meter Earthstone (he calls it "Starstone") mirrors. This is similar to David Rowe's design for the support of a cellular mirror by an 18 point wiffletree. You can see 12 support points on an outer ring and 6 on an inner ring.

One of the first batches of Earthstone from the New Mexico kiln will be shipped to Andrew to use in making lightweight meter-class mirrors (estimated at about 50 lb) faced with thin plate glass. Alt-Az Initiative members have shown interest in testing and using the first large mirrors produced by Andrew.

Foamglas Mirrors in Italy

While following Andrew and David's work, Russ searched around for other interesting, similar activities. He found a doctoral student in Italy, Rodolfo Canestrari, who was developing a lightweight 1 meter diameter mirror composed of thin (2 to 3 mm) glass plates adhered to each side of a Pittsburgh Corning Foamglas core. One surface of the Foamglas core is machined to the precise concave shape desired for the approximately *f*/17 focal ratio for a Cherenkov radiation telescope (CRT). The surface of the machined Foamglas is coated with a thin layer of very low shrinkage epoxy, and the top glass plate is pulled via a vacuum into contact with the Foamglas. Once the epoxy sets, a thin, flat bottom glass plate is adhered to the Foamglas core. The mirror is then aluminized and overcoated and is ready for use without any optical grinding, polishing, or figuring.

It might be noted that the optical resolution requirements of CRTs are modest, as the pixel sizes of their cameras are typically an inch across—their cameras being formed from an array of end-on photomultiplier tubes. It might also be noted that while the overall *f*/ratio of CRTs is very fast, the *f*/ratio of the individual mirrors is very slow—*f*/17 in this case. Thus the thin plate glass does not need to be bent appreciably and, as a result, bending forces do not shatter the glass.

Acknowledgments

We are pleased to acknowledge the support and materials provided by Steve Badger and his associates at Pittsburgh Corning and also by George Morandin and his associates at Earthstone. We are also grateful for the borosilicate blank provided by United Lens. The members of the Alt-Az Initiative have provided encouragement and occasional helpful suggestions throughout the development process. Finally, our thanks to Rodolfo Canestrari and the Italian development team for sharing their experiences with us.

References

Manly, Peter. 1995. *Unusual Telescopes*. Cambridge, UK: Cambridge University Press.

Chapter 21

Graphite Fiber Nano-Composite Mirror Technology

Peter C. Chen

Lightweight Telescopes, Inc.

Introduction

Graphite fiber composites have long been recognized as promising candidate materials for the development of large lightweight telescopes in space and on the ground (Meinel 1985, Swanson 1986). The major advantages are:

- High stiffness to weight ratio – Mirrors can be made with very low ratio of aperture to mass. This quantity, usually expressed as the areal density or mass per unit area, is in the range 1 to 10 kg/m^2 for composite mirrors. An illustration of this capability is shown in Figure 1.

Figure 1: A 0.9 m composite mirror being held by two girls. The mirror weighs about the same as the cat looking through the center hole (the cat is held by a third girl standing behind the mirror).

- Low cost – Composites fabrication is a well established industrial process. It does not involve handling toxic materials or specialized high temperature/high vacuum equipment. The composite replication process is especially suited to the rapid fabrication of large optics in multiple identical units since only a few days are required to lay up and cure each mirror.
- Large aperture capability – Unlike other traditional optical substrates, such as glass or metals, there is essentially no limit to the maximum size of the substrate materials (mirror blanks). Graphite fiber plies or tows are available from multiple vendors within a wide industrial base, in lengths ranging from meters to kilometers.
- Ruggedness – Graphite fiber composite materials are tough and do not shatter regardless of how thin they are. These qualities make composites the most promising candidates for large aperture adaptive optics, which are crucial components for the next generation of extremely large telescopes.
- Versatility – Depending on the application, graphite fiber composite mirrors can be made thin or thick, stiff or pliable, and heavy or lightweight. All types of mirrors (passive, active, or adaptive) in just about any shape or size are possible.

The advantages of composite laminate reflectors have long been recognized in radio astronomy. Multi-meter radio, submillimeter and infrared telescope mirror panels, and support structures have been in use for many years on the ground and in space (Abt et al. 1990). However, only recently has composite technology advanced to the stage of making precision optical mirrors with smooth surfaces and figure accuracy < 1λ rms visible (Chen et al. 1998, 2000). The present article reports on progress made in the technology during the past three years.

Basic Structure

Figure 2 shows the basic components of a composite mirror. It is, in its simplest form, a three-layer structure consisting of a reflective coating, an optical surface, and a support structure.

Figure 2: Basic three-layered structure of a fiber composite mirror.

The substrate is a graphite fiber reinforced polymer laminate. It supports the optical surface, which is usually a layer of neat resin. The optical surface is, in turn, covered with a thin reflective layer such as aluminum or silver. The structure as shown in Figure 2 can in practice incorporate many variations, depending on applications. The reflective layer, for example, can range from simple vacuum deposited aluminum to complex multilayer coatings and nanoparticle coatings. The optical surface can consist of several thin resin layers not necessarily of the same chemical composition. Sometimes additional barrier layers are used to isolate print-through between the support structure and the reflective coating. The substrate can be fiber composite laminates made up of different fibers (graphite, Kevlar, fiberglass, etc.), bonded by a variety of resins (epoxy, cyanate ester, polyester, polyimides, etc.). The laminates may consist of different number of plies ('layups') with different angular orientations. And finally the support structure can be made of two laminate layers with a core structure in between. The core can be made of different materials (fiber composite, phenolic, Nomex, aluminum, etc.) in different shapes (honeycomb, rectangular, flexible composite core, etc.).

Achievement of Diffraction Limited Performance

After much effort over many years, our group at NASA's Goddard Space Flight Center (GSFC) succeeded in realizing optical telescope quality composite mirrors with diffraction limited performance. The critical components were found to be a resin surface with low cure shrinkage, a barrier layer to reduce print through, and a stiff, thick laminate substrate. The last quantity was especially important. We found that, for example, the best surface accuracy achievable with a single laminate substrate (thickness 3 mm, areal density 5 kg/m^2) was ~0.17λ rms (λ=632.8 nm). By changing to a double laminate structure with a core sandwich (thickness 2.2 cm, areal density 12 kg/m^2), the surface accuracy was improved to 0.037λ rms. An interferometric measurement of the thick mirror, made at the GSFC Optics Branch facility, is shown in Figure 3 (next page).

To the best of our knowledge this was the first time that diffraction limited optical performance had been achieved using graphite fiber polymer composite material.

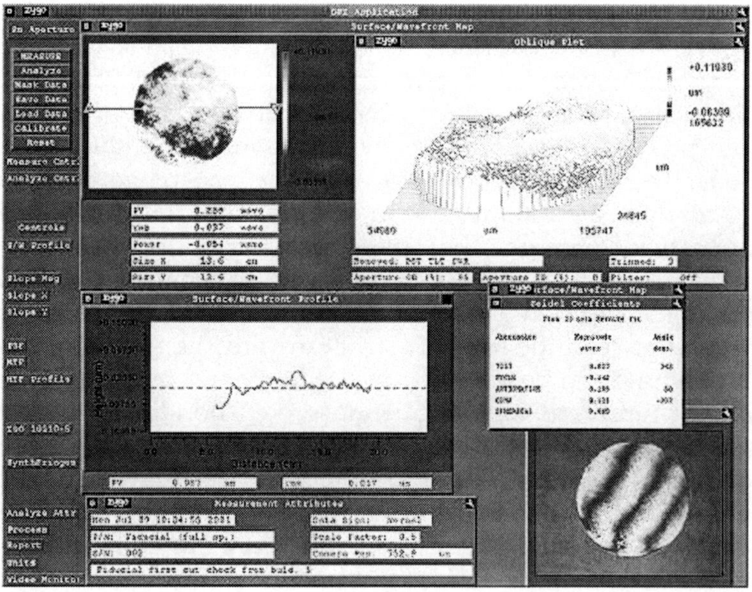

Figure 3: Zygo interferometer print-out of a 15 cm fiber composite mirror, showing diffraction limited optical performance (0.037λ rms, or λ/27). Courtesy NASA GSFC Optics Branch.

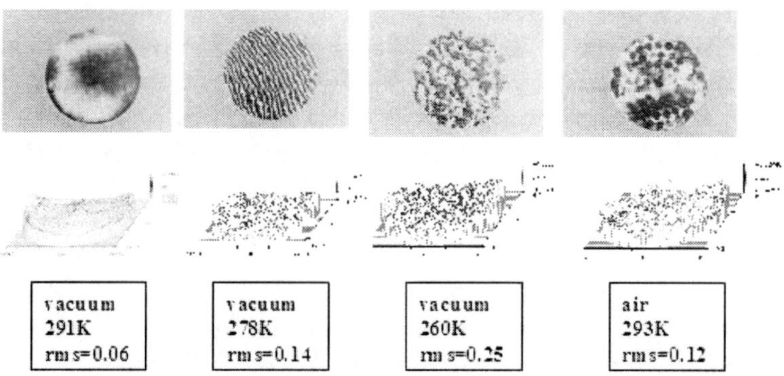

Figure 4: Behavior of fiber composite mirror subjected to vacuum and cooling. Top row: Zygo interferometer printout of interference fringe patterns. Middle row: Zygo interferometer printout of surface profile. Bottom row: conditions of test (air/vacuum, temperature, and rms wavefront error @ 632.8nm). All data taken with a NIST-calibrated Zygo Mark V Phase Shifting Interferometer at the NASA GSFC Optics Branch facility.

Thermal-Optical Testing

Having achieved the desired optical performance, work proceeded at GSFC to performance space qualification testing in vacuum and at cryogenic temperatures. However, the behavior of the fiber composite mirrors with temperature changes, in and out of vacuum, were surprising and totally unexpected. A large data set was generated on six mirrors with different constructions. A succinct précis of the most important results is shown in Figure 4 (opposite page).

In Figure 4, starting from the left, the first column shows the mirror at room temperature. In the second column, the mirror surface had developed a series of regularly spaced 'bumps' or high spots at T=278°K. In the third column, the mirror surface was starting to show a pattern of regularly spaced large diameter annuli. In the fourth column, the large round annuli pattern was found to persist after warming to room temperature. This indicated a permanent deformation of the mirror surface.

Analysis of Thermal-Optical Data

Without going into great detail, the observed behavior of fiber composite mirrors can be understood as follows:

- A fiber composite mirror consists of two major components, a surface resin layer and a support graphite fiber laminate. The surface usually must be a thin layer of pure resin in order to obtain a smooth optical surface. The laminate supports the surface layer and maintains its shape.
- The two layers differ greatly in thickness, rigidity, and thermal expansion coefficients (CTE). The resin layer is <0.3 mm thick, has a modulus of ~3 GPa, and high CTE typically ~60 ppm/°K (Peters, 1998). The laminate layer, on the other hand, is much thicker (~2.5 cm), much stiffer—modulus ~80 GPa (Abt et al. 1990)—and has CTE <1 ppm/°K. This gross mismatch is the cause of the thermal stability problem.
- When first cooled, the surface layer tries to contract but is prevented by the laminate from doing so. An internal stress develops within the resin, which relieves itself in the direction perpendicular to its surface by a process known as 'wrinkling' (Mei et al. 2007). This action results in the pattern of surface bumps seen in Figure 4, column 2.
- On going to lower temperatures, the stress in the resin layer increases. Beyond a certain point, the stress overcomes either the adhesion of the resin to the laminate, or, in our case, the adhesion of the laminate to the core. The resin and front laminate become detached from the core (delamination).

- The delaminated resin and front-side laminate layers reveal stress in areas where adhesives have been used to bond the two to the core. The pattern of circular annuli seen in Figure 4, columns 3 and 4, corresponds exactly to where the core used to be attached to the front-side laminate.
- Delamination from the core persists after warming to room temperature (Figure 4, column 4). This indicates permanent damage to the fiber composite mirror assembly.

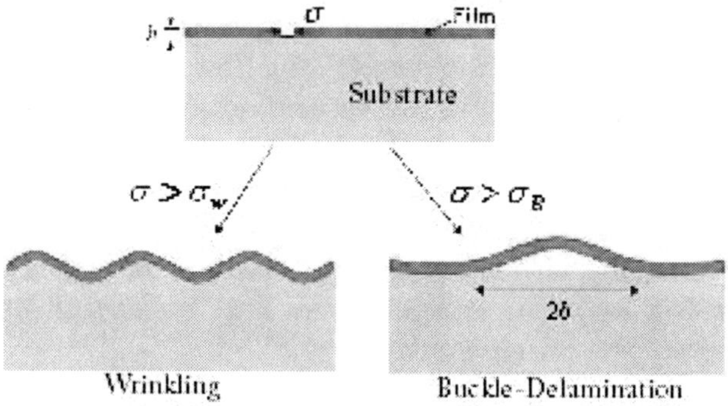

Figure 5: Wrinkling and buckle-delamination of a thin film bonded to a rigid substrate (from Mei et al. 2007).

A New Generation of Graphite-Fiber Nano-Composite Mirrors

Development of fiber composite mirrors suffered interruption between 2004 and 2005. In 2005 work started anew in a new organization with funding from commercial partners and the support of colleagues at NASA GSFC. New materials and new fabrication processes had since been developed to overcome the shortcomings found in the earlier design. Finally, realization of the full potential of fiber composite mirror technology is at hand.

Major developments since 2005 have been finding new materials, developing new processes, making improvements in the mirror structure, and the introduction of nanotechnology. The crucial points are:

- Experimenting with and finding new polymer systems that can withstand cryogenic temperatures, possess low vacuum outgassing, and are qualified for space flight applications.
- Developing new procedures to cure the polymers to obtain maximum short and long term stability.
- Decreasing the stiffness of the substrate by using softer fibers. Also, make thinner laminates with fewer layers. A proprietary layup and cure procedure has been developed that yields substrates with practically zero astigmatism.

- Reduce the CTE mismatch between the surface resin layer and the substrate by interspersing one or more layers with graduated CTE values that are intermediate between the resin surface and the support substrate.
- To compensate somewhat for the decrease in laminate stiffness, carbon nanotubes (CNTs), carbon nanofibers, and other ingredients have been introduced into the system. Carbon nanotubes are distinguished by their very high stiffness (modulus ~1 TPa), excellent electrical conductivity, and good thermal conductivity (Li et al. 2004).

Using the new structure and formulation, it has been possible to make a new generation of graphite fiber nano-composite mirrors (GFNC) with greatly improved characteristics. The optical figures of GFNC mirrors are much less sensitive to temperature changes. Tests have shown that they can be cooled to liquid nitrogen temperature (77°K) without suffering permanent damage. The GFNC substrates are more pliant, which permit the use of low force, low cost actuators for active figure control when needed. Figure 6 below show test images of two GFNC mirrors.

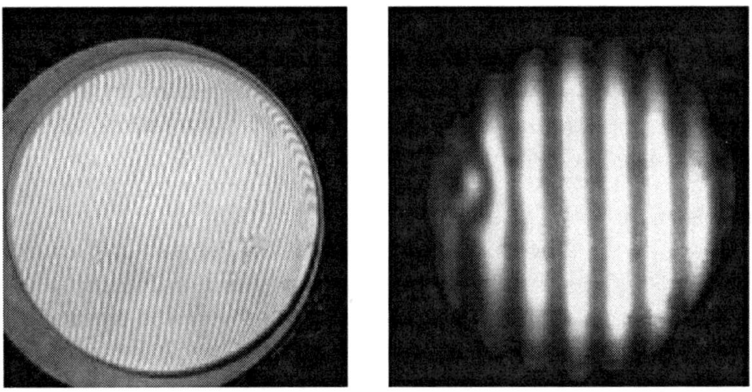

Figure 6: Optical figures of two graphite fiber nano-composite mirrors. Left: interferogram of a flat replicated mirror. Right: Ronchi fringe pattern of a spherical mirror taken at 100 lines per inch.

Work In Progress

At present, work is focused on the development of larger aperture GFNC mirrors. Based on the experience of ourselves and others, we believe that direct replication, the process which has been employed thus far, is ultimately limited to apertures of 1 meter or less. This comes about for two reasons:

- Cost – The cost of precision convex parabolic glass mandrels increases almost exponentially with size. At 1 meter the estimated cost is ~$\$10^5$. At ~2 meters they are simply not available at any cost, as a government funded project found out some time ago.
- Durability of mandrels – In addition to the cost factor, precision mandrels are also very fragile. Glass as a material suffers from a low threshold against damage by shear. As replicating polymers shrink while curing against the mandrel, they not infrequently tear off a chunk of glass in the process. The situation is clearly not desirable nor cost effective.

To overcome these obstacles, we are developing a method for direct generation of concave parabolic surfaces by spinning a liquid polymer while it cures, a process known as spin-casting. Spin-casting by itself is not expected to yield a precise optical surface (Richardson 2001). One or more additional steps are required to fine figure the surface resin layer. The fine figuring can be accomplished either by contact or non-contact methods. We have done work to demonstrate the validity of these concepts. Parabolic mirrors 0.5m and larger have been generated by the spin-casting process. Mirrors of 0.3m diameter and having supersmooth surfaces (rms microroughness <4 Å) have been made without the use of mandrels. The optical surface of a 15 cm mirror, with a rough generated and initially visibly bumpy surface, has been fine figured to almost diffraction limited performance. These and other results will be reported in separate publications elsewhere.

Conclusion

We have shown that diffraction limited optical performance can be achieved using graphite fiber polymer composite materials. However, the traditional structure of a rigid laminate substrate supporting a resin surface has been found to be unstable against changes in temperature. With the use of new polymer systems, development of a modified structure, and the incorporation of nanotechnology, these obstacles can be overcome. New powerful processes are currently being developed to make large aperture, lightweight, and low-cost GFNC mirrors for a wide range of applications in space and on the ground.

Acknowledgements

The author acknowledges helpful discussions and support from D. Rabin, J. Brasunas, P. Mirel, M. Van Steenberg (NASA GSFC) and R. Hale (KU).

References

Abt, B., Helwig, G., and Scheulen, D. 1990. Composite technology for light-weight optical mirrors. *Proc. SPIE* 1236, 696-705.

Chen, P.C., Saha, T.T., Smith, A.M., and Romeo, R. 1998. Progress in very lightweight optics using graphite fiber composite materials. *Opt. Eng.* 37(2), 666-676.

Chen, P.C., Bowers, C.W., Content, D.A., Marzouk, M., and Romeo, R.B. 2000. Advances in very lightweight composite mirror technology. *Opt. Eng.* 39(9), 2320-2329.

Li, X, Gao, H., Scrivens, W., Fei, D., Xu X, Sutton, M., Reynolds, A., and Myrick, M. 2004. Nanomechanical characterization of single-walled carbon nanotube reinforced epoxy composites. *Nanotechnology* 15, 1416-1423.

Mei, H. and Huang, R. 2007. Buckling modes of elastic thin films on elastic substrates. *Appl. Phys. Lett. 90,* 151902.

Meinel, A.B. and Meinel, M.P. 1985. A view in the mirror – or through the looking glass. *Proc. SPIE* 571, 2-6.

Peters, S.T. 1998. *Handbook of Composites.* London: Chapman and Hall.

Rapp, D. 1992. Precision segmented reflector program, precision composite mirror development task final report for FY1988-1991. *JPL Report D-9204,* NASA Jet Propulsion Lab.

Richardson, R. and Griffiths, P. 2001. Generation of front-surface low-mass epoxy-composite mirrors by spin-casting. *Opt. Eng.* 40(2), 252-258.

Swanson, P.N., Breckinridge, J.B., Diner, A., Freeland, R.E., Irace, W.R., McElroy, P.M., Meinel, A.B., and Tolivar, A.F. 1986. System concept for a moderate cost Large Deployable Reflector (LDR). *Opt. Eng.* 25(9), 1045-1054.

Chapter 22

Replica Processes for Mirror Manufacturing

Glass Slumping and Foamglas Backing

Rodolfo Canestrari, Mauro Ghigo, and Giovanni Pareschi

INAF–Osservatorio Astronomico di Brera, Italy

Introduction

During the last decade in the field of professional astronomy, the need grew to have larger optical telescopes, surpassing the present day 10 meter class telescopes (i.e. KECK I and II, VLT, Subaru, GranTeCan). An initial study to consider these large telescopes was accomplished by ESO (European Southern Observatory) with a project called OWL (Over-Whelming Large telescope): a 100 meter class primary mirror telescope. This sort of telescope is well beyond the technological capabilities that are now in the market or even the new ones under development. This project would require so much effort and development time as not to be affordable, if not impossible, today. Nevertheless, this study was an inspiration in both the U.S. and Europe. Consortia of research institutes began feasibility studies to address the problems related to telescopes with 20 to 40 meters of aperture. This new class of ground-based optical telescopes now under investigation is called ELT (Extremely Large Telescopes). There are three major telescopes that may be in operation around 2016-18: TMT (Thirty Meter Telescope), GMT (Giant Magellan Telescope), and E-ELT (European-Extremely Large Telescope).

Increasing telescope aperture permits astronomers to collect photons coming from the faintest sources located at the edges of the universe, to reduce the time needed for imaging, photometry, and spectroscopy, to improve knowledge in cosmology, and to study the early moments of the universe's birth just after the Big Bang. Another very interesting and challenging research topic that could be carried out with such telescopes will be the search for Earth-like exoplanets.

From the technological point of view, these ELTs are a considerable challenge from many aspects. Let's take the example of E-ELT and let us focus on the optics, in particular the primary and the secondary mirrors.

The primary mirror will have a diameter of 42 meters, about half a soccer field! It will be manufactured in many mirror segments that will be mounted and aligned together with an active system to form a large monolithic-like astronomical mirror. The active system will also manage and minimize, during the observations, the shape distortions due to gravitational force. Each segment will be of hexagonal shape, about 1.3 meters face-to-face: this implies a total number of segments of about one thousand! To produce such a large number of mirror segments, it will be mandatory to develop and use a technology that will be able to provide many mirrors with high optical quality that are lightweight in a short delivery schedule. All these requirements are in conflict with each other. The traditional optical manufacturing processes simply will not suffice. An attractive alternative is offered by processes that accurately replicate the shape of a master mold. In this way it is possible to reach a high production rate of identical mirror segments from each mold. Thus, with a limited number of optically figured molds, it is possible to achieve mass production of mirror segments.

The secondary mirror of E-ELT is also a sizable challenge. In the initial design of the telescope it was a 4 to 5 meter diameter mirror 2 mm thick having a fully integrated adaptive support. Flexures are induced by a matrix of actuators placed on the backface of the mirror that deforms its shape to compensate for the aberration introduced by atmospheric turbulence. The use of adaptive optics is fundamental to producing top level science with the present generation of telescopes, and it will be mandatory to realize the best performance of the ELTs.

This chapter will discuss an innovative technology based on a glass slumping process. The study was aimed at the secondary adaptive mirror for E-ELT, but with fascinating possibilities to translate it into the production of lightweight, stiff, and low cost primary mirrors for modest-aperture telescopes.

The Hot Press Direct Slumping Approach

Currently, adaptive optics are systems added to telescopes as an upgrade to the original instrument. This implies one more reflection surface in the optical path; more reflections mean less efficiency of the overall instrument. A key breakthrough in this field is the Large Binocular Telescope (LBT), where both secondary mirrors of the two twin 8.4 meter telescopes will be fully adaptive from the very beginning. The secondary mirrors are two meniscuses 0.9 meters in diameter, each with 672 electromechanical actuators placed on their backface. Both mirrors are floppy enough to accept deformations; their thickness is only 1.6 mm.

The LBT mirrors were produced starting from a blank disk of material a few centimeters thick (Figure 1). Both surfaces of the blank disk were machined: the one that was the backface needed a surface precision of a few microns, while the front optical surface needed to be figured and polished as a classical astronomical mirror. During this operation, the meniscus was created from a thicker disk and the initial thickness was reduced to the desired one. These tasks took a long time to complete and they introduced a severe risk of breaking the mirror during the thinning phase. The costs for the production of such a mirror are very high, on the order of some hundreds of thousands of US dollars! It is clear this technology cannot be used in projects like the ELTs that are composed of many mirror segments, both for cost and production-time reasons.

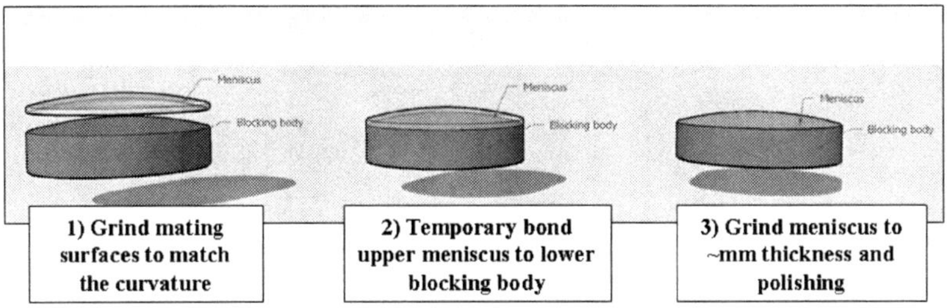

Figure 1: Conventional manufacturing technique for thin mirrors.

The slumping process

A process able to deliver a thin mirror starting from a blank disk with the desired thickness (on the order of 1 to 2 mm) and based on the concept of a replicated master mold is a very attractive possibility to strongly reduce costs, production time, and the risk of catastrophic mirror damage. With these aims, a research team located in Milan, Italy, proposed and initiated a feasibility study based on the concept of glass hot slumping (Ghigo 2006). A thin, flat glass sheet is placed above a master mold previously optically figured and polished to the desired shape; then a suitable thermal cycle is applied. When sufficiently heated, the glass will soften enough to slump onto the mold surface and adapt to the mold's shape (Figure 2). When the system is cooled to room temperature, the slumped glass shell is released from the mold and coated with the proper reflecting layer.

Let's go into more detail. A mold having the complementary optical shape and the same surface micro-roughness quality desired for the optical segment is manufactured from a suitable material. In a clean-room environment, both mold and glass sheet are thoroughly cleaned of any residual grease from handling and dust; then they are placed into a muffle with the glass above the mold. The glass' surface touching the mold becomes the optical surface of the mirror.

The muffle is a stainless steel box designed to maintain a vacuum seal at high temperature. The air is removed, making use of mechanical vacuum pumps. The use of this muffle helps the process in many ways. It reduces the air convection: heating the system just through irradiation makes it easier to reach a greater thermal homogeneity. This is essential to avoid creating stresses in the glass shell. Also, the muffle protects the mold and the glass from the dusty environment of the oven. In fact, dust particles can be named as one of the possible show stoppers for this technology: a dust grain trapped between the mold and glass surfaces during the slumping process prevents a correct copy of the mold's shape in an area of some square centimeters.

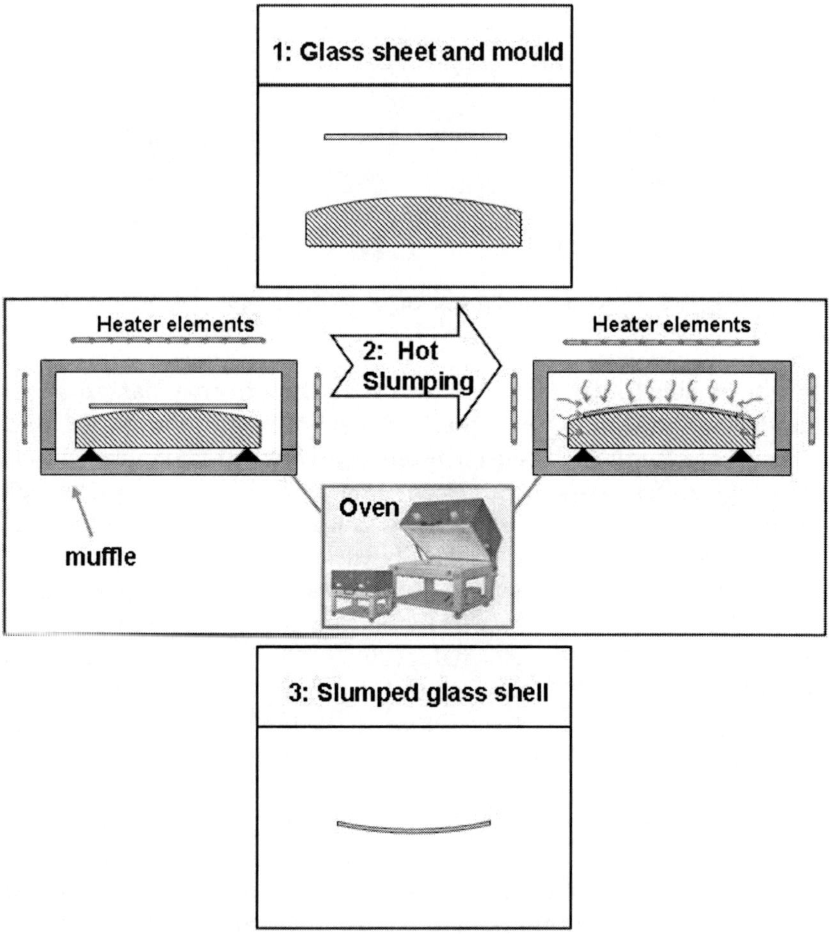

Figure 2: The hot glass slumping process studied in Italy.

The muffle is then placed inside the oven and a suitable thermal cycle is applied. The maximum temperature reached during the process is about 650° C. The speed in heating and cooling of the overall system is not simply limited by the power input to the oven used, but much more by the thermal conductivity of the materials used for mold and glass. The presence of thermal gradients during the slumping process can introduce shape errors on the glass optical surface. This is due to stresses arising mainly during the cooling-down phase of the process, and in particular when the glass approaches its transformation temperature, T_g. This kind of error is typically in the low spatial frequencies range affecting the optical figure of the slumped glass shell.

Bending a surface from a flat geometry to a spherical one is quite a complex task and it requires a long time if the slumping is assisted simply by gravity (it is the same for parabolic and hyperbolic shapes). To ensure the full (optical) contact of the glass against the mold, a pressure is applied during the slumping process when it reaches the maximum temperature. It forces the glass against the mold, helping both to bend the initially flat glass sheet and to speed up the procedure. This gives the process the capability to copy the overall geometry and part of the surface micro-roughness of the mold. For this last reason it is important to have a mold with a micro-roughness level consistent with a surface for optical application, otherwise the mirror will mostly scatter the light instead of reflecting it.

At the end of the thermal cycle, the thin glass shell will ideally have the same shape and micro-roughness of the mold. Its optical quality can be measured with an interferometer. Due to the floppy nature of the slumped shell (the thickness is only a few mm), it is necessary to use an astatic support to measure its shape. This device is able to support the shell, simulating a weightless condition in order to remove the gravity induced deformations.

Some related problems

One critical aspect is defining the best material for the mold and the glass type for the segments. To this end, it is useful to compare the theoretical thermo-mechanical parameters of suitable materials for the mold and, in parallel, to gain experience regarding the problems related to the slumping of different types of glass onto different types of molds (i.e. using test samples). The critical parameters we have identified can be summarized and grouped as follows in Table 1:

Table 1: Parameters for mold material and glass type.

Mechanical	Physical	Structural	Fabrication	General
Young's modulus	CTE and CTE homogeneity	Voids, inclusions	Machinability	Availability
Hardness	Thermal conductivity	Max appl. temp.	Polishability	Scalability
	Density	High temp. stability	Optical microroughness	Costs
	Glass adhesion Transparency		Mould charact.	

A consideration that must be kept in mind is that the choice of a mold material cannot be considered only in terms of thermo-mechanical material properties. Both the slumping process under development and the final application for which the technology is addressed could in principle affect or be affected by the mold and/or glass properties. Indeed, it must also be considered part of the selection. For example, if the sample for the study is a mirror segment of X size with a requirement on the scalability up to 10X, the choice of the mold must take into account the possibility for larger dimensions.

The selection of the type of glass is close to some of the characteristics of the mold itself. For example, finding a good match in CTE between glass and mold helps reach the desired shape on the final glass segment; the temperature necessary to slump the glass must be in the range of application of the mold; the chemical composition of the glass results in different behavior in glass-mold adhesion.

The thermo-mechanical properties of the glass itself are also important. In particular they are related to the final application of the slumped segments. For example, the actuators for the adaptive support heat the segment during the operation, hence the glass shells must be able to dissipate the heat with sufficient efficiency. Different glass types are produced/sold in different size ranges (depending on the technology used and the market they are addressed to), so the linear dimensions and the thickness of the glass sheets must be compatible with the application and possible scaling-up.

To choose the best material for the mold and the glass type, all these considerations must be properly weighted in a merit function to reach an acceptable trade-off between the many input parameters in the effort. The materials tested for the mold were Alumina, Silicon Carbide, Technical Quartz, and ZERODUR K20. A detailed analysis of the materials investigated has been presented Ghigo, 2007. The choice for the mold was the ZERODUR K20 material, while Borofloat was chosen for the glass. Both materials are produced by Schott.

Another crucial point is the heating and cooling rates for the thermal cycle. The slumping is done at temperatures between the annealing point and the softening point in order for the glass to slump without loosing its original good surface properties. The graph in Figure 3 shows an example of an optimized thermal cycle. However it differs, it always includes ramps and plateaus. The exact timing is strongly dependent on the type and size of the material used as well as the oven. In general, the heating can be relatively fast up to the required slumping temperature; then a holding time ensures that no temperature gradients are present and the glass has time to slump against the mold. The cooling down is slower to allow the glass to reach the annealing point without internal stresses, since below this point the glass is no longer able to relieve stresses.

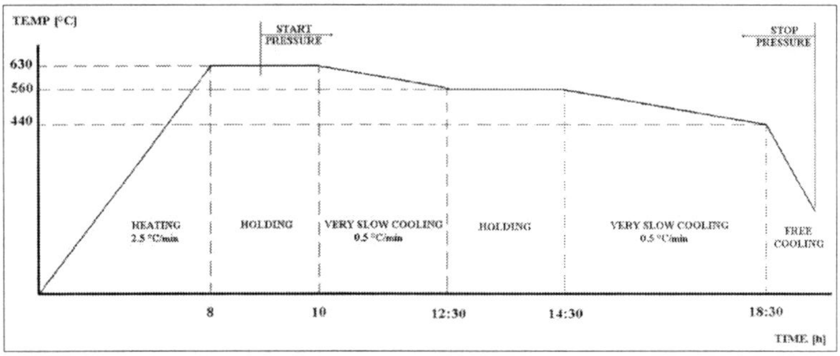

Figure 3: Optimized thermal cycle for slumping process.

State-of-the-art

Initial tests were performed on a small mold with a diameter of 150 mm and a convex spherical shape of about 4 meters radius of curvature. A number of glass disks having a thickness of 2 mm were slumped.

A representative measure of these shells is shown in Figure 4. On the left side, the shell is still in place on the mold just after the cooling down of the oven and is illuminated with sodium light. The periodicity in light and dark coronas is the interference fringes between glass and mold due to their shape difference. This evaluation is qualitative, but it permits us very quickly to assess the results of the experiment. The fringes are well rounded, smooth, and centered, meaning a good copy of the mold. There is some shape difference due to thermal effects (CTE variation from room to slumping temperature). On the right side, an interferometric measure of the entire surface is visible. The values range from $\lambda/11$ to $\lambda/3$ (where λ = 632.8 nm typical of the HeNe laser). The slumped shells so far have shown values of optical accuracy very similar to each other. This indicates a good repeatability for the process. Moreover, the surface features' pattern visible in the measures is present in all the segments slumped so far. These features are very likely present on the mold and are simply replicated onto the segment surface; on these tests the mold shape is limiting the performances obtainable with the slumping technique (Ghigo 2007).

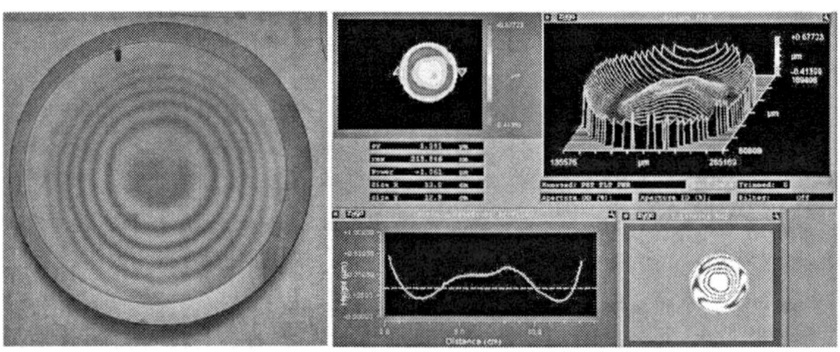

Figure 4: Left: Slumped shell still on the mold; Right: measure of surface.

Recently, a scaled-up process has been performed to slump glass mirror shells with more representative dimensions. The larger ZERODUR K20 mold with a diameter of 700 mm and thickness of 110 mm was used. The radius of curvature was about 10 meters, while the surface finish was on the order of 2.5-4.5 nm rms. This mold has a constant thickness for thermal reasons, and hence its bottom surface is concave. The optical figuring of the upper surface was performed following the specifications related to the project that require an overall figure of about 4λ rms, but more stringent values at higher spatial frequencies. The slumped glass disks have a diameter of 500 mm and a thickness of 1.7 mm.

From the thermal point of view, the time to reach the slumping temperature and to cool down the system minimizing the thermal gradients is longer than those shown in Figure 3. A complete slumping cycle takes up to four days (Canestrari 2008a).

In Figure 5, the small circular and concentric paths of fringes are due to dust grains or particles that prevent the glass from having full contact with the mold. Only a few number of dust grains can be counted on the whole surface, thanks to the cleaning tools and the meticulous procedure used. On the right side is the measurement of the shell: only the central part of the glass has a shape close enough to a sphere to be interferometrically measurable. This central part is about 220 mm wide (half diameter). It shows an error deviation from a mathematical sphere of about $\lambda/2$.

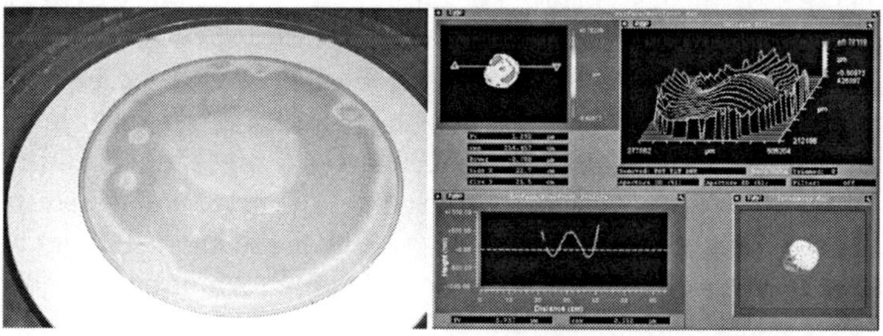

Figure 5: More representative (larger) slumped shell and measures.

The Foamglas backing

The experience gained from telescopes for Cherenkov light

INAF-Osservatorio Astronomico di Brera, in collaboration with Media Lario Technology, has developed a new technique for the production of mirrors that specifically addresses the production of segmented primary mirrors for the Cherenkov telescopes (Figure 6). It is based on the replication from a convex aluminum master mold to impart a concave shape to a thin sheet of glass (1 to 2 mm thick). This concept works in the regime in which the glass is elastic and it accepts small deformations without introducing distortions of higher orders. At room temperature, by means of a vacuum suction, the initially flat glass sheet is bent and forced into contact with the mold. A honeycomb structure is then bonded with glue to the back face of the glass and sandwiched using another glass sheet on top of it. At the end of these steps, we have a good replica of the master mold in a stiff and lightweight mirror panel ready to be coated with a suitable reflecting film. The implementation of a symmetric sandwich-like structure helps achieve the desired stiffness of the panel and improves the thermal behavior of the mirror segments. This approach is very attractive since mass production with short manufacturing times is possible (e.g. one mirror per day per master mold), maintaining very low costs and weights, but with only modest angular resolution (Pareschi 2008, Vernani 2008). Cost for a 1 square meter mirror is about 2 to 3 thousand Euros; weight for a 1 square meter is about 10 kg.

Figure 6: The MAGIC Cherenkov telescopes sited at the Canary Islands

Stiffening a floppy glass slumped shell

To permanently freeze the shape of a curved slumped glass, it is forced into full contact with a mold through air suction; then a suitable foamed substrate is bonded with glue to the mirror shell, forming a sandwich-like structure. One side of the foam substrate must be machined in advance to accept the curvature of the slumped glass segment. This also maintains as constant as possible the thickness of the glue (Figure 7) (Canestrari 2008b). The foams typically show good performance in stiffness and weight when compared with bulk materials of the same density.

Attention must be paid to the choice of the foam and glue. In particular, it is important to minimize the CTE mismatch between the foam and the glass shell to avoid distortions due to environmental thermal gradients. For the glue, a very low-percentage shrinkage is mandatory to minimize local stresses during the curing phase.

With these steps, the optical shape of the shell is frozen and the final mirror panel has a low weight while maintaining the good optical quality obtained from the slumping process. In addition, with this approach the cost of a finished segment can be quite low and the production rate can be fast.

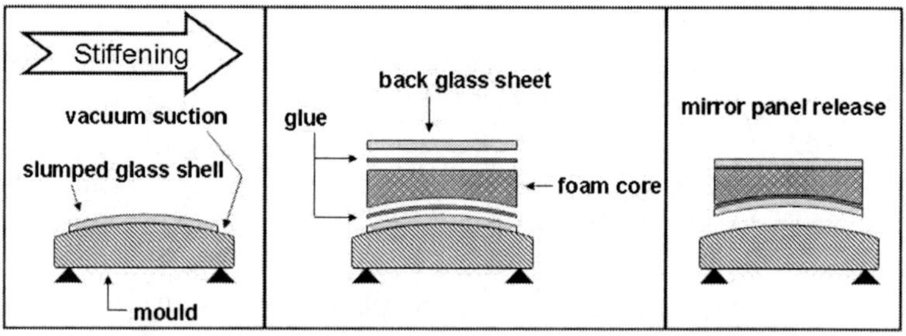

Figure 7: Bonding to maintain shape of slumped shell.

Preliminary results of the investigation
The substrate used for this development was Foamglas from Pittsburgh Corning. The CTE of this foam is different from that of the glass sheet. To minimize the effect of the CTEs mismatch, the curing of the glue and the interferometric measurement were made at the same temperature. The use of foam with a better match in CTE (with respect to the glass shell) is a step that will be implemented in future development. The glue used is a two-component epoxy product that has very low per cent shrinkage, on the order of .037%, and the same quantity of epoxy was used in both of the foam substrate faces. We also tested for the capability of the foam to withstand a vacuum without breaking. This permits the panels to be vacuum coated.

The left side of Figure 8 shows the stack during the curing of the glue. Visible is the mold with the slumped glass kept in full contact by vacuum suction. The foam substrate, with a thickness of 40 mm, is placed above the slumped glass with the glue in between. At the top of the stack is visible a weight used to spread the glue over the entire surface. The right side of Figure 8 is a picture of the final mirror panel after the aluminization of the optical surface.

After curing, the mirror panel was measured interferometrically. Figure 8 (bottom) shows the results. Compared with the measurement of a standard slumped shell (Figure 4), it might be noted that the overall surface shows essentially the same optical quality of about $\lambda/2$. This confirms that the implementation of the sandwich structure does not appreciably degrade the overall optical quality (Canestrari 2008b).

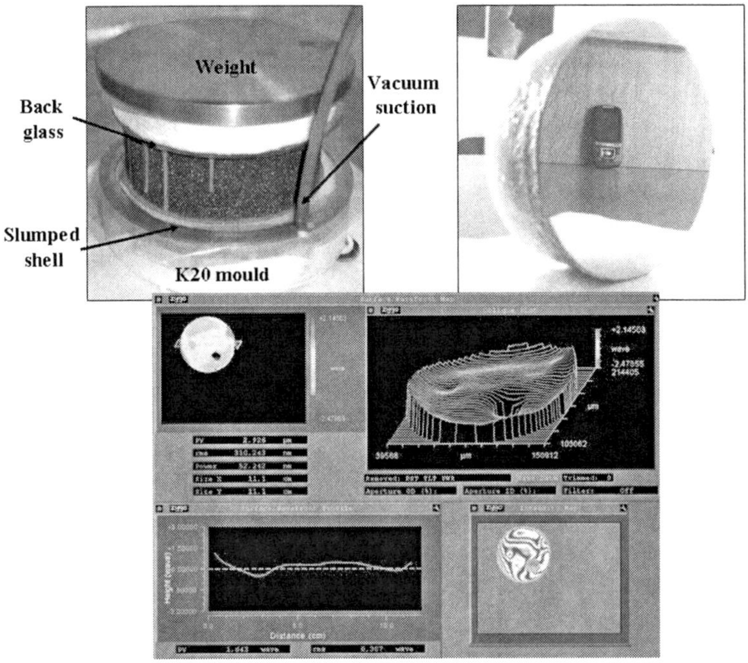

Figure 8: A mirror panel formed through slumping and stiffening, with measures.

Author's Note

Contact: rodolfo.canestrari@brera.inaf.it
Address: via Emilio Bianchi 46, Merate 23807 (LC) Italys
Office phone: +390395971044

References

Canestrari, R., Ghigo, M., Basso, S., Spiga, D., and Prosperpio, L. 2008a. Lightweight optical segment prototype for adaptive optics manufactured by hot slumped. *Proc. SPIE*, 70153S.*

Canestrari R., Ghigo, M., Pareschi, G., Busso, S., and Prosperpio, L. 2008b. Investigation of a novel slumping technique for the manufacturing of stiff and lightweight optical mirrors. *Proc. SPIE*, 70180D.*

Ghigo M., Basso, S., Citterio, O., Mazzoleni, F., and Vernani, D. 2006. Manufacturing of lightweight glass segments for adaptive optics. *Proc. SPIE*, 62720X.*

Ghigo M.. Canestrari, R., Basso, S., and Spiga, D. 2007. Development of lightweight optical segments for adaptive optics. *Proc. SPIE,* 66910K.*

Pareschi G., Giro, E., Banham, R., Basso, S., Bastieri, D., Canestrari, R., Ceppatelli, G., Citterio, O., Doro, M., Ghigo, M., Marioni, F., Mariotti, M., Salvati, M., Sanvito, F., and Vernani, D. 2008. Glass panels by cold slumping to cover 100 m^2 of the MAGIC II Cherenkov telescope reflecting surface. *Proc. SPIE,* 70180W.*

Vernani D., Banham, R., Citterio, O., Sanvito, F., Valsecchi, G., Pareschi, G., Ghigo, M., Giro, E., Doro, M., and Mariotti, M. 2008. Development of cold-slumping glass mirrors for imaging Cherenkov telescopes. *Proc. SPIE,* 70180V.*

**Proc. SPIE* (Proceedings of the Society of Photo-optical Instrumentation Engineers) can be accessed at http://spiedigitallibrary.org/

Chapter 23

Replica Polyurethane Foam Mirrors for Photometric Applications

Kiran Shah

Chroma Systems, India

Introduction

While telescopes with apertures in the range of 0.5 to 2.0 meters are regularly produced, their costs are high, especially for telescopes 1.0 meter and larger. If the costs of such telescopes could be reduced, many institutions (and even individuals) could afford a telescope in this aperture range, and both scientific research and student education would be advanced.

The cost of these telescopes is high for two primary reasons. First, as mirror diameters increase, the weights of the mirrors go up very rapidly. Heavy mirrors require heavy mechanical support structures, and the entire telescope becomes very expensive, with many requiring a crane for installation. For these telescopes, transportability, let alone portability, is out of the question. Second, the cost of grinding, polishing, and, especially, optically figuring mirrors is high, even for smaller mirrors, and this cost rapidly increases as mirror diameters go up. Telescope costs would be greatly reduced if a very lightweight mirror could be mass-produced that did not require each mirror to be ground, polished, and figured.

This chapter describes the methodology of an attempt to determine whether a replication process using a urethane foam substrate and a solid urethane reflective surface can create a lightweight mirror. This method, if it can be demonstrated to work, will have three advantages: a much lower cost, a lightweight optical surface, and a shorter time to produce the mirror. The ideas presented here are admittedly experimental in nature and need further work to evaluate their viability.

The goals of this developmental project are purposely modest. Specifically the intent is to reach large "effective" apertures by producing modest aperture mirrors in quantity at very low cost. A number of these "small" mirrors (perhaps eventually 0.25 to 0.5 meters or so) could be used together in a multiple mirror telescope. Also, at least initially, the goal is to produce lightweight urethane replica mirrors which are spherical and of only modest optical quality—good enough for some photometry or low-resolution spectroscopy tasks, but not for demanding astrophotography applications.

It may turn out that urethane replica mirrors are too flexible to even provide optical performance suitable for some photometry and low-resolution spectroscopy tasks. However, the flexible nature of the mirror substrate might be overcome when used in conjunction with an active mirror support system to deliver the performance required for a photometric or low-resolution spectrographic application.

The replication process needs a complimentary master of the part to be replicated. In our case that would mean a convex master of the shape and size that we require. Before we start on this path, the idea needs to be checked for viability, as the cost of making a convex master is high.

What is proposed is to create a flat mirror using polyurethane foam as the backing and multiple layers of clear polyurethane as the main reflecting surface. This mirror will then be placed into an auto collimation test setup to characterize its surface under a Ronchi test.

Making the flat mirror

A mold of the diameter and the thickness required is created. It is coated with a release agent (car wax polish). The foam is a two-part mixture of the resin (with additives for foaming and strength) and hardener mixed in a proportion of 1:1. This mixture is stirred for a minute and poured into the mold as soon as the foaming starts.

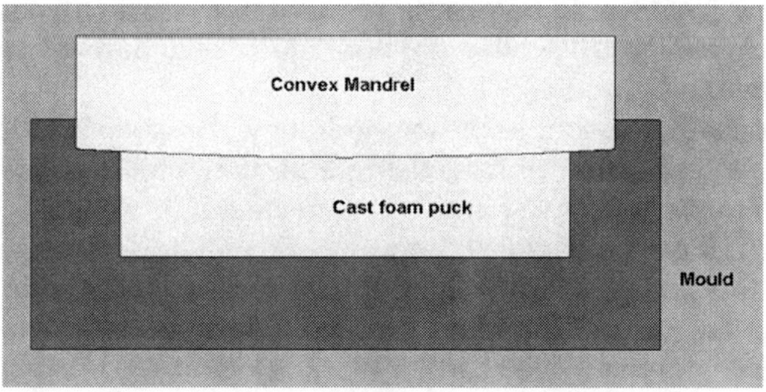

Figure 1: Photo of mold with foam poured in and glass surface making contact with foam.

As the foam begins to rise in the mold, the glass surface with a very thin layer of car wax as a release agent is placed on top of the mold. This gives the working flat surface. The foam is tack free in 10 minutes and has 80% of its structural strength in 2 hours. It reaches its maximum strength in about a week. The foam puck is allowed to cure for two days in the mold after which it is released from the mold and allowed to cure for a week. The first surface, which will be used for the next coat of non-foaming solid polyurethane, is cleaned with solvent, and sanded, and blown clean with dry compressed air.The working glass surface (the replication "master") is cleaned again and coated with a very thin layer of release agent and placed face up on a synthetic grass mat or a shaggy carpet to minimize deflection of the glass optical surface. A mixture of non-foaming polyurethane with additives for strength mixed with hardener is poured in one continuous movement on the glass surface. The foam puck is placed on this poured mixture and worked around till a thin layer of the polyurethane mixture remains on the surface. This setup is cured for a week and released by cooling the glass and foam in a freezer.

This process is repeated two (or more) times until a smooth surface is obtained. The idea is that the solid polyurethane layers progressively fill up any voids or tendency to warp due to stresses being relieved during the cure of the foam puck, and finally result in a smooth surface.

Some work on the thickness of the foam back plane is necessary so that if an active support system is used, the thickness is just right for the deflecting forces being used without being too floppy.

This surface under an auto collimation test should reveal the usefulness of this method of replicating mirrors.

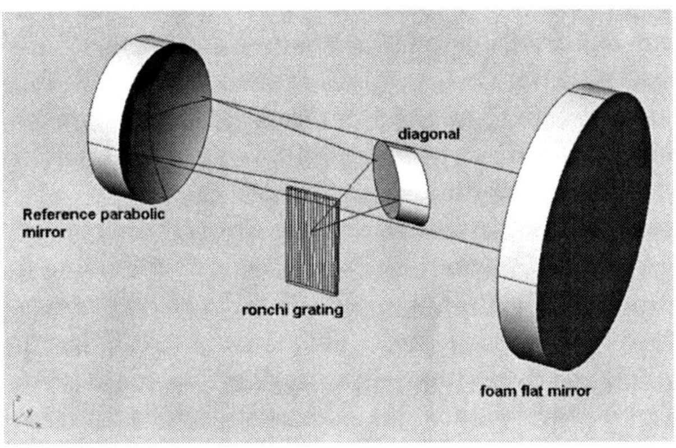

Figure 2: Proposed auto collimation setup with Ronchi grating.

Vacuum coating of polyurethane foam is an unknown that needs to be investigated. Alternatives could be silvering by reduction of silver nitrate.

For Concave Mirrors

Once this method is established by casting larger flat mirrors, say in the 10 to 20 inch range, we could use it to replicate concave mirrors that are capable of photometry. An important development would be tackling the problem of an active support system for larger floppy mirrors.

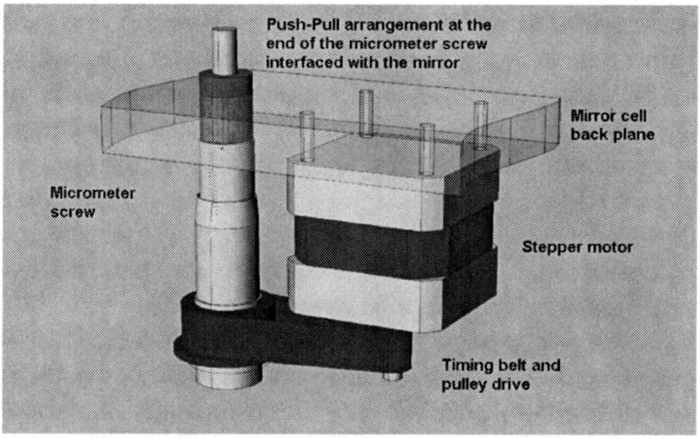

Figure 3: An active support system.

The Convex Glass (or Metal) Master

Testing convex surfaces is a challenge. The Hindle test requires a concave spherical mirror typically larger than the convex mirror under test and is a null test for hyperboloid mirrors.

What is proposed is to place the convex master on a precision turntable that can also be adjusted in X Y with micrometers. This will enable good centering of the convex glass master. Then a 'scanning' Hindle test could be carried out to check for gross imperfections on the convex glass master. An iterative process should lead to the glass master being figured as a sphere without gross zonal errors.

The mold will be similar to the one created for the flat foam puck and the convex glass master will be pressed into the rising foam surface. Multiple layers of polyurethane will be molded off the master onto the polyurethane foam mirror puck. This should result in a smooth, relatively zone free concave spherical surface on the foam puck.

It is hoped that this surface, after an aluminum vacuum coating, perhaps combined with an active support system, should be able to provide imaging quality suitable for photometric observations. It is also hoped that a number of these spherical mirrors used in a multiple mirror configuration with a corrector before the prime focus could provide a large aperture 'light bucket' with a combined aperture on the order of 1.5 to 2 meters.

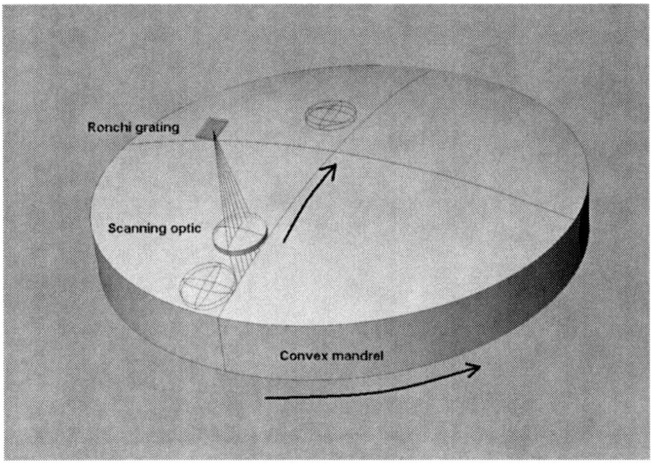

Figure 4: The 'scanning' Hindle test.

Chapter 24

Experiments with Pneumatically-Formed Metalized Polyester Mirrors

Bruce D. Holenstein[1], Richard J. Mitchell[1], Dylan R. Holenstein[1], Kevin A. Iott[2] and Robert H. Koch[3]

[1]Gravic, Inc., Malvern, PA; [2]Iott Precision Instruments, Petersburg, MI; [3]Department of Physics and Astronomy,University of Pennsylvania, Philadelphia

Introduction

In a contribution earlier in this volume (Holenstein, Mitchell, & Koch 2010) (HMK), some of us presented an analysis for quantifying the imperfections of unconventional optical systems and correcting for them. Our primary motivation is credible and reproducible astronomical observational work and not lab applications. In this presentation, we lay out our developmental efforts toward that intended purpose and apply some of the earlier results to our developments and accomplishments since 1991.

Background

In 1991 Peter Waddell visited the Laboratory for Research on the Structure of Matter (LRSM) at the University of Pennsylvania. A member of the Department of Mechanical and Process Engineering of the University of Strathclyde, Glasgow, UK, Waddell delivered a lecture about John Logie Baird (1888-1946) and his sadly neglected (according to at least some Scots) claim to be the inventor of television. Waddell's baggage included an 8 inch pneumatic mirror which he drew into a focusing figure by sucking on a plastic tube and co-author Robert Koch witnessed the image of an outside scene displayed on an office wall with this device. Some results and interpretations from Strathclyde had already appeared in, e.g., King et al. (1985), and Manly (1991) has reviewed a part of Waddell's results. Gavin (1979), a British amateur astronomer, briefly reports on and illustrates a pneumatic telescope that he had designed and constructed.

The imaging stunt was reported to Kenneth Lande, then Chairman of Astronomy and Astrophysics, who authorized spending limited departmental funds to investigate the theory and realization of like mirrors. Over the next 5 years, a few people participated in fabrication and tests of them: Robert Hee as professional machinist and the late Samuel

Seeleman and Robert Koch as amateur ones and opticians; Richard Mitchell as hardware and electronics specialist; and an assortment of graduate students. Effort was dedicated to design, construction, and testing of 3 pneumatic imaging systems, and a few in-house demos and talks were given about them. Robert Koch visited the Waddell lab in 1994 studying a 24 in pneumatic mirror that had been kept under a substantial pressure difference for more than a year and, to visual inspection, preserved a good figure. Robert Koch's retirement in 1996 brought the opportunity to try a different design which he pursued until about 2000. Work on the project was re-engaged in 2008 when Bruce Holenstein pulled together the surviving collaborators to start anew, bringing Kevin Iott on board as designer and machinist and Dylan Holenstein as lab technician.

Hardware In Summary

There follows a summary of the items and processes that we have experimented with:

Reflecting materials: 0.2, 0.4, 0.6, 1.5, 1.7, 2.0, 4.5, 6.3, 6.8, and 7.2 mil aluminized or silvered polyester films (either one- or two-sided metalizing), some of which were Mylar® marketed under different brand names from U.S. manufacturers; 1 mil stainless steel shimstock.

Structural materials: 7-ply, 0.75 inch sealed marine plywood; 0.75 inch primed, painted, and sealed oak; 0.25 inch aluminum; 0.25 and 1.00 inch acrylic.

Retaining and clamping materials: threaded acrylic with O-rings; aluminum standoff posts with O-rings; hot-melt glue with cellular foam; discrete clamped spreader bars with cellular foam; bolted aluminum ring cells; rolled and welded aluminum sheet.

Optical diameters: 2 each 7.25 inch; 8 inch; 2 each 12 inch; 2 each 42 inch; and 70 inch.

Focal ratios: *f*/0.5 – *f*/10.

Pressure differences with respect to atmospheric: positive in a bladder with a transparent film or acrylic window; negative without a transparent window.

Pressure difference control: mouth; hand pump; electrical vacuum/pressure pump; laptop control of mini-valves.

The reader should not imagine that all combinations of these characteristics were tried experimentally nor should one suppose that the chronology of change has been monotonically from small to large or from naïve to complex. The reader will also notice in the following a mixture of metric and English units according to the measuring devices that we actually used.

Motivation

World-wide there have been several investigators of deformable optics and a certain number of them (including Waddell's radiant heating of some hardware for a bench experiment) had non-astronomical purposes in mind for their designs. Until very recently, we had not looked at the patent literature but were content to be guided by our own efforts. Our review of that literature does not indicate that we have significantly duplicated earlier work.

The ultimate purpose of our efforts has been to develop a pneumatic mirror that would be useful for astronomical photometric programs with a feedback control against ambient changes so as to sustain a constant focal length. A mirror surface that will satisfy our observing program will collect photons from a stellar program object and deliver them to the detector surface, yet exclude enough background photons in order to yield an acceptable signal-to-noise ratio.

Test Results

In the following we summarize our experience with significant features of our devices:

Film Stability

The film thicknesses given above are nominal and we did not make enough measures to determine the uniformity of thickness over an extended area of any sample. Thickness of the metalizing layers ranged from about 0.2 mil to about 1.2 mil but this too was not tested extensively.

No matter what the film thickness and no matter the manufacturer, reflecting film can be degraded easily as it is peeled from a roll. The width of our rolls was typically 48 inches but wider dimensions can be bought up to at least 12 feet. Unless film is restrained as the roll is unwound and unless the metalized surface is protected from abrasion, linear, curvilinear, or cornered folds are inevitable. The same care must be looked to when moving a cut film sample, for it may fold if caught in an airstream. Even with all precautions observed, an area of about 2 feet^2 or larger will not lie flat under its own weight as it comes off a roll; it will have obvious ripples with semi-amplitudes of the order of 0.125 inch.

At the LRSM we cycled a sample film from room temperature to -70 C a few times to see if a glass transition would occur in the film. No such transition happened in these repetitions. Eventually, the metal and polymer delaminated, presumably because of the difference of the coefficients of thermal expansion.

Repeated cycling to a pressure differential of -0.5 inch of Hg and back to atmospheric led to an unchanged figure for the film and we were unable to detect any passing of the elastic limit. In order to look further at the elasticity of the film, we cut a sample 0.25 inch x 33.5 inch x 0.005 inch of 3M silvered film and loaded it statically along its long dimension with successively greater weights in a test that lasted 3 days. There were repeated changes, repetitions, and reversals of the loading during that interval. Over a considerable range of load the film performed according to:

$$Elongation = 0.006(4) - 0.000269(6)(\boldsymbol{Load}). \quad (1)$$

The parenthesized numbers (here as well as later) are the 1σ-errors of the last decimal place of the foregoing coefficients. This linear fit to the data shows that the sample elongated by about 0.4% up to (1200 gm < *Load* < 1500 gm). The result is consistent with the published elasticity of Mylar film and leads to a Young's modulus of 4 GPa, also consistent with the specs of the sample. Beyond a load of 1500 gm, elongation was faster.

Structural Materials

Because of pressure-difference losses, the oak frame needed many repetitions of interior and exterior fiberglass and epoxy sealing before all pores were sealed. None of the other materials suffered from this defect.

Retaining and Clamping Mechanisms

For a criterion of retaining a substantial pressure difference for many hours, the threaded 12 inch acrylic ring cell furnished with 0.125 inch O-rings and with contacting surfaces coated with stopcock grease would be considered adequate. (This device appears in Figure 2.) The small aluminum bolted cells were superior; they retained pressure differences of -5 inches of Hg against atmospheric for more than a month. The other mechanisms were failures. Insofar as discrete clamping points were used, their adjoining locations could not be separated more widely than about 1 inch on the cell circumference if ripple structure was to be invisible to the naked eye on the reflecting surface.

Control Against Ambient Changes

With an A. H. Thomas pump and a water manometer that we built, we could control the calibrated pressure difference to ±0.002 bar. The pump, with a capacity of 2 psi, moved 3.6 liters/minute and required of the order of 60 seconds to attain a good figure on the film mounted in a 12 inch cell. For film mounted in the 42 inch cells, pumping time was of the order of a few minutes using a Brey G12/02-8LC vacuum pump monitored with a Honeywell 142PC05D pressure sensor. There are numerous temperature and pressure sensors commercially available which can be suitably interfaced for computer control.

Figures and Focal Ratios
Early tests used a depth micrometer to determine gross figures for both of the 12 inch mirrors. These tests were run across a number of different diameters and corrected for the shape and size of the micrometer probe. One sample is shown in Figure 1. Polynomial curve-fitting to order 2 by the Levenberg-Marquart (L-M) method led to the representation of the figure:

$$y = 0.0003(7) + 0.0003(1)x + 0.00733(4)x^2 \quad (2)$$

The linear term is due to an asymmetry in mounting the film in the frame. Curve fitting to order 3 recovered the three coefficients above within their errors and showed the cubic term not to be significant. In principle, the quadratic term signifies an f/2.70 beam with a focus at F = 32.34 inch.

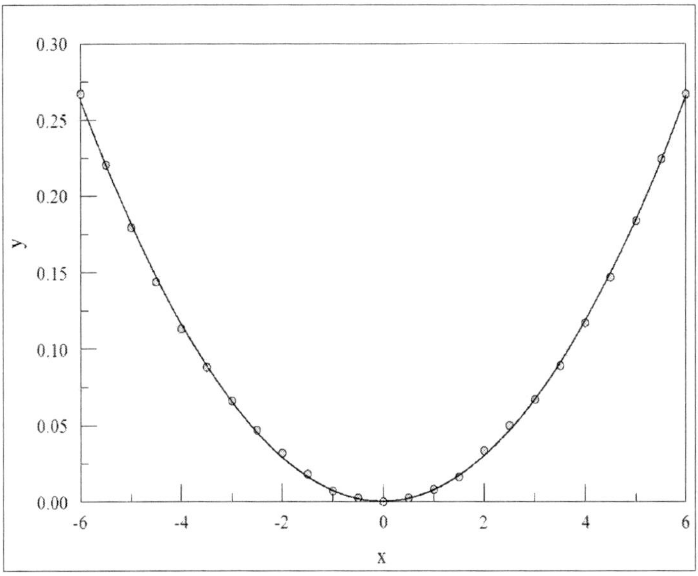

Figure 1: A mechanical determination, made in 1992, of the figure of a silvered Mylar membrane mounted in one of the 12 inch frames. The continuous curve is that of equation (2) of the text. A ripple structure can be seen from about |5.25| inch to about |1.5| inch.

However, as can be seen in the illustration, higher even-order terms are likely and curve fitting indeed established 4th- and 6th-order terms at close to the 3σ level. These fittings diminished the sum of the squared residuals by factors of 1.5 and 2.3, respectively, compared to the value found by the quadratic fit. The coefficients of the 4th- and 6th-order terms correspond to 9,140 Å and 2,290 Å, respectively, on the film surface. No matter whether this "fine" structure is confined to the sampled diameter or is limited to a circular sector or is circularly symmetric, it emphasizes the need for uniform tensioning of the film into a plane before the cell is closed. A tensioning frame must obviously be larger than the cell diameter.

Curve fitting, again by the L-M procedure, converged to a circular fit for a radius of 68.3(3) inches but with a sum of squared residuals about 20% worse than the simple parabolic fit.

There are published suggestions that the figure of a pressurized film is a catenary – the usual example being that of a cable restrained at its endpoints and hanging under its own weight. In a formal mathematical way, the L-M method cannot fit the data of Figure 1 by a catenary, but physically there is a more forceful reason why this geometry should not be suitable for a pressurized film. One lays aside, first of all, consideration of a textile string or rope, which characteristically has little weight and feels aerodynamical forces and, as an example, concentrates instead on a braided steel cable as an analogue to the pressurized film. This familiar element of a suspension bridge becomes configured as a catenary under its own mass but it lacks one characteristic of a pressurized film – the film deforms under the pressure difference because of its relatively large elasticity and this does not happen with construction steel. For this reason, further consideration of a catenary was considered unrealistic.

Our mechanical experiments, therefore, led us to believe that imaging defects would be traceable to zonal errors and classical astigmatism of the stretched film due to defects in tensioning, spherical aberration, and the bi-axial structure of the polymer itself.

Quickly thereafter and very recently as well, we ran numerous optical tests of figure by constructing the Right-angle Bath interferometer (mentioned above) fed by a He-Ne laser. The setup is shown in Figure 2 with the beam aimed at a 1 mil aluminized film mounted in one of the 12 inch cells. The same figure also shows typical results for a 1 inch radius of the film (a choice made to keep the number of fringes to be analyzed reasonably small) offset by 4.5 inches from the center with a pressure difference of -5 inches of Hg against atmospheric.

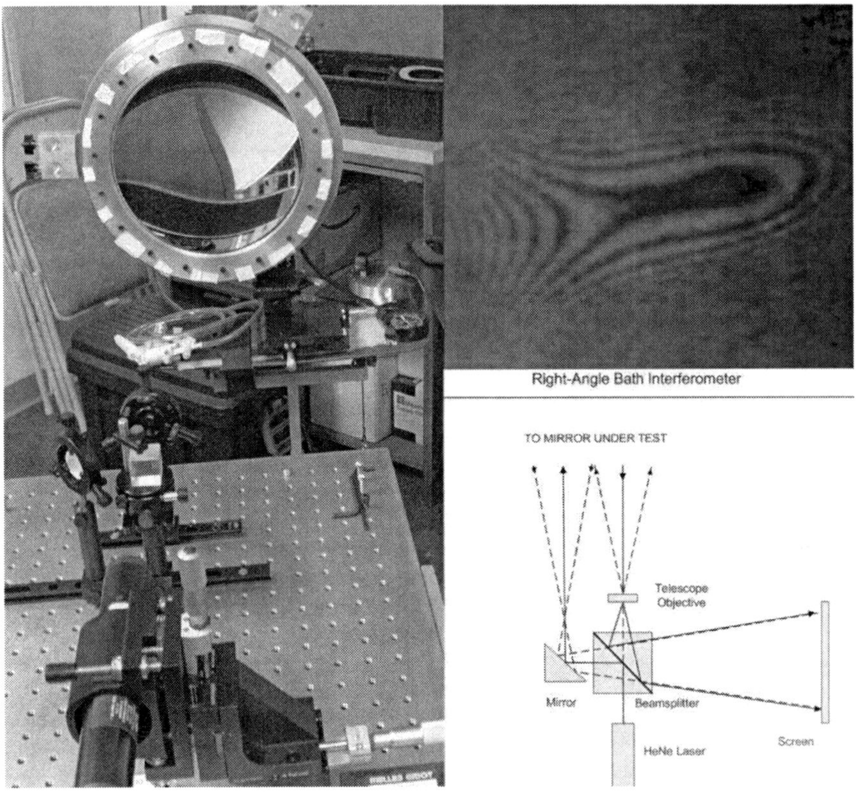

Figure 2: The tabletop optical bench mounting the interferometer, laser, and sample 12 inch film mirror, which is not under tension. The bolting pattern on the table is a 1 inch grid. In the lower right there is a schematic of the optical components and paths and above this sketch a sample interferogram.

Fitting of all interferograms was accomplished by *FringeXP* (Rowe 2003) and resulted in the sample Zernike coefficients listed in Table 1.

Table 1

Zernike Coefficients for Example Interferogram					
		(n,m)	**Coefficient**		**Type**
	Z_0	(0,0)	7.993		Piston
Z_1	Z_2	(1,1)	5.114	2.131	Tilt
	Z_3	(1,0)	2.147		Defocus (power)
Z_4	Z_5	(2,2)	-5.955	2.089	Astigmatism
Z_6	Z_7	(2,1)	0.9333	0.5471	Coma
	Z_8	(2,0)	-1.547		Principal Spherical
Z_9	Z_{10}	(3,3)	-0.9508	-0.08976	Trefoil
Z_{11}	Z_{12}	(3,2)	4.443	-2.009	Secondary Astigmatism
Z_{13}	Z_{14}	(3,1)	-1.305	-0.3269	Secondary Coma
	Z_{15}	(3,0)	0.1612		Secondary Spherical
Z_{16}	Z_{17}	(4,4)	-2.242	2.23	Tetrafoil
Z_{18}	Z_{19}	(4,3)	0.09358	-0.304	Sccondary Trcfoil
	Z_{21}	(4,2)	0.4233	-0.2649	Tertiary Astigmatism
Z_{22}	Z_{23}	(4,1)	-0.4218	-0.3383	Tertiary Coma
	Z_{24}	(4,0)	0.1182		Tertiary Spherical
					RMS fit Error: 0.05889 waves

The first three terms (piston, tilt, and defocus) are not related to the mirror figure for interferograms centered on the mirror. For off-axis interferograms, the terms must be understood as offsets from the values derived on centered and neighboring (partially overlapping) interferograms and must be used in a mirror analysis. Column 3 gives the conventional notation for the coefficients and columns 4 and 5 the fitting evaluations in waves. Terms 4 and higher are characterized simply by a large degree of astigmatism caused by the Young's modulus of

Mylar being bi-axial by about 5%, where the axes are along the rolling machine direction and in the transverse direction, and by our lack of tensioning control. This condition may be reduced markedly in future cells provided with mechanical or electrical control elements. The test configuration itself introduced about one wave of astigmatism because the right-angle Bath interferometer is not a common path interferometer. This instrumental astigmatism can be easily reduced by moving the right angle mirror closer to the beamsplitter. The tetrafoil aberration is due to the mutually-opposing four bolts that in this particular experiment were the first ones secured on the membrane clamp ring, again a detail that may be easily corrected.

Contour and graphical representations for a sample interferogram are shown in Figure 3.

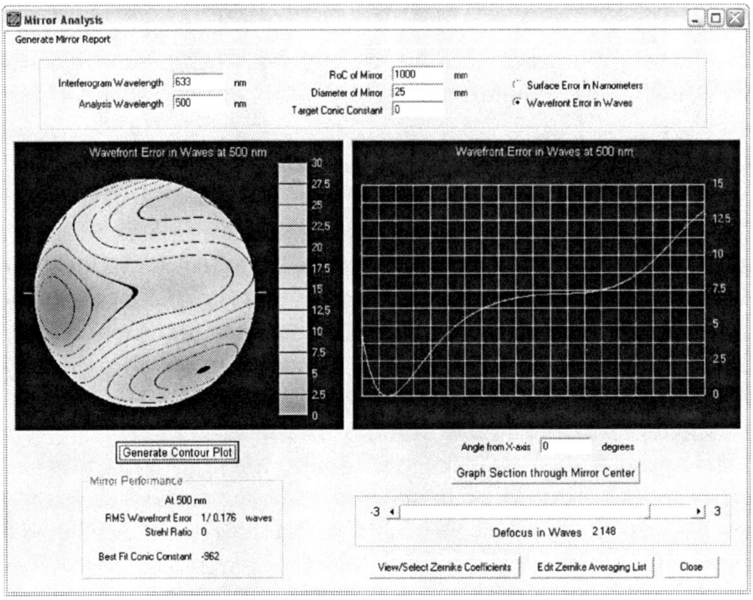

Figure 3: The Zernike coefficients are summed into contour and graphical displays in these panels.

The target conic section in the sample fit is a sphere and the mirror analysis was done at 500 nm. Because of the significant astigmatism, the PV aberration averages about 30 waves but is only about 10 waves along the best axis, with an rms aberration of about 10 waves for this 1 inch radius section. This result holds up for other areas measured across the mirror surface as well and we estimate, using the Holenstein, Mitchell, Koch (HMK) equation (23) from chapter 16 in this book, that the total waveform aberration for this surface is around 200 waves. Except for gross edge defects such as wrinkles which should be masked off, one should expect less than 1000 waves rms aberration on a 1.6 meter mirror.

Reflectivity

The transmissions of unpressurized film samples were measured with beams from white photodiode and He-Ne laser sources with a precision of about ±2%. Film thinner than 2 mil was insufficiently metalized. Thicker film, on the other hand, showed essentially 0% transmission and this value remained stable for months. Reflectivities were much harder to measure. Diffuse reflection from microripple and microroughness of the film surface were probed by bouncing a laser beam off various spots on stretched, unpressurized film. As an example, a typical 5 mil sample showed a divergence due to diffuse reflection of 0.59(3) milliradian.

We have not checked the possibility that transmission and reflectivity change significantly when a pressure difference is applied to the film. This might happen because the metal domains themselves can become stretched and attenuated with a significant pressure difference.

Selected Imaging Results

Four of the mirror devices were made into skeletal telescopes with carbon fiber or aluminum structural members; one of these was intended to function as a broken-Cassegrain reflector, another worked as a prime-focus instrument and two as Newtonian telescopes. For the Newtonian configuration, the secondaries were aluminized glass flats while a pneumatic mirror with a positive pressure difference against atmospheric was used for the broken-Cassegrain. Only two of these instruments were attached to suitably stable mounts.

The newer 42 inch cell was loaded with the same 5 mil film for more than 6 months during which interval it had been pressure-cycled scores of times. By June, 2009 it was loose in its cell but was not re-tensioned before being taken outside where daytime ambient temperature was about 80° F. The appearance of the assembly is shown in the main part of Figure 4.

A pressure difference was established and the mirror aimed at a communications tower visible to the naked eye after being stabilized on an IPI393 tripod and GEM mount. The target was about 10,000 feet distant and was seen over the convection patterns from two intervening office parks of sunlit brick buildings and asphalt parking lots and a 4-lane highway at an elevation of about +5° above the mirror. A handheld digital camera made the image of the target shown as an insert in Figure 4 with the telescope working at approximately *f*/0.7. The resolution of the image is of the order of 10″.

Somewhat later, one of the 12 inch cells was loaded with 1 mil film and was mounted on an undriven optical table to obtain sample images of bright stars. A column sum profile of a CCD image of Vega is shown in Figure 5. (Opposite)

Figure 4: A 42 inch cell and OTA mounted on an IPI393 GEM mount. A daytime image showing a near field of trees and a distant one of a structural tower is shown in the upper-right insert. The camera in automatic mode had difficulty deciding which fraction of the scene to focus on. The piecrust distortions around the edge of the image result from failing to re-tension the film in its cell. Stratified inversion layers above manmade structures account for the ragged tower image. In order to facilitate development testing, the inside cables on the truss members are not in place here and this condition caused the truss poles to bend inwards.

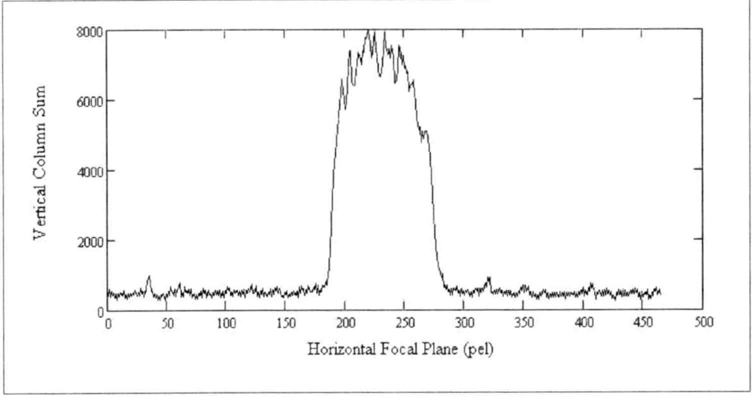

Figure 5: A CoC profile of Vega from a CCD image made with a 12 inch cell operating at *f*/4 using a factor of 0.5x focal reduction to *f*/2. The scale is 5.0 μm/pixel at the focal plane.

We find that a focal reducer lens placed prior to the prime focus of the telescope and in front of the detector markedly improves the character of the point spread function. The diameter of the CoC decreases by the focal reduction factor making the focal reducer act as a light concentrator. Also, some aberrations become less pronounced with the longer focal distances made possible by the use of the reducer, the dominant one being spherical aberration since it decreases as the cube of the focal ratio.

However, the diameter of the CoC encircling substantially all of the flux is about 0.5 mm in the figure. The prediction, using HMK equation (3), again from chapter 16 of this book, and the measurement of 200 rms waves aberration on the surface, is just 0.08 mm. The spherical aberration in this experiment was uncorrected and according to HMK equation (2) produces a spot 0.07 mm in diameter after the focal reducer. The local slope aberrations using HMK equations (15) and (17) with data from Table 1 amount to 0.32 mm. However, the interferogram sampled only a small area of the mirror and it is not clear (to the authors at least) that local slope aberrations in this case add in quadrature.

Co-added local slope aberrations, diffuse reflection from microripples, and non-random systemic surface defects are believed to be responsible for the discrepancy. The sample of 1 mil film used on Vega was not sampled for microripples, but the diffuse reflection of the 5 mil sample mentioned above would account for a 0.73 mm CoC. Understanding and diminishing this gap between prediction and experiment is an active area of research. Published vendor literature on commercial Mylar films indicate that microscopic particles are purposefully deposited on the film during manufacture to prevent the layers from sticking once the film is rolled. We expect that a solution to this problem may be found with vacuum aluminizing of low-surface-defect membrane materials.

Future

Figure 6 is a detail from a sketch for a 1.6 m mirror on an Alt-Az mount with control of the gross astigmatism of the primary mirror using push-pull screws placed on the mirror cell back and sides. If needed, secondary correction for significant Zernike coefficients may be done pseudo-statically or dynamically on a small deformable secondary feeding the detector. This design is expected to produce a CoC well under 1 mm. The mirror cell and tube assembly are expected to weigh less than 500 lbs and can be mounted on commercial off-the-shelf equatorial or Alt-Az mounts. The mirror cell may be made of cast aluminum with a vacuum-sealed epoxy finish to reduce costs and prevent air leakage. The entire structure is projected to cost under USD $65,000.

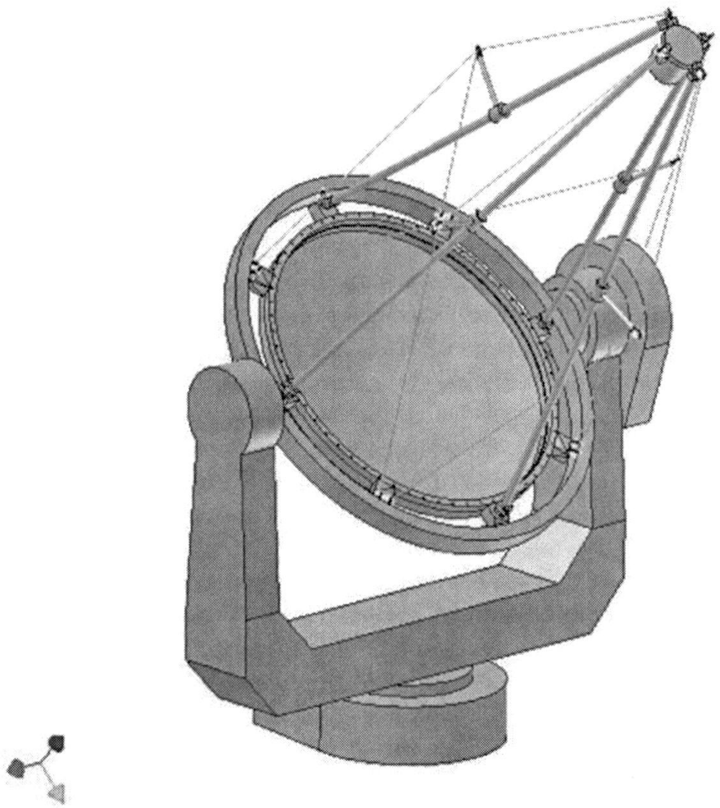

Figure 6: A sketch of one possible next-generation mounted trial mirror.

Conclusions

• We have been able to develop two varieties of cells that will maintain a pressure difference for the long term. Mylar permeability is not a problem. Bare Mylar against machined surfaces forms an airtight seal. There is no need to add rubber gasket or glue to hold the vacuum for hundreds of hours. Active air pressure stabilization is not needed for observing runs of less than an hour when the atmosphere is relatively stable. Fine focusing or vacuum adjustment is sufficient and can be automated if needed between observations.

• Edge clamping and pre-tensioning of Mylar can make wrinkle-free mirrors. Folding the Mylar before clamping causes wrinkles.

• Film thickness should be at least 4 mils for metalizing to be sufficiently reflective. Sources of aluminized film with low levels of microripple must be identified.

• Mirror cell and structural members should be made of metal. Wood and fiberglass were subject to flexure from the cycling of the modest air pressure differences needed to form a suitable surface and they suffered numerous leaks. For instance, a 3% atm pressure difference against atmospheric (commonly used by us) on the 42 inch cell (at *f*/2) results in about 600 lbs load on each of the cell rear and mirror faces.

• Membrane mirrors without astigmatism correction can produce figures with about 10 waves rms aberration peak-to-peak per inch. Our construction of a 1.6 m cell, however, is pending perfection of large-amplitude aberration control. This happens because rms aberrations of considerably less than 300 waves rms across the membrane surface, with rms gradient norms of just 4 waves, are needed in order to allow detectors to work adequately with 1 mm apertures. Our current plans for future research emphasize static and dynamic control of the primary mirror Zernike coefficients and secondary mirror waveform detection and control.

• We project that mirrors with large degrees of smooth aberrations may be used for astronomical research as light buckets in the not-too-distant future. Also, we expect that a pneumatic 1.6 m mirror cell and optical tube assembly, with adjustable figure control, will be transportable and be affordable for small research institutions. We project that the upper limit for pneumatic mirror telescopes will exceed 3 m using commercial films.

Authors' Note

Bruce D. Holenstein can be reached at Gravic, Inc., 301 Lindenwood Drive, Suite 100, Malvern, Pennsylvania 19355 USA.

References

Gavin, M. 1979. Aluminized mylar as a flux collector. *Sky and Telescope*, 57, 490.

Holenstein, B. D., Mitchell, R. J., and Koch, R. H. 2010. Light bucket mirror figures of merit. In *Lightweight Alt-Az Telescope Developments*, eds. Genet, R., Johnson, J., and Wallen, V. Santa Margarita, CA: Collins Foundation Press.

King, W., Waddell, P., and Raptodimos, T. 1985. Optics in engineering measurement. *Proc. of SPIE*, 599, 80.

Manly, Peter L. 1991. *Unusual Telescopes*, 15. Cambridge UK: Cambridge U. Press.

Rowe, David. 2003. FringeXP: A Fringe Analysis Program, Ver. 2. Available at http://www.ceravolo.com/fringeXP.html.

Chapter 25

Spinning the Way to a Perfect Figure
Spin-Cast Polymer Mirrors

Terry R. Richardson, College of Charleston, SC
K. Lisa Brodhacker, Lander University, SC

Introduction

Imagine a telescope mirror manufacturing process in which a two-meter mirror blank is produced, ground to a paraboloid, polished, figured, and readied for aluminizing in a matter of weeks. Such is the allure of spin-cast polymer mirrors. Decades before Borra's progress with liquid mercury mirrors began in the 1980's (Borra 1982), experiments with the spin-casting of plastic mirrors are reported (Hass 1954, Archibald 1957, Schmidt 1966). Investigations with this process report mirrors produced with epoxies as the most likely material for desirable results (Ninomiya 1979).

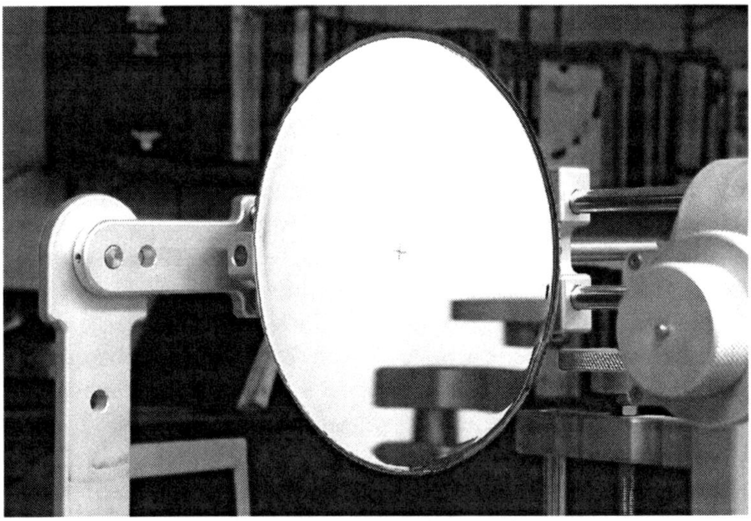

Figure 1: A 0.2 meter spin-cast epoxy mirror in the optics lab for testing.

Borra's liquid mirror telescopes (LMTs) have demonstrated stable optical surface on spinning containers of liquid mercury (Borra 1995), and studies have shown carefully chosen epoxy mixtures solidify with

RMS surface irregularities less than a few percent the wavelength of visible light (Mollenhauer 2002). Given those results, it should be possible for spin-casting to produce optical-quality polymer mirrors without any polishing or figuring being necessary. Given the flexibility of polymeric solids, even mirrors with relatively small figure errors could be corrected by an adjustable mirror support, and, just as with LMTs, telescopes with epoxy mirrors would provide significant light gathering power at low cost yet, unlike the LMTs, they could study any part of the sky.

In the last several decades groups working in the infrared have used spin-casting to produce mirrors used in research. These efforts include the 62 inch resin mirror for a two-micron survey of the sky (Olsson-Steel 1986) and a 1.8 meter off-axis epoxy paraboloid for the submillimeter cosmic microwave background survey (Dragovan 1994).

Epoxy Compared to Glass

A comparison of epoxy mirrors to conventional mirrors composed of borosilicate glass or ceramics reveals the advantages and limitations of the epoxy mirror approach and indicates areas where improvements in polymer characteristics might be directed. The clear advantage in the epoxy mirror approach is both the low cost and ease and speed of production. A 4 meter spin-cast epoxy mirror could be completed in the same amount of time as a 40 centimeter mirror.

Ever since the silvering of glass was perfected, some form of glass has dominated mirror making whether large or small. The chief concern of using glass, especially in the case of large mirrors, is its fragility. A single accident with an 8 meter mirror could idle a major observatory for years. The next generation of telescopes makes this potential liability even greater. Such concerns have revived interest in metal mirrors that are seen as more forgiving than glass. The VLT was first planned with metal mirrors for this very reason (Bely 2003). Epoxy mirrors are robust in this regard and it is easy to imagine a mirror accident that would shatter glass or bend metal yet would leave an epoxy mirror relatively undamaged.

Epoxy, just like metal or glass, is not the ideal material for telescope mirrors primarily because of the thermal behavior of unmodified epoxy. Its thermal diffusivity, a measure of how rapidly the material adjusts to ambient temperatures, and its thermal figure of merit that takes into account the coefficient of thermal expansion (CTE) are both low, and higher values are desired (Bely 2003). This circumstance arises because of the large CTE for epoxy coupled with a large heat capacity that counteracts the advantage provided by the relatively low density of epoxy.

However several methods are available to reduce these undesirable properties. The thermal conductivity of epoxy can be substantially elevated with a conductive additive without significantly affecting other desirable properties. Additionally, studies underway by our group have already demonstrated a reduction in the CTE by the inclusion of components with negative CTE. How far this approach can be taken and whether the method adversely affects other desirable properties is yet to be ascertained. Furthermore the lightweighting of mirrors discussed below also reduces the negative effects of thermal properties by reducing the mass. What seems clear is that the thermal properties of epoxy require some modification if the material is to be viable for mirrors 2 meters and larger.

The density of epoxy is less than half that of other mirror materials, yet owing to a relatively low Young's modulus, unmodified epoxy's structural performance is comparable to glass. Its specific stiffness, the ratio of Young's modulus to the density, is half that of borosilicate glass; however, composite materials utilizing epoxy are two orders of magnitude better in specific stiffness than glass or simple epoxy (Bely 2003). It is possible that a mirror of epoxy built up in layers with the first layer a composite with appropriate reinforcing could produce a strong yet lightweight design.

Without composites it is clear that the lightweighting of epoxy mirrors is necessary to obtain structural rigidity without unnecessary mass; and even composite designs would benefit by a lightweight approach. At first it seems that lightweighting these mirrors could utilize the same method applied to the spin-casting of borosilicate mirror blanks. In that work a system of voids is created with ceramic fiber core boxes that protrude into the glass during casting (Hill 1998). Calculations show that differential thermal shrinkage in epoxy mirrors during cooling from the cure temperature to the operating temperature would result in a print through of the ribbing pattern of the cores. Spin-cast glass mirror blanks are still ground and polished so that any print through is unimportant. Not so with epoxy where the objective is to avoid any polishing. Thus casting a ribbed backing with an epoxy mirror seems problematic.

Two other methods of lightweighting show promise, though either would make the manufacturing process more complex. A lightweight spin-cast facesheet could be glued to a ribbed core and bottom plate formed of fiber reinforced epoxy or other suitable material. This method was adopted for the production of the Hubble Space Telescope primary, though the components were of ultra low expansion glass. An alternative would be to spin-cast a mirror with the appropriate thickness and then remove material by machining from the rear of the mirror, producing a ribbed support in that manner. Given the soft nature of cured epoxy, removing material would be relatively easy and, if there are no significant stresses cured into the body of the mirror, the method should retain the surface figure, though this outcome is yet to be tested. With epoxy, the facesheet of the mirror can be thinner than for glass mirrors. For glass the front must be thick enough to support the grinding and polishing tools without deformation, otherwise the pattern of ribbing would show through on the surface. Since epoxy will not be subjected to the stresses and heating of grinding and polishing, the face of the mirror need be only as thick as necessary to provide sufficient support for the optical surface.

Polymer System Selection Criteria

There is no perfect polymer system available for optical quality spin-cast mirrors. A good system should have as many desirable characteristics as possible. Those characteristics that have been identified as important for epoxy mirrors include a low viscosity and surface tension, a relatively high glass transition temperature, high strength and hardness, low shrinkage on curing, a slow reaction rate, and the lowest possible CTE given the relatively high values associated with most epoxies.

Viscosity and Surface Tension

Viscosity and surface tension act to counter the parabola-forming forces of the spinning process, so that keeping their values as low as possible are necessary for the best surface figure. In addition, complete mixing of the epoxy components is essential for uniform properties throughout the final mirror and low values for viscosity and surface tension aid in mixing. Since epoxy components are relatively larger molecules, careful and complete mixing is difficult but absolutely necessary. Given the viscosity of even the most fluid of epoxies, the addition of a diluent is indicated. Since surface tension usually increases proportionately with viscosity (Richardson 2001), the diluent aids in lowering surface tension as well. The diluent could be a simple solvent or an agent that reacts with and becomes an integral part of the final material. A solvent would create outgassing problems during and after curing and increase shrinkage. For those reasons, a reactive diluent is indicated for use here.

The specific polymer chosen and its curing agent affect the viscosity and surface tension in several ways. Not only is the viscosity of the mixture important, but also the cure temperature. If the system requires an elevated temperature to cure in a reasonable period of time, the process benefits from the lower viscosity that heating provides. There are epoxy systems that require days to gel at 70 °C and their viscosity is significantly lower at the elevated temperature and this situation aids both in mixing and in achieving a satisfactory surface figure.

Figure 2: A 0.6 meter mirror with a central hole cast into the mirror. The edge distortion shows in the reflection of the wall.

Glass Transition Temperature

In a polymerization reaction there are important transformations that occur. The point at which the viscous liquid changes to an elastic gel is the gel point. This gelation is somewhat sudden and irreversible and marks the point at which a network of infinite molecular weight first appears. Sometime following gelation comes vitrification, the point at which the gel becomes a glassy solid and has a maximum crosslink density. If the glassy solid is heated above the glass transition temperature, T_g, the polymer will lose its glasslike properties and assume those similar to a rubber until its temperature is reduced below T_g. This temperature is one of the most important characteristics of a polymer (Stevens 1999) and those polymers with a high T_g have high crosslink density (Nunez 2001) and good strength.

Shrinkage and Polymerization Rate

As a polymerization reaction proceeds there are volume changes that occur in the matrix and these changes induce stresses in the final product that can cause warping and cracking (Schoch 2004, Lange 1997).

It is important to note that any shrinkage in volume prior to gelation induces no stress in the final material (Schoch 2004), and studies by Russell and Lee (1997) have proven that tuning the cure process can eliminate most shrinkage in the thermosetting resin.

During the curing process, the mixture undergoes a thermosetting reaction in which the monomers are converted into a cross-linked matrix. A cure study we performed has revealed that the gel point of the system can be postponed for days, allowing the matrix to undergo shrinkage without inducing stress. If the gel point can be avoided long enough, only a small amount of residual stress will be present in the final material. Given these considerations, a low Polymerization Rate was given the highest priority in our selection criteria.

Coefficient of Thermal Expansion

As discussed previously, thermal qualities are a serious potential liability of epoxy mirrors; however if the thermal conductivity can be enhanced by a highly conductive fill agent, the CTE of the polymer system alone has little significance and therefore the CTE was given minor importance in resin selection.

A Specific Polymer System

Given the considerations just discussed, the resin system consisting of diglycidyl ether of bisphenol A (DGEBA) epoxy with 4,4-methylen-bis-3-chloro-2,6-diethylaniline (MCDEA) as the curing agent is a reasonable selection. Most commercial resins are a mixture of monomers. For consistent and uniform mirror properties we purchase purified reagents with a more uniform molecular weight. In addition, the reactive diluent ED757 (cyclohexane dimethanol diglycidyl ether) is used to lower the viscosity of the mix. This polymer system requires up to 100 hours at 70 °C to reach gelation, thus allowing enough time for thorough mixing before pouring a layer of the mirror. The T_g for this combination is 137.9 °C.

Mirror Casting

Spin-casting an epoxy mirror is relatively simple in principle but the details are important. Our method employs equipment modeled after the setup used for LMTs (Borra 2003). There is an air bearing supporting a steel platter. The air bearing is driven by a stepper motor and controller and is held by a triangular frame resting on three cast concrete supports. Leveling mechanisms are installed where the triangular frame meets the supports.

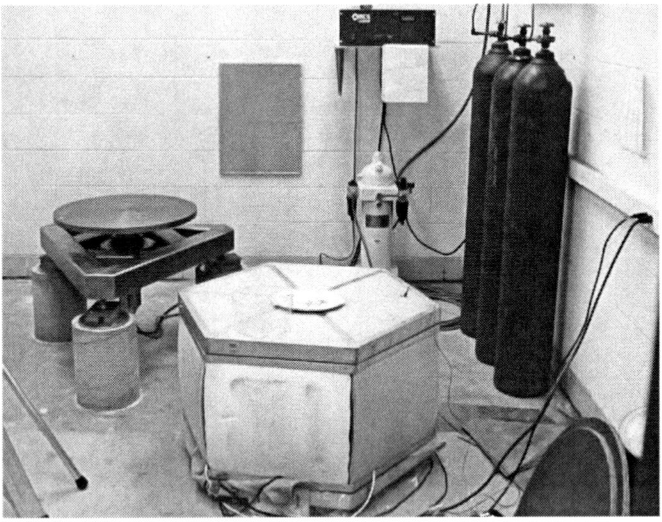

Figure 3: Lab for producing 0.6 meter mirrors showing (clockwise from left) the stainless steel platter supported by a precision air bearing and held by a triangular frame, the nitrogen tanks in case of air loss from the compressor and on the floor the oven that has been removed from the spinner for maintenance.

Because gelation and vitrification occur above ambient temperatures, the platter and container of spinning epoxy are enclosed in an oven heated by electrical resistance and modulated by thermostatic controls. The mixing is done at the gelation temperature and since the mirror is built up from successively smaller pours of epoxy, the oven has a small opening in the top for adding epoxy while maintaining the temperature.

The mixing must be thorough and can require some time for chemicals such as those used in mirrors. The polymer system chosen permits a mixing time of 25 hours or longer. While mixing proceeds, degassing is accomplished by reducing the pressure above the liquid with a vacuum pump. After mixing, each batch of resin is added to the spinning container and allowed to reach gelation as determined by the amount of time at 70 °C. Then the oven is raised to 120 °C until complete vitrification occurs. After that the temperature is raised above the glass transition temperature and held there for a short time to relieve chemically induced stresses that occur during the reaction. The temperature is then reduced to 70 °C and allowed to stabilize in preparation for pouring the next layer.

Figure 4: Triangular frame for a 2 meter spinner resting on concrete supports.

Since there are forces on a liquid at the point of contact with its container, the gravity-centripetal acceleration balance that gives rise to the paraboloid shape of the surface is disturbed at the edge of the container. Though we have experimented with slanting the container sides to reduce the effect, very little remedy is achieved with this approach and we have concluded the solution of masking off the outer 2 centimeters of the mirror radius to be the most practical solution.

Epoxy has a relatively large CTE and that quality comes in to play in casting a mirror. Since the polymer system used needs a temperature of about 120 °C for vitrification and since the material is a rigid solid of fixed shape when below that temperature, there is considerable thermal shrinkage as the mirror is brought to room temperature. If the shape of the disk includes a flat bottom, the center, because of the sagitta of the mirror, will contract less on cooling than the edge. In even modest size mirrors this differential shrinkage amounts to tens of wavelengths of visible light. Therefore it is clear to minimize this effect, the bottom must have a curvature similar to the top.

To this end the container could have an insert of the correct curvature or be filled with a liquid of greater density than the polymer to provide a paraboloid bottom. A combination of these two approaches seems even better. A paraboloid bottom of epoxy could be spin-cast into the container. After curing, a relatively thin layer of liquid such as an ionic liquid, mercury or a bismuth casting alloy which melts below the gelation temperature could serve a uniform surface for spin-casting the mirror. Experiments with metal have shown that the thermal expansion of the metal as the oven is raised to T_g is a problem. The expansion of the metal can damage the figure of the mirror, so that a means of relieving this effect must be provided. Research into this approach is ongoing.

Figure 5: One of the authors attaching the support arms for the spinner while the panels for the oven enclosure can be seen leaning against the wall on the left.

We have conducted a number of experiments with custom containers of various materials from metal lined with a plastic coating to containers machined from a solid block of polymer material. With the idea of producing mirrors of a meter or more, off the shelf containers of polyethylene, polypropylene, and Teflon FEP now seem the most practical.

Conclusion

Though there is yet no optically perfect mirror for visible light, one seems possible in the next few years. We have upgraded our existing air bearing and platter to cast 0.6 meter mirrors and are planning a telescope to demonstrate the quality of a mirror this size. Also, we are presently completing a lab with a second larger air bearing and platter that will permit the manufacture of mirrors in the two meter class. At the present time epoxy mirror production is an active area of research with some challenges to be overcome before the production of a 2 meter optical mirror can be realized. Given the results of our investigations to date, we believe the future of these mirrors is promising.

Author's Note

We would like to acknowledge the generous support this work has received from the following sources: College of Charleston Faculty Research and Development Fund; Lander Foundation Faculty Grant Fund; NASA, Jet Propulsion Laboratory; South Carolina Space Grant Consortium.

References

Archibald, P. B. 1957. A method for manufacturing parabolic mirrors. *Journal of Solar Energy Science and Engineering*, 1, 102.

Bely, P. Y. 2003. *The Design and Construction of Large Optical Telescopes*, New York: Springer.

Borra, E. F. 1995. Liquid mirrors. *Can. J. Phys.* 73, 109.

Borra, E. F. 1982. The liquid-mirror telescope as a viable astronomical tool. *JRASC*, 76, 245.

Dragovan, M. and Alvarez, D. 1994. Making a mirror by spinning a liquid. *Sci. Am.*, 270, 116.

Hass, E. and Erbe, W. 1954. Method for producing replica mirrors with high quality surfaces. *JOSA*, 44, 669.

Hill, J. M., Angel, J. R. P., Lutz, R. D., Olbert, B. H., and Strittmatter, P. A. 1998. Casting the first 8.4 meter borosilicate honeycomb mirror for the large binocular telescope. *SPIE Proc. On Advanced Technology Optical/IR Telescopes IV*, ed. L. M. Strepp, 3352, 172.

Lange, J., Toll, S., Manson, J. A. E., and Hult, A. 1997. Residual stress build-up in thermoset films cured below their ultimate glass transition temperature. *Polymer*, 38, 809

Mollenhauer, D. and Camping, J. 2002. Multilayered polymer mirror experiment. *Journal of Spacecraft and Rockets*, 39, 691-694.

Ninomiya, Y. 1979. Parabolic mirror made by the rotation method: its fabrication and defects. *Applied Optics*, 18, 1835.

Nunez, L., Fraga, F., Castro, A., Nunez, M. R., and Villanueva, M. 2001. TTT cure diagram for an epoxy system diglycidyl ether of bisphenol A/1,2 diamine cyclohexane/calcium carbonate filler. *Polymer*, 42, 3581.

Olsson-Steel, D. 1986. A note on the history of the liquid mirror telescope. *JRASA*, 80, 128.

Richardson, R. L. and Griffiths, P. R. 2001. Generation of front-surface low-mass epoxy-composite mirrors by spin-casting. *Opt. Eng.*, 40 (2), 252.

Russell, J. D. and Lee, A. Y., 1997 Effect of cure cycle on the specific volume of 3501-6 epoxy. *29th International SAMPE Technical Conference*. October28-November 1, 289.

Schmidt, F. J. 1966. Electroforming of large mirrors. *Applied Optics*, 5, 719.

Schoch, Jr. K. F., Panackal, P. A., and Frank, P. P. 2004. Real-time measurement of resin shrinkage during cure. *Thermochimica Acta, 417,* 115.
Stevens M. P. 1999. *Polymer Chemistry*. New York: Oxford University Press.

Lightweight Mirrors

Allan Keller, Floyd Fiechtner, Rick Hedrick, Russ Genet, and Jo Haberman stand beside the 1 meter mirror from DOTI Optical. The "marks" on the mirror are primarily a reflection and some dust from unpacking the mirror.

Joe Haberman evaluates the performance of the DOTI Optical 1 meter spherical mirror at its radius of curvature some 27 feet from the mirror. Russ smiles for the camera. The Ronchi and knife edge tests were made from the second story balcony in the open bay telescope assembly area at PlaneWave Instruments.

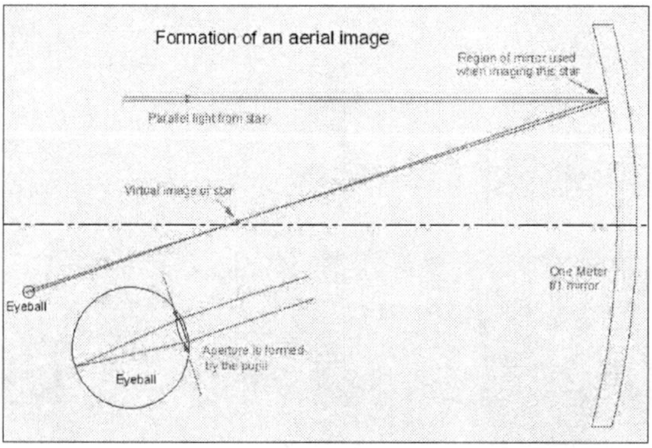

Dave Rowe made this diagram to illustrate why stars can appear sharp to the eye (or a camera with a lens) when one looks at their reflection from the sky. They will appear sharp as long as the "local" areas on the mirror are smooth even if the overall figure of the mirror is very poor.

The Instron is a computer controlled testing machine that measures the deflection of materials through a range of pressures. The results provide valuable information on the performance of the materials. A piece of Pittsburgh Corning Foamglas is under test here with a glass top plate and sealant material on the bottom. The tests were made at California Polytechnic State University in San Luis Obispo under the supervision of John Ridgely, a professor in the Mechanical Engineering Department.

Andrew Aurigema's vacuum chamber. He can aluminize and overcoat mirrors up to 60 inches in diameter. A very fast mirror, just coated, can be seen at the left fastened to the chamber's door. The vacuum chamber resides in Andrew's airplane hanger about 25 miles north of Cape Kennedy in Florida.

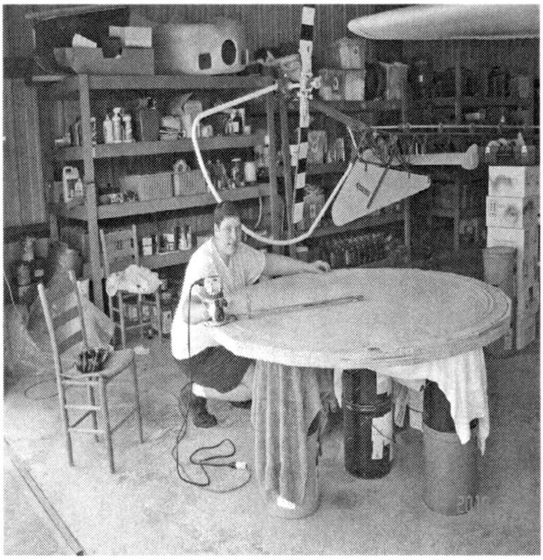

Andrew Aurigema working on the turntable for his large automatic mirror making machine he will be using to grind and polish mirrors up to 60 inches in diameter. When this picture was taken, he was also putting the finishing touches on a large CNC machine for milling out foam glass mirror backs. Directly behind Andrew is the tail rotor of his helicopter, and upper right is the right wing of his Cessna 172 aircraft. His telescopic activities are gradually crowding out his aviation toys.

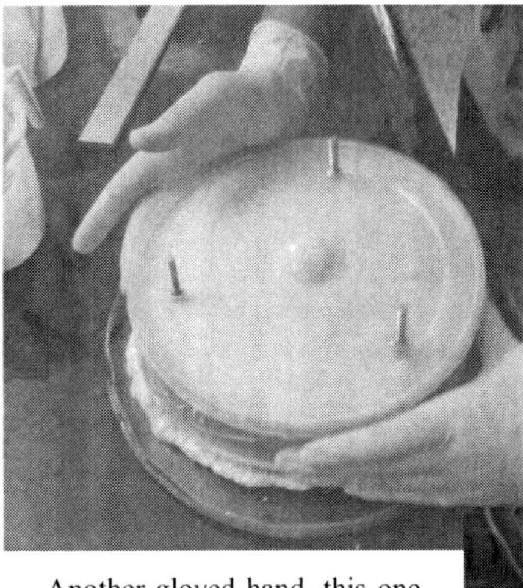

Kiran Shah's gloved hands hold a small polyurethane replica mirror. The mirror is upside down, with the bolts to fasten down the mirror sticking up. The front surface of the mirror (on the bottom) has been pressed up against a convex master.

Another gloved hand, this one Andrew Aurigema's, holds one of his milled mirror backers. Andrew's mirrors are made from "Starstone," a lightweight foam glass material. The front glass plate is fused to the foam glass in a kiln. Large Starstone mirrors up to 60 inches in diameter are in the works.

Terry Richardson holds a small mirror made by spinning an epoxy solution at a constant speed. The epoxy hardens into a parabolic surface which can be aluminized and over-coated without any grinding, polishing, or figuring. Lisa Brodhacker, a chemist with a keen interest in astronomy, developed the special epoxy material. Terry and Lisa are putting the finishing touches on an air bearing machine to spin-cast 2-meter epoxy mirrors. The reflected camera, which took the picture, is held by Russ Genet.

Part IV

INSTRUMENTS, TECHNIQUES, AND SOFTWARE

Chapter 26

The Bath Interferometer

Vladimir Galogaza

Physicist

Introduction

The Bath interferometer is one of many possible interferometer variations. It is special for its simplicity, which qualifies it as a basic metrology tool in evaluation of telescope optics for amateur telescope makers. While this field is still dominated by the Foucault test, recent developments in data analysis methods available to amateurs could change that trend. Interferometry provides full quantitative data about primary mirrors, and the Bath interferometer is well within the financial and craftsmanship capabilities of amateurs.

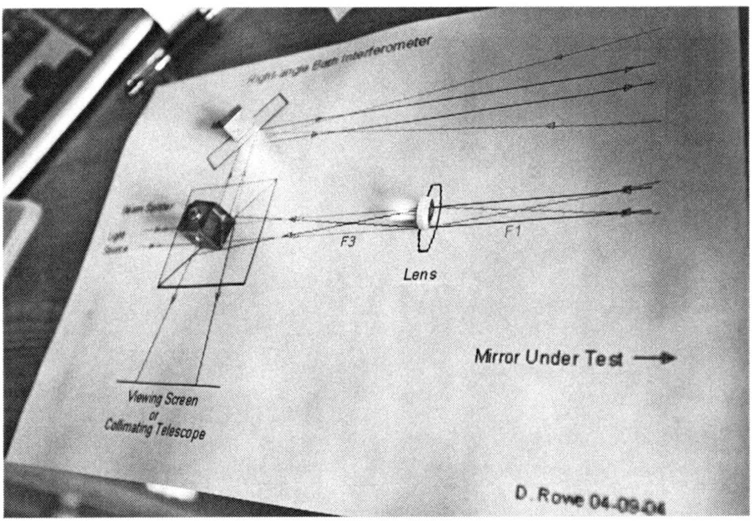

Figure 1: Essential optical parts of the Bath interferometer are a non-polarizing 50:50 ratio (cubic) beam splitter, biconvex diverging lens, and front surface flat mirror, here placed over a Bath interferometer drawing from Interferometry wiki: http://starryridge.com/mediawiki-1.9.1/index.php?title=Main_Page.

Figure 2: This figure shows a complete working conventional Bath interferometer with light source (red diode laser pointer) and camera, in this case a webcam. Squares on the graph paper are 1 cm to stress its small size and compactness.

Figure 3: The most advanced version of the Bath interferometer, the PSI-Bath, uses the Phase Shift Interferometry technique (PSI). To the conventional Bath configuration are added two linear polarizers, the first one between the laser and beam splitter cube, and the second one between the quarter wave plate and camera (this one must be rotated to several discrete angles). The beam splitter cube of the PSI-Bath interferometer is of the polarizing variety, i.e., the two output beams are mutually orthogonally polarized.

Sample interferograms, obtained from a Bath interferometer, are presented in the following pictures.

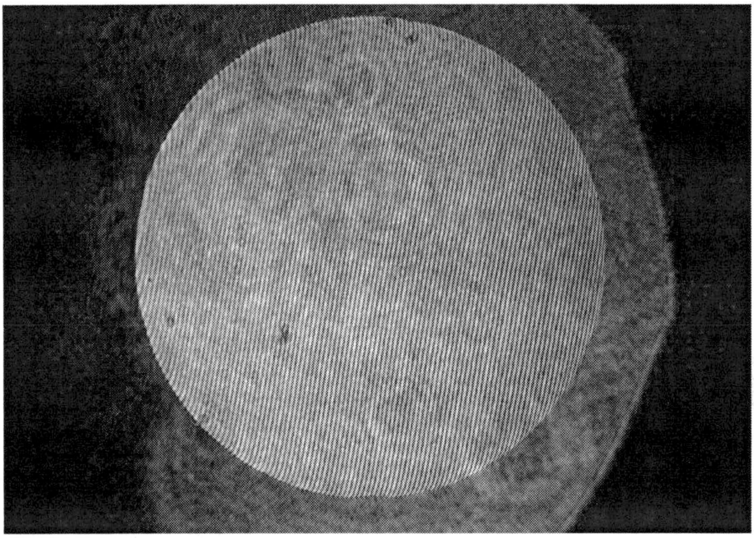

Figure 4: This multi-fringe interferogram with a high degree of tilt was made with a Bath interferometer (DSLR camera, 6 Mpixels, green diode laser pointer).

Figure 5: Another shot with few fringes (DSLR camera, 6 Mpixels, green diode laser pointer).

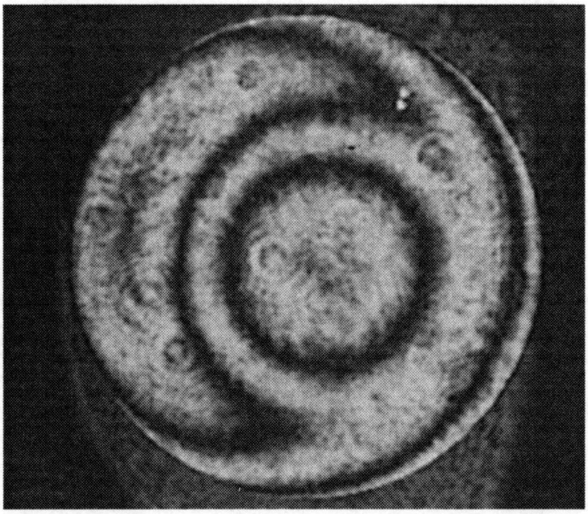

Figure 6: This interferogram has only several fringes. Note the noisy background, from dust, scratches and imperfections in light source, and other components. The presence of such large amounts of noise does not prevent the analysis of this interferogram (it was analyzed using the PSI technique), demonstrating the robustness of the analysis method and how forgiving the Bath Interferometer is to disturbances of this kind (Webcam, 0.5Mpixels, red diode laser pointer).

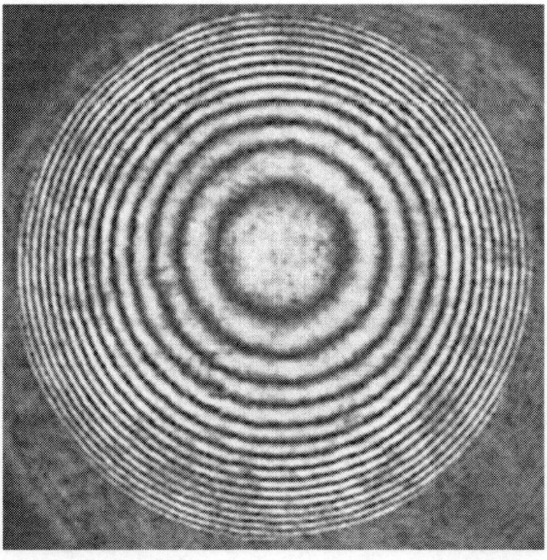

Figure 7: This interferogram has negligible tilt and many fringes of defocus (Webcam, 0.5 Mpixels, red diode laser pointer).

Data Analysis Example

When interferograms are analyzed, the results are similar to the following, where *Wyant* indicates the number of the Zernike polynomial. According to Wyant, *n* is the highest order of the radial part of the polynomial, *m* is frequency of the angular part, *t* indicates that the angular part is sine, cosine or neither, and *Name* is the verbal description of the aberration represented by an individual polynomial. In this presentation the actual aberration of the mirror under test is the weighted sum of Zernike polynomials. Weights are calculated in the analysis and tabulated as *NETCOEFF*, the Zernike coefficient of the ortho-normal polynomial.

Table 1: The results of analysis of a typical interferogram are presented as a list of the first 15 Zernike coefficients, followed by the wavefront P-V and RMS errors, and the Strehl ratio. Analysis provides 64 coefficients by default, but the number can be adjusted.

	Wyant	nmt	NETCOEFF	Name
1	Z0	00 n	1.98704	piston
2	Z1	11 c	-1.54267	x-tilt
3	Z2	11 s	3.60639	y-tilt
4	Z3	20 n	0.08835	defocus
5	Z4	22 c	-0.00395	x-astig. (corrected)
6	Z5	22 s	-0.02659	y-astig.
7	Z6	31 c	0.00148	x-coma
8	Z7	31 s	0.01242	y-coma
9	Z8	40 n	0.04142	prim. SA (corrected)
10	Z9	33 c	0.00387	x-trefoil
11	Z10	33 s	0.00720	y-trefoil
12	Z11	42 c	-0.00332	x-sec. ast.
13	Z12	42 s	0.00519	y-sec. ast.
14	Z13	51 c	0.00170	x-sec. coma
15	Z14	51 s	-0.00235	y-sec. coma
16	Z15	60 n	-0.05769	sec. SA (corrected)
17	- -		532	wavelength [nm]
18	- -		946.8	radius [pixels]
19	- -		0.0879	rms [waves]
20	- -		0.737	Strehl ratio
21	- -		0.722	P-V [waves]

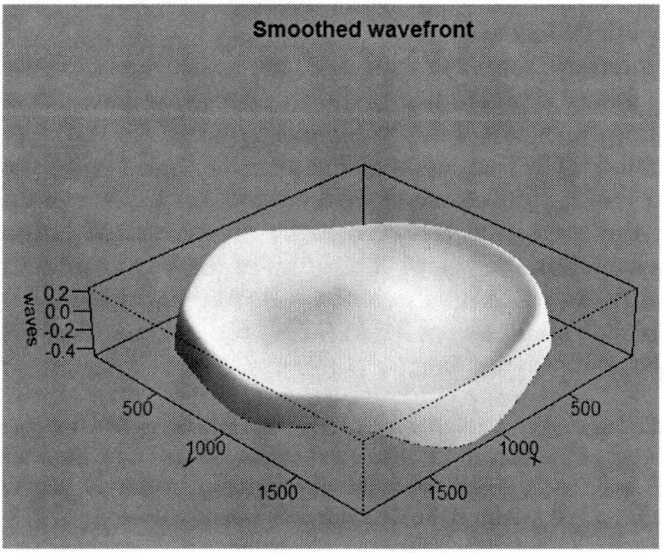

Figure 8: Another way to present analysis results is the 3D graph of the aberrated wavefront of the tested mirror.

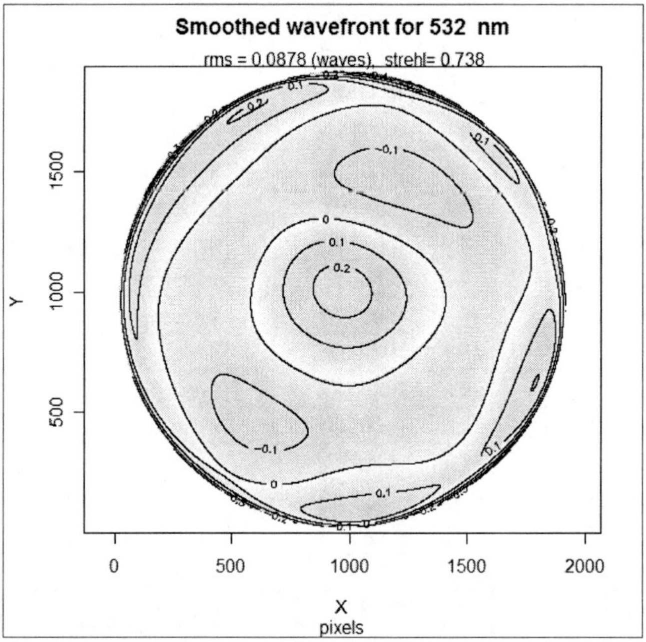

Figure 9: This figure shows a wavefront map with contours. On a computer screen, it would appear color coded.

Most Bath interferometers use laser pointers (green or red) or a HeNe laser as a light source. Although white light can be used, lasers are today widely available, acceptably priced, stable, handy, and with sufficient power. A 1 mW laser is more than bright enough for the Bath, and a 5 mW green laser pointer enables wall projection of interferograms in a lighted room.

The Bath interferometer position must be adjusted relative to the mirror under test so that light reflected from the mirror re-enters the interferometer. Common practice is to mount the interferometer on an x-, y-, z- translational stage. Convention is that the z axis is the optical axis of the mirror under test. Moving the interferometer along the z-axis is equivalent to the defocus adjustment, while movement along x- and y-axes adjusts the tilt. By adjusting tilt and defocus, the number and shape of fringes are easily controlled.

The Bath interferometer belongs to so-called common path interferometers. Reference and test light beams trace approximately the same path. This makes the interferometer less sensitive to vibrations. Images of interferograms (Figure 4 to Figure 7) were made in an apartment in a multistory building during the day. The only precaution against vibrations was placing the test rig on 3 inner tubes from the tires of a wheel cart. Thermal air currents were reduced by a simple shroud.

For basic Bath interferogram analysis, a program, FringeXP, was developed by David Rowe, available at Peter Ceravolo's site: http://www.ceravolo.com/fringe_analysis.php. Building on FringeXP, Dale Eason wrote the analysis program OpenFringe available at http://sourceforge.net/projects/openfringe. It incorporates several analysis methods. The simplest is manual fringe tracing, and the most sophisticated is the Vortex method by Kieran Larkin and Stephen Koehler. The OpenFringe program leaves little to be desired. It has interferogram simulation, many different graphical outputs, and much more.

Analysis of PSI-Bath data is done by a program written by Michael Peck, available at: http://home.earthlink.net/~mlpeck54/astro/astro.html and also at http://wildlife-pix.com/rpackages/.

Some common characteristics of the aforementioned programs are that all are free, and their authors are alive, well, available, and willing to help interferometry newcomers.

With a working Bath interferometer, data acquisition, reduction, and presentation of results require only a couple of minutes. This makes feasible the enhancement of accuracy, as needed, by averaging the results of many individual interferograms. At this time there is no other measuring method suitable for ATMs providing 2D quantitative test results of comparable quality.

There is also a Wiki on interferometry containing detailed information for the practical construction and use of interferometers, especially the Bath: http://starryridge.com/mediawiki-1.9.1/index.php?title=Main_Page

Principles of Operation of the Bath Interferometer

Interferometry is the technique of measuring the optical path difference of two wavefronts, one of known shape, i.e., the reference, and the other with the imprinted aberration of the optics under test. The two wavefronts are brought together, and the resulting interference fringes are recorded and analyzed to extract the aberration in the optics. To enable recording, interference must be sufficiently long lasting and spatially fixed. This is easily achieved if interfering wavefronts originate from a source of highly coherent light and if the path length difference is small. In 1973, Karl-Ludwig Bath invented a method to do this in a particularly simple way.

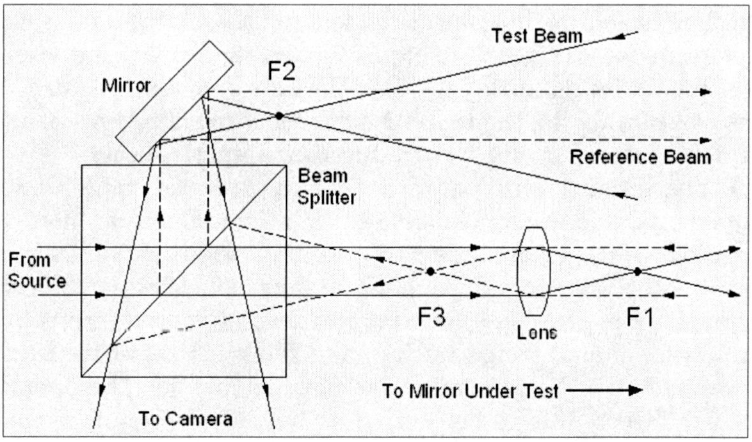

Figure 10: The layout of the right-angle Bath Interferometer.

The light from the light source, entering from the left in Figure 10, is divided into two beams of approximately equal intensity by the beam splitter. One beam, called the test beam (solid lines), is directed through the biconvex lens. The light focused at F1 spreads into a diverging cone, illuminating the mirror under test. For this to happen, a simple design parameter must be satisfied, i.e., the laser beam width and diverging lens focal length must be such to ensure the complete illumination of the mirror under test. The mirror under test is placed at a distance of its radius of curvature (ROC) from the diverging lens focus. Thereby, reflected light comes to focus near the diverging lens (at F2) and is reflected from the flat mirror, redirecting it into the splitter and out to the camera. The other beam, the reference beam (dashed lines), is redirected by the same flat mirror toward the mirror under test, reflected back toward the diverging lens, coming to a focus at F3, from where it continues as a diverging light cone, representing a wavefront of a known shape, i.e., a spherical wave. It enters the splitter and is reflected to the camera. At the camera sensor, the two beams interfere and an interference pattern, like those in Figures 4-7, is recorded.

Fringes are locations of constant path differences. Centers of dark fringes are where the two waves are out of phase. Aberrations of the test wavefront are thus encoded in the fringe shapes. The Bath interferometer, as described herein, inherently suffers from two instrumental errors: instrumental astigmatism because of the lateral shift of incoming and reflected beams, and instrumental spherical aberration caused by measuring conics at the ROC. They are easily calculated and numerically subtracted. The beam cross-section at the interferometer is small (typically one or two millimeters), consequently the necessary components are of a similar small size. Beam splitter cube size need not be larger than 5 mm on a side, but the typical size used is 10 to 15 mm.

The Phase Shifting Bath Interferometer (PSI-Bath)

The original, conventional Bath interferometer provides one interferogram for analysis, and all the required information has to be mined from it. The information is encoded in the shape of the fringes and the most basic analysis method looks for fringe peaks and/or valleys (obtained by fringe tracing). Therefore, the analysis results describe the surface of the mirror under test at a limited number of measured points along the fringe centers. Between these points, the surface is estimated based on the assumption that the surface between measured points does not deviate from expectations; in other words, it is smooth in a mathematical sense. Steps, bumps, and other irregularities will remain unobserved.

Another method is Phase Shift Interferometry (PSI), where each imaged surface point of the mirror is considered. This is a huge advantage, and modern commercial interferometers use this method. Fortunately, it is possible to convert the Bath interferometer to this mode of operation without excessive investment or difficult requirements. PSI interferometers use several interferograms for a single analysis, each interferogram having a different phase shift between reference and test wavefronts. With these multiple images at each point, the wavefront phase can be computed at each location on the mirror. The required phase shift is usually obtained by changing the optical path of one beam, for example, by translating the reference mirror. This can be done only if the reference and test beams traverse different paths or components. The Bath interferometer is a common path type and moving any part of it is sensed equally by both beams. Therefore, there will be no relative (net) phase shift between them. Another method to introduce phase shift, suitable for common path interferometers, is to decouple the test and reference beams by making them orthogonally polarized. In this case the beams travel along the same path, but are orthogonal and noninteracting. In other words, they do not "see" each other. At the end, they have to interact (interfere) with each other, so the polarizations of both beams must be made the same. The polarization diversity and interferometer geometry is shown in the following drawing (see Figure 11).

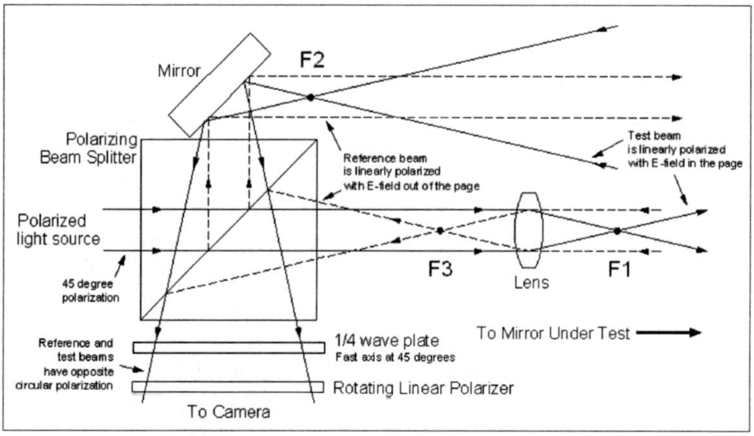

Figure 11: The layout of the PSI-Bath Interferometer.

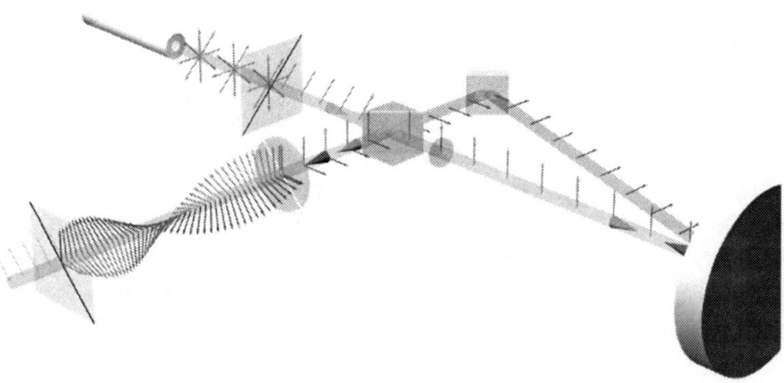

Figure 12: A 3D view of the PSI-Bath.

A diode laser light source is only partially polarized. To increase its polarization, the light is first passed through the polarizer, usually Polaroid film, oriented so that output polarization is at 45 degrees with respect to the cube beam splitter (in other words, along the diagonal of the input cube face). The beam splitter cube must be of the polarizing kind, where the light passing straight through is polarized along one cube edge (vertically) and the light reflected at 90 degrees is polarized along the other edge direction (horizontally). The two emerging beams are mutually orthogonally polarized. In vector parlance, each beam has only a very small, residual component of light in the polarization direction of the other beam. The reflection of the beams in the flat mirror, from the mirror under test, and refraction through the diverging lens do not change the state of polarization, so they emerge from the beam splitter in the direction of the camera, orthogonally polarized.

From the splitter, the beams are transmitted through a birefringent plate. The speed of the light through the plate is different for the two directions of polarization. If the light is polarized along the "fast axis" of the plate, it has greater speed than if it is polarized along the "slow axis", which is 90 degrees relative to the fast axis. Light traveling along the slow axis is retarded. Therefore, the plate is also called a retarder plate.

If the polarization direction of the incoming light is between the fast and slow axes of the retarder plate, the vector component of the electric field along the fast axis emerges from the plate sooner than the component along the slow axis (which is retarded). When the two components of the sinusoidally oscillating electric fields, mutually phase shifted in proportion to the amount of retardation, emerge from the plate, the resulting polarization vector is changed. The light is said to be circularly polarized in the special case when the phase of one component of polarization is 90 degrees relative to the other, which is equivalent to one quarter of the wavelength. In this case, the birefringent plate is called a Quarter Wave Plate (QWP).

As discussed, the reference and test beams enter the QWP with orthogonal polarizations. The QWP is oriented so that its fast axis is at 45 degrees to both input beam polarizations. Because of this, the reference beam emerging from the QWP is circularly polarized in one direction, for example clockwise, while the test beam is circularly polarized in the opposite direction.

All this was done to enable the introduction of a variable phase shift between them, which is accomplished by the insertion of the (final) linear polarizer into the beam path. The linear polarizer will pass only the components of the electric vectors of both rotating beams which are aligned with its polarization direction.

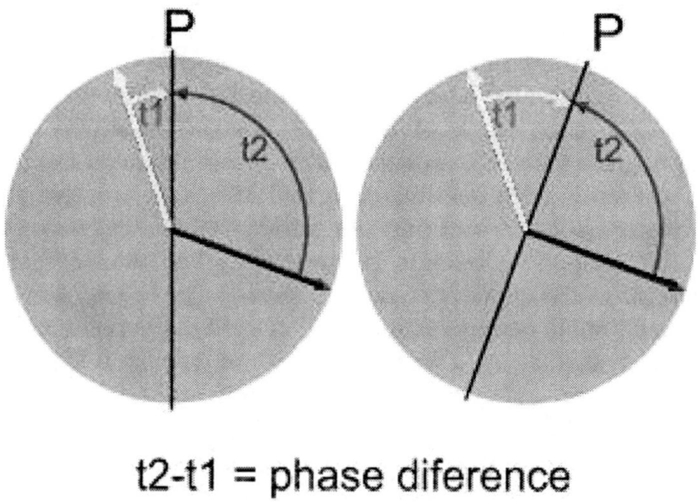

Figure 13: In the two examples shown, the reference vector is counter-clockwise from P and the test vector is clockwise from P. The final polarization direction is marked by P.

If for clarity we observe only the maximum values of the electric field (as all other values in a sinusoidal oscillation behave in the same way), then in the left part of Figure 13, it will take the reference vector rotating vector times t1 to reach the polarization direction P, when it can emerge from the polarizer. The test beam vector will take longer, t2, and the two emerging sinusoidal waves will emerge from the polarizer mutually phase shifted by an amount that is proportional to t2-t1. If the final polarizer is rotated to another direction, as in the right-hand image of Figure 13, t1 will be longer in proportion to this angle, and t2 shorter by the same amount. The resulting phase difference introduced by rotating the final polarizer by angle phi will be 2*phi. Beams leaving the final polarizer are now aligned in the same polarization direction and will interfere at the detector, usually the sensor in a digital camera, producing the desired interferogram.

Phase shift interferometry requires several interferograms, each one phase shifted relative to the others, so that the total phase shifted range for the collection of interferograms is 2*pi or 360 degrees. If, for example, we decide to phase shift each interferogram 90 degrees, then four interferograms will suffice to cover the full 360 degrees, or one full cycle. The final polarizer should be rotated by 45 degrees for each step. After each step, the fringes will be displaced on the interferogram by a certain amount proportional to the phase shift. In the previous example of 90 degree steps, the fringes will move a quarter of the distance between two successive fringes.

The intensity distribution across one fringe is sinusoidal. Consequently, the intensity change of each pixel in the interferogram is also sinusoidal. This fact is used by the PSI algorithm in fitting the intensities of all pixels in order to deduce the shape of the test wavefront relative to reference wavefront. The reference wavefront is usually constructed to be spherical.

This was a reduced and simplified description. In practice, we are dealing with practical approximations of theoretical assumptions. In fact, the intensity of light on each pixel is given not only by the interference of pure wavefronts. There is additional noise in the form of shadows from dust, scratches on the optics, vibrations, thermal air currents, imperfect polarizers and splitter cubes, etc. The beauty of PSI analysis is that in the process of calculation, many of the mentioned effects are eliminated or reduced. Consequently, the interferometer components need not be of exceptional quality, allowing amateurs to build high-accuracy interferometers and use powerful interferometric analysis methods and programs.

There are some limitations to the application of Bath and PSI-Bath interferometers. At the time of writing this chapter, they continue to be investigated, analyzed, and improved. One limitation is the size of the mirror and its F-ratio. A specific problem for PSI-Bath, relative to the conventional non-PSI Bath interferometer, is in the use of an HeNe laser. This gas laser, especially its inexpensive versions, has a problem with beam intensity and polarization direction stability. Since PSI-Bath is sensitive to input beam polarization, inexpensive HeNe lasers may not be suitable. However, common diode lasers, available in various wavelengths and output powers, are more than adequate and easier to use than gas lasers.

Those interested in ATM Interferometry, or aspects of it, can look for more information and ongoing discussions at the following URL: http://tech.groups.yahoo.com/group/interferometry/messages.

Chapter 27

Lucky Imaging for High Resolution Measurements of Double Stars

Rainer Anton

Altenholz/Kiel, Germany

Introduction

Lucky imaging is an alternative (or even supplemental) method to adaptive optics that increases image quality by reducing seeing effects. By careful selection and superposition of the "lucky" frames recorded with highly sensitive and fast CCD cameras, the resolution can be pushed to near the theoretical limit even under non-optimum seeing conditions. It is demonstrated, with some representative examples, that diffraction limited images can be almost routinely obtained with modest amateur telescopes with error margins of position measurements well below 0.05 arc seconds. Besides the seeing, this depends more on the resolution in the image than of the telescope. Another benefit of direct imaging is that brightness differences of the components can be measured, which is especially interesting for pairs with contrasting colors.

The term "lucky imaging" refers to moments of good seeing, which can be captured in short exposure recordings. A number of the best frames out of a longer series are registered and stacked, which results in a significant improvement of image quality and resolution over single exposures. Not only are seeing effects reduced, but so is image noise. This technique is widely used by amateurs and professionals on all kinds of celestial objects, including planets and double stars. Because the seeing usually varies at high frequency, exposure times have to be short, typically in the range of milliseconds. However, even under good seeing conditions, the "lucky" image is virtually never perfect. Therefore, one has to carefully select the better ones out of many bad ones, and stack them in order to average out residual image distortions.

The effect is a significant reduction of the seeing disc compared with that resulting from longer exposures, which is rarely smaller than 1 arc second—larger than the Airy diffraction disc of modest amateur telescopes. As an example, for an aperture of 20 cm (~ 8 inches), the full-width at half-maximum (FWHM) is 0.57 arc seconds, at a wavelength of 550 nm. For resolving double stars, Rayleigh's or Dawes limits of the separation, with dips of 2.5% or 3% between the peaks, would be 0.69 or 0.58 arc seconds, respectively. These limits are somewhat arbitrary, and can in principle be exceeded, both by visual observers and by imaging techniques.

As a double star observer, one has to discern between the splitting power of a telescope and the accuracy of position measurements. For the latter, most important is the definition of the centroids of the seeing discs of the components. Provided that both are symmetric, or at least similar, this is given by the resolution in the image, i.e. the size of the pixels of the CCD camera in relation to the focal length of the telescope. In this respect, relatively long exposure times help in averaging image distortions by seeing effects, in order to obtain well defined peak shapes. The improvement by lucky imaging is given by smaller errors in position measurements, due to inherently sharper peaks. In particular, measurements of close pairs benefit from this, as the separation limits by peak overlapping will be reduced. It will be shown in the following that the accuracy can be better than expected from average seeing conditions, as well as better than the theoretical resolution limits.

Equipment

Telescopes

So far, I have used telescopes with apertures from 20 cm, up to 50 cm. My Newtonian telescope at home is a 10 inch *f*/6. During several stays in Namibia, I have used SCT's from 20 cm to 35 cm and, more recently, two Cassegrain telescopes with apertures 40 cm and 50 cm, and focal lengths of 6.3 m and 4.5 m, respectively. The latter are located at the International Amateur Sternwarte (IAS) in Namibia, with generally good seeing conditions. In most cases, I use a Barlow lens to double the focal length, and increase the resolution of the image. This, however, requires insertion of a band filter to reduce the chromatic aberration of the lens. As a by-product, the sensitivity is reduced by about 2 magnitudes. Four filters (R, G, B, and NIR) are mounted on a wheel in a homemade combination with the Barlow. The R-G-B filters are of dichroic type (Edmund Scientific Co.), transmitting above 600 nm (red), between 500 and 575 nm (green), and below 500 nm (blue), respectively. The range of the NIR filter was above 700 nm. The filters are also used to produce RGB composite images. As a small benefit, red or NIR filters help in reducing seeing effects.

Camera

In 1995 I started recording double stars with a low cost CCD video module on tape. This camera was designed for surveillance at infrared wavelengths, and was very sensitive. With my 10 inch Newtonian, I could detect stars up to 10th magnitude. Disadvantages were the fixed image frequency of 25 Hz, and the relatively poor image quality as compared with today's standards. Image noise, however, could be partly compensated by stacking many frames.

This module was replaced in 2000 by the STV CCD camera from SBIG (Santa Barbara Instrument Group), which offered, besides many other interesting features, variable exposure times between 1 millisecond and 10 minutes, and was well suited for lucky imaging. It is also very sensitive, with pixel size of 7.4 μm square. However, fast recording was only possible via the video output. Double star measurements obtained with this camera were described in articles in Sky & Telescope and in the Journal of Double Star Observing (Anton 2002, Anton 2004, Anton 2006).

Meanwhile, fully digital and affordable CCD cameras appeared on the market, such as the popular webcams, which, with ever increasing storage capacities of computers, allow for almost loss-free transfer of huge amounts of image data at high speed. A very reasonable compromise between sensitivity, resolution, and price is the DMK camera from The Imaging Source (TIS). I am using the black and white version DMK21AF04, which is connected via a firewire interface to my notebook. Images are stored as uncompressed bitmaps at frequencies up to 30/sec. Although the internal grey level resolution is 10 bits, it is reduced in the output to 8 bits. Exposure times can be set by the accompanying software between 0.1 milliseconds and 10 minutes, although the chip is not cooled. The chip contains 640x480 square 5.6 μm pixels. The field of view can be reduced by setting a region of interest, which saves a lot of hard disc space and increases the image transfer rate. As the pixels are smaller than in the STV camera, the sensitivity and dynamic range may be somewhat less, but this seems to be more than compensated for by the electronics. Also, the smaller pixel size increases the resolution. This is especially beneficial when using telescopes with short focal lengths. In Table 1, resolution values are listed for the combination of the DMK camera with my Newtonian at home, a Schmidt-Cassegrain, and two Cassegrain telescopes, which I have recently used in Namibia. These values are not calculated from the focal lengths, as this would not be accurate enough. Rather, more exact calibrations were obtained in an iterative way by measuring systems with well known and predictable separations. This procedure has been described in earlier articles in the Journal of Double Star Observations, and is mentioned below in Section III (Anton 2004, Anton 2006, Anton 2008). Error limits are estimated to +/- 1 percent. The Barlow lens approximately doubles the focal length of the telescope. However, the exact value may differ because it depends on the distance to the camera, which should be fixed.

Telescope	FL (m)	without Barlow	with Barlow
10 inch Newt	1.5	0.770	0.386*
11 inch SCT	2.8	0.424	0.220
50 cm Cass.	4.5	0.254	0.132
40 cm Cass.	6.3	0.187	0.097

Table 1: Calibration factors (resolution) in arc seconds/pixel for combinations of the DMK camera with different telescopes, as determined from reference double stars. *the distance between Barlow and camera was slightly different.

Exposure times are chosen as a compromise between signal-to-noise ratio and pixel saturation, which should be avoided. As an example, for the system STF 3050 AB shown in the next section, with magnitudes of the components of about 6.5, I found a value of 12 milliseconds adequate with my 10 inch Newtonian with a Barlow and red filter. In contrast, the pair alpha Crucis with magnitudes 1.3 and 1.6, was overexposed at 0.1 milliseconds with the 50 cm Cassegrain without a Barlow and filter. High-contrast pairs may cause problems by blooming effects from the bright star. In such a case, diffraction spikes, produced by the secondary mirror spider, are quite useful for determining the peak centroid.

Selection of images

Typically, sequences of 1000 up to 3000 frames are recorded within a couple of minutes, out of which the best are selected by visual inspection for further evaluation. Image quality changes rapidly. Often, there are periods of several seconds with only bad frames, and sometimes relatively good frames appear in bunches, but not necessarily consecutively. An example of the seeing statistics is illustrated in Figure 1 for the binary pair STF 3050 AB in Andromeda (Mason, WDS23595+3343), with the brightness of the components of 6.5 and 6.7 mag, and a current separation of about 2.2 arc seconds. Out of 1500 frames, the best 22, plus the second best 26—in total 48 frames—were selected for registering and stacking. This and the resulting measurements are shown in Section III and Figure 3, below.

The selection of good frames can be done automatically by image processing programs. Excellent results have been obtained for planets, with up to several thousand frames being registered and stacked. A threshold can be set for the image quality, which is estimated from the local distribution of bright pixels. This can also work fine with double stars when the star images are more or less well defined and separated, as in the case shown in Figure 1. However, problems arise, for example, for close pairs at high magnification, with more or less distorted profiles of the star images, including the diffraction rings, which may fluctuate in shape and intensity, or for systems with very different brightness of the components, or when anisoplanatic effects occur. In such cases, a simple measurement of something like the Strehl ratio can be misleading, and I would not trust automatic software. It is better to select the useful frames by visual inspection. A good strategy is to prefer good images when they come in bunches, like the ones at around 20 sec in Figure 1. This indicates moments of steadier seeing than average, with increased probability that the images are less distorted. In contrast, a single, apparently good frame between distorted ones may show a transient, but probably also a distorted state. When dealing with anisoplanatic effects, I first register and stack a large number of frames using an automatic program, such that distortions average out, which, however, results in increased sizes of the intensity profiles. In the next step, I select frames of better quality with reference to this template.

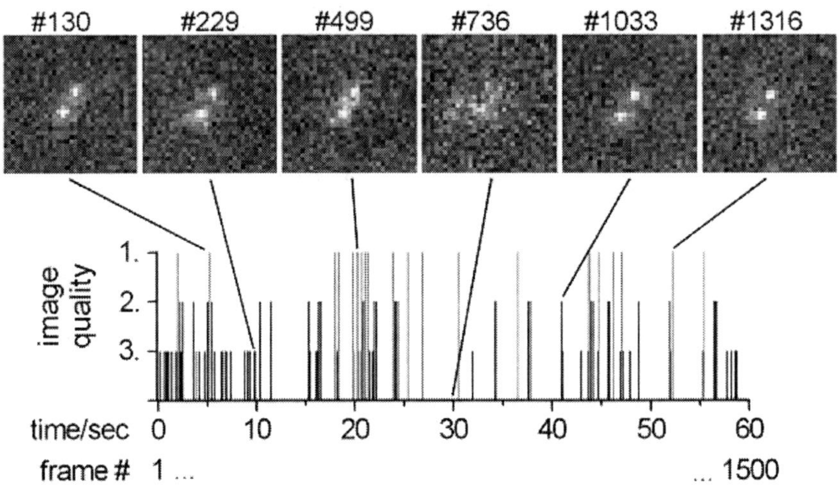

Figure 1: Seeing statistics during recording of the system STF 3050 AB (DMK camera, red filter, exposure time 12 milliseconds, 10 inch Newtonian, f/12, date 2007.797). Out of 1500 frames obtained within 60 seconds, some are selected (top row, with numbers, enlarged, not resampled) and classified into four categories, which are marked in the plot by vertical lines with different lengths. Category 1 represents the best quality, like nos. 130, 499, and 1316, whereas frames with quality below 3, like no. 736, were not useable. The result of superposition is shown in Figure 3 below.

This results in better definition of the profiles and their centroids. Figure 2 illustrates the effects of anisoplanatic distortions for the wide pair psi Piscium, with components of 5.3 and 5.5 mag, separated by 29.5 arc seconds. This is much greater than the typical isoplanatic field size of the order of 5 arc seconds (which somewhat increases at longer wavelengths). Although some single frames exhibit good quality, relative displacements of star images of about +/-0.5 arc seconds are observed. Such frames were discarded, when noticed. The measurements resulted in 159.3 degrees for the position angle and 29.5 arc seconds for the separation. This compares well with the last entry of 2007 in the Washington Double Star Catalog (WDS) with 159 degrees and 29.6 arc seconds (Mason).

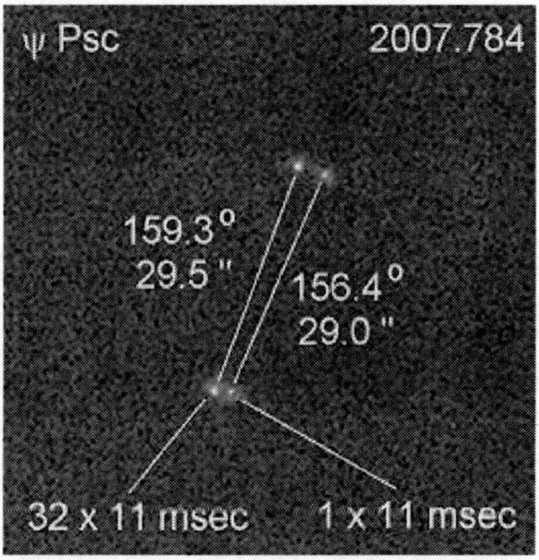

Figure 2: Demonstration of anisoplanatic effects during recording of the system psi Piscium (same telescope and configuration as in Figure 1). The pair is shown twice, as indicated by connecting lines. The image of the pair at left is the result of superposing 32 best frames, while the image at right is a single frame, which, although good looking, exhibits an extreme displacement as indicated by the position data. Those at left agree well with the last entry in the WDS (see text). (Note: All images have been re-sampled during processing.) North is down, east is right, as in all other star images.

Image Processing

As the yield of useful images is generally of the order of only a few percent, the number of images to be processed is limited, say, to between some tens and a hundred. It will be illustrated in the next section that this is usually sufficient to obtain well defined star images when they are suitably processed, and does not impair the results of measurements. All frames are re-sampled, i.e. the images are recalculated by doubling the number of pixels in x- and y-direction, up to three times, and stacked with sub-pixel accuracy, referred to the original pixel size. The first re-sampling is usually done with bicubic interpolation. This procedure results in rather smooth intensity profiles and in accordingly more accurate definitions of the peak centroids. Sometimes, the grey levels are linearly stretched in order to fill the range of 255 steps between black and white. I normally do not apply non-linear manipulations, such as gamma or logarithmic corrections, as this would not increase the accuracy of measurements. Only in cases of very different magnitudes of the components does gamma correction help to enhance the visibility of the dimmer star. Measurements of star positions are straightforward. Distances are calculated from the pixel coordinates of the peak centroids, and from the calibration factors. For measuring the position angle, the east-west direction is determined from superposed images recorded while the telescope drive is temporarily switched off. I do this routinely for each individual recording of any system. The error margin is about +/-0.1 degrees.

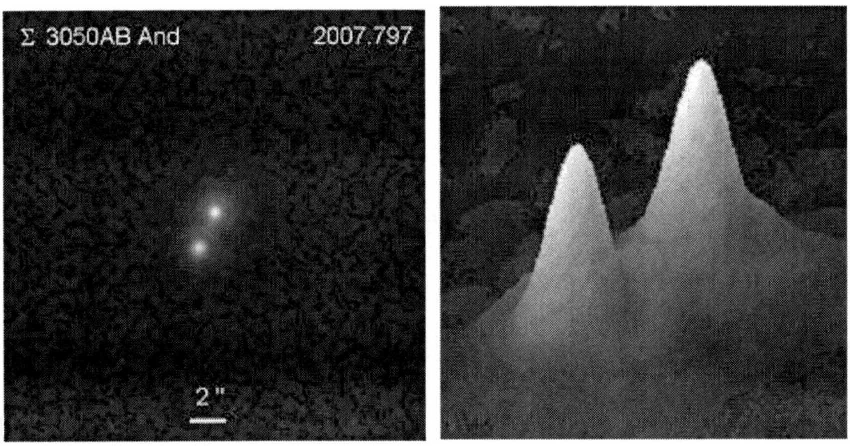

Figure 3: Left: Superposition of the 48 best images of the binary STF 3050 AB recorded with my 10 inch Newtonian at *f*/12 as described in the text. Note the diffraction rings around the star images. North is down, east at right. Right: 3-D view of the peak profiles. No non-linear manipulation of the grey levels was done, as in all other images.

Results

In the foregoing section, the selection of lucky images was illustrated for the system STF 3050 AB. Figure 3 shows the result of superposing the 48 best frames. Due to double re-sampling before stacking, the intensity profiles are smooth and well defined, and the Airy diffraction rings are clearly resolved, as can also be seen in the 3-D plot at right in Figure 3. The pair is a binary with assumed period about 800 years. The result of my measurement in 2007 was 335.6 degrees and 2.18 arc seconds, with estimated error margins of +/-0.5 degrees and +/-0.05 arc seconds. This is compared with recent interferometric data as well as with the currently assumed ephemeris in Figure 4. Deviations of my data from extrapolated, linear fits are +0.3 degrees and +0.01 arc seconds, respectively, although it is not quite clear whether a linear extrapolation is justified. In any case, the data of the last 5 years or so deviate from the calculated orbit, which is of grade 3, by more than the error limits.

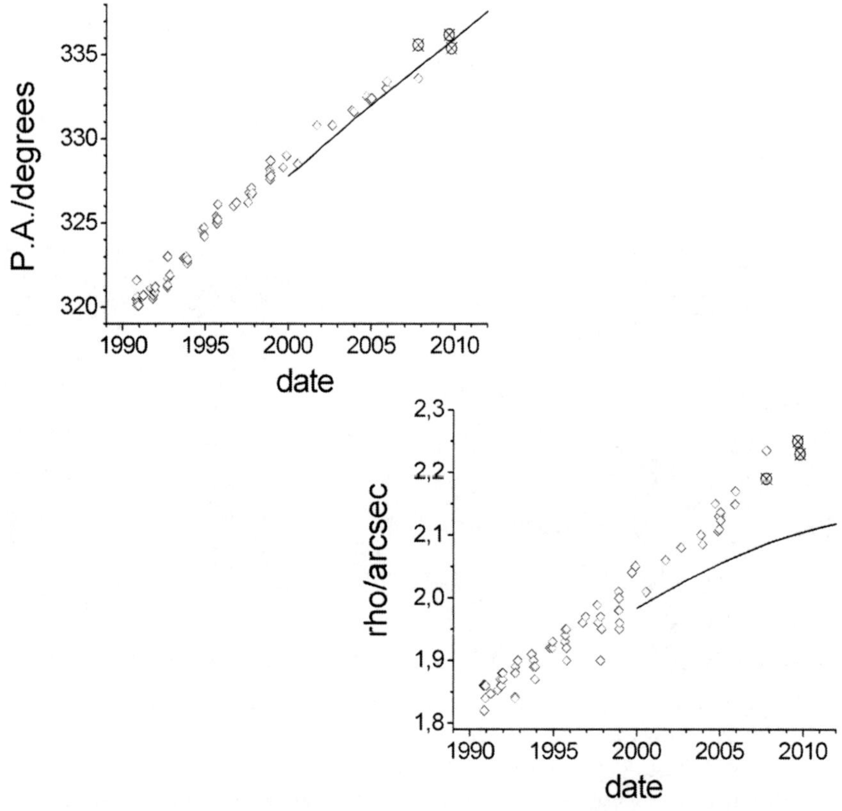

Figure 4: Plots of recently measured position angles P.A. (left) and separations rho (right) of STF 3050 AB versus time. Data shown as diamonds are taken from the 4th Catalog of Interferometric Measurements of Binary Stars (USNO) (Hartkopf *Fourth Catalog*). Solid lines represent the ephemeris of the assumed orbit, taken from the 6th Catalog of Orbits of Visual Binary Stars (USNO) (Hartkopf 2006). Circles with crosses are my own measurements.

During a recent stay in Namibia, I tested the accuracy and reproducibility of measuring double stars from lucky images by six recordings each of two binary systems, ζ Aquarii (STF 2909 AB, WDS22288-0001) and β Phoenicis (SLR 1 AB, WDS01061-4643), with current separations of about 2 arc seconds and 0.4 arc seconds, respectively, using the two Cassegrain telescopes at the IAS observatory, as mentioned above, on different nights. A representative image of ζ Aqr, obtained with a 50 cm Cassegrain, is shown in Figure 5.

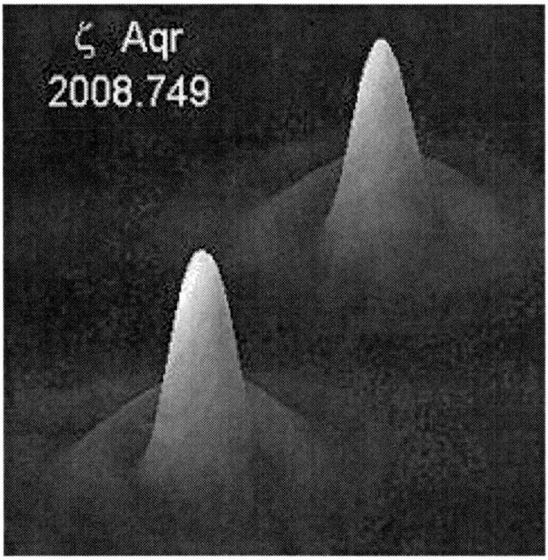

Figure 5: 3-d view of the peak profiles in an image of ζ Aqr AB, obtained with a 50 cm Cassegrain from the 48 best frames of 2 millisecond exposures.

Note the increased resolution, when compared with the image of STF3050 in Figure 3, with about the same separation, due to better seeing, and to the larger telescope. Separations were measured between 2.076 and 2.113 arc seconds, with a mean value of 2.094 arc seconds, and a standard deviation of 0.014 arc seconds. Addition of the possible error of the calibration constant of 1% results in a total error margin of about +/- 0.03 arc seconds. The mean value of the position angle was 170.5 degrees with a range of +/- 1.5 degrees, and a standard deviation of 0.6 degrees. This data compares well with measurements by other authors, as can be seen in the plot of the orbital motion in Figure 6. Note the deviations from the currently assumed orbit. This is believed to be caused by an unseen companion orbiting around B with a period of about 25 years.

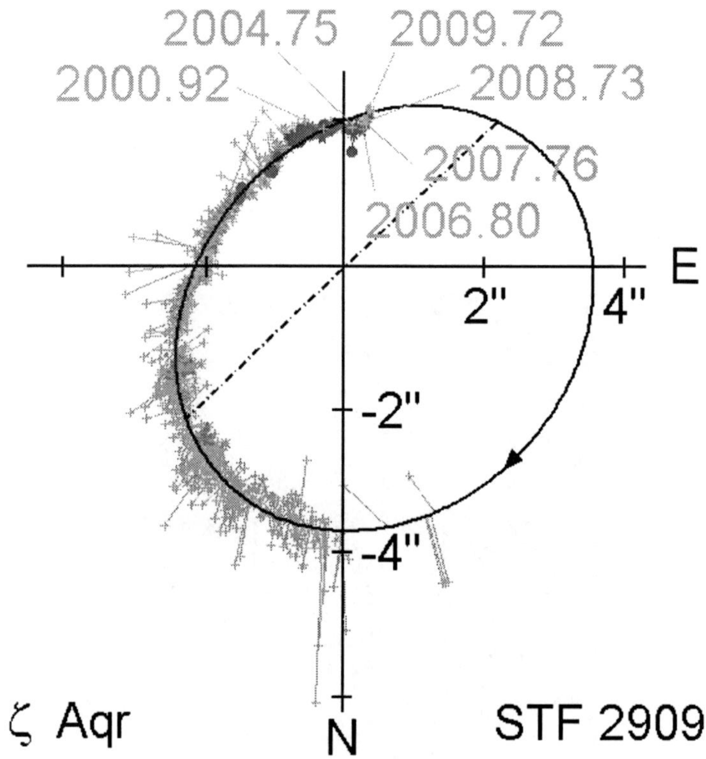

Figure 6: Plot of the orbit of ζ Aqr, adopted from the 6th Catalog of Orbits of Binary Stars, with markings of my own measurements in the last eight years. Dates 2008.73 and 2009.92 refer to six measurements each. Error margins are of about the symbol size. Note temporary deviations from the calculated orbit.

The image of β Phe in Figure 7 is a superposition of 64 frames with 0.67 millisecond exposures, obtained with the 50 cm telescope. At a separation of about 0.4 arc seconds, the peaks are clearly resolved, as is also shown in the line scan in the plot at the right. Again, six recordings of the pair were made under different conditions. The measurements resulted in a mean value of the position angle of 119.4 degrees, within a range of +/- 1.7 degrees, and a standard deviation of 1.1 degrees. The range of separation values was 0.014 arc seconds, at a mean value of 0.390 arc seconds, and a standard deviation of 0.007 arc seconds. While this value is surprisingly small, the total error margin is somewhat greater for two reasons. First, the error of the calibration constant of about 1 percent has to be added. Second, the peak profiles are slightly distorted by partly overlapping. This effect was analyzed by decomposition and revealed an apparent inward shift of the centroids by about 0.01 arc seconds in an average. Thus, the end result of the separation is shifted to 0.40 arc seconds, with an estimated total error limit of +/- 0.02 arc seconds.

The additional broad profile shown in Figure 7, at right, is produced from a superposition of 500 unselected frames, and represents an exposure of 1/3 of a second. This demonstrates the gain in resolution by lucky imaging. An estimate of the limit, according to Rayleigh's criterion, which could have been reached under the present seeing conditions, resulted in a value of 0.35 arc seconds, which should be compared with the theoretical limit of 0.28 arc seconds. In principle, even smaller dips of the intensity between the peaks than 25% could be detected, and the resolution could be further increased. However, it becomes increasingly difficult to recognize the lucky images in the original frames, which suffer not only from seeing effects but also from noise. Thus, for practical purposes, and with regard to the accuracy of measurements, I would estimate the limit of useful double star recordings with a 50 cm scope and lucky imaging techniques to not be below 0.4 arc seconds. For my 10 inch Newtonian, I set my limit to 0.8 arc seconds. These limits go further up with increasing differences of the brightness of the components.

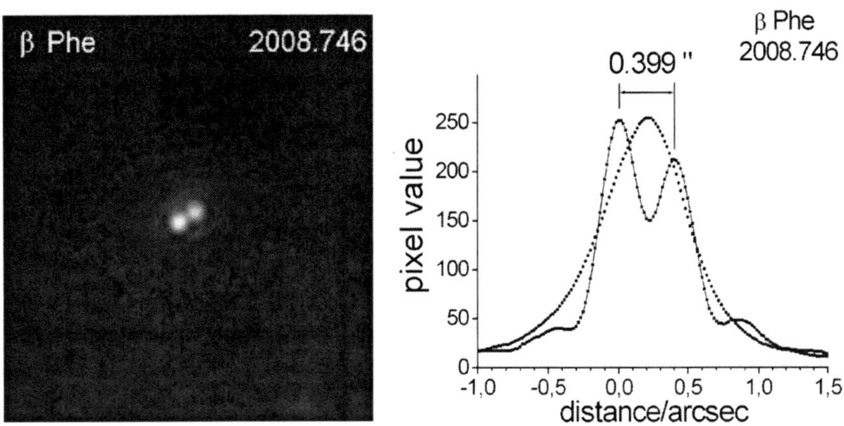

Figure 7: Left: superposition of 64 selected frames (exposure time 0.67 milliseconds) of the pair β Phoenicis, obtained with a 50 cm Cassegrain at f/9. Right: intensity profile across the components (data points and solid line), taken from the image at left. The additional broad profile was produced by superposition of 500 unselected frames, and represents a seeing-limited exposure of 0.33 seconds.

In the Equipment section above, the iterative calibration of the image scale by known double star separations was mentioned. This is illustrated here in Table 2 and in Figure 8, as it also delivers an estimate of the absolute accuracy of the measurements. Table 2 is extracted from an earlier paper in JDSO (Anton 2008), and lists measurements (using an 11 inch SCT) of position angles and separations for a number of systems in the southern sky which are well documented in the literature either as so-called "relfix", or with trustworthy predictable movements. By analyzing the differences of my separation measurements to extrapolated literature data, I fine-tuned the image scale such that the average deviation and the standard deviation assumed minimum values. The scatter of the residuals is shown in Figure 8, together with those of the position angles. The resulting calibration factor of 0.220 arc seconds /pixel is listed in Table 1. The residuals may have various origins, i.e. error limits of the literature data and/or ambiguous extrapolation. Nevertheless, the scatter of the deviations indicates the accuracy of the measurements. Separations deviate at maximum by +/- 0.1 arc seconds, and their standard deviation is 0.06 arc seconds. Position angles deviate within +/- 0.8 degrees, with standard deviation 0.5 degrees. The average values were in both cases virtually zero.

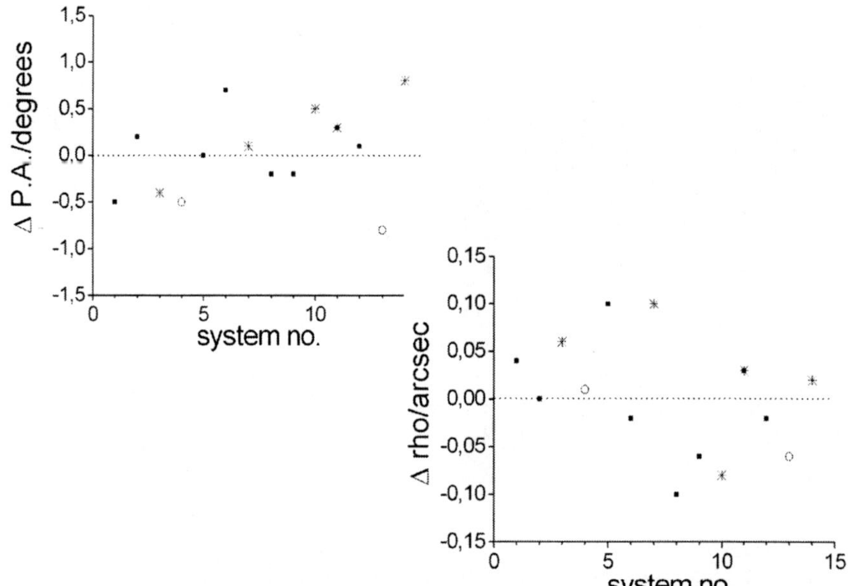

Figure 8: Residuals of position angles (left) and separations (right) plotted vs. the system number from Table 3. Star symbols denote "relfix" systems, open circles mark binaries, and squares are other systems.

PAIR	name or constellation	RA + DEC (WDS)	MAGS	P.A. meas.	rho meas.	DATE	no.
DUN 252 AB	α Cru	12 26.6 -63 06	1.25 1.55	112.5	4.04	2007.364	1
DUN 252 AC	α Cru	12 26.6 -63 06	1.25 4.8	202.2	90.0	2007.364	2
RMK 16 AB *	Θ Mus	13 08.1 -65 18	5.65 7.55	186.6	5.36	2007.373	3
RHD 1 AB	α Cen	14 39.6 -60 50	0.14 1.24	235.7	8.78	2007.367	4
CPO 62	Lup	14 58.5 -47 26	7.37 8.36	164.0	24.6	2007.373	5
HJ 4728	π Lup	15 05.1 -47 03	4.56 4.60	66.7	1.68	2007.367	6
CPO 415 *	Lup	15 10.7 -43 44	7.07 7.66	20.5	49.9	2007.373	7
DUN 176 *	ζ Lup	15 12.3 -52 06	3.50 6.74	248.8	71.7	2007.373	8
HJ 4753 AB	μ Lup	15 18.5 -47 53	4.99 4.93	307.8	0.94	2007.373	9
PZ 4 *	ξ Lup	15 56.9 -33 58	5.09 5.56	49.5	10.3	2007.373	10
BU 245 *	Sgr	18 10.1 -30 44	5.77 7.99	352.2	4.10	2007.378	11
BU 132	Sgr	18 11.2 -19 51	7.01 7.13	189.1	1.38	2007.378	12
HN 126	Sgr	19 04.3 -21 32	7.87 8.06	188.5	1.18	2007.375	13
S 715 *	Sgr	19 17.7 -15 58	7.07 7.90	16.8	8.42	2007.373	14

Table 2: List of double star measurements with a C11 SCT (with Barlow) at the southern sky in 2007. All systems are either "relatively fixed," or exhibit reasonably well predictable motions, and were used for calibration of the image scale (extracted from Anton 2008).

Direct imaging of double stars can also be used for measuring brightness differences of the components and, when using color filters, at different wavelengths (although I haven't done this very systematically). Intensities are measured by integrating the pixel values in the peak, and subtraction of the background in the surroundings. Differences of the magnitudes are calculated from the intensity ratio IA/IB according to $\Delta mag = -2.5 \log(IA/IB)$. An example is illustrated in Figure 9 for the pair STF 644 AB (WDS05103+3718) in Auriga, which shows a nice orange-blue color contrast, with spectral classes K3 and B2II of the components. Their intensities are about equal in the "visible" (about green), but differ in opposite directions in the red and in the blue. Magnitudes in the visible are listed in the WDS as 6.78 mag and 6.96 mag. The difference of 0.18 mag compares well with my own measurement of 0.11 mag in the green. As a general remark, it should be noted that color imaging may become difficult when the components exhibit strong contrast. In such cases, one of the components may be overexposed such that its color will be shifted to white in the RGB composite.

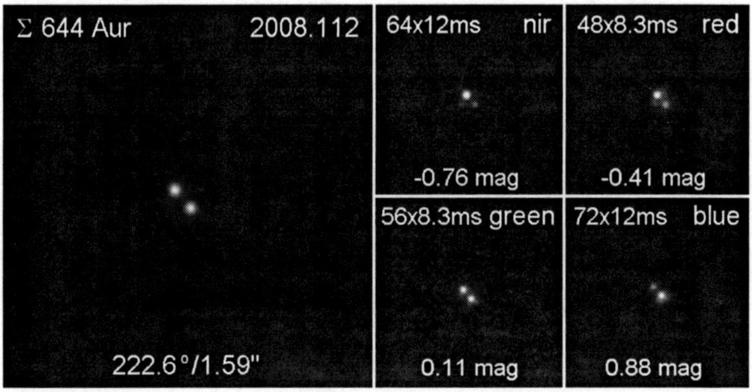

Figure 9: Filtered images (NIR, R, G, B, at right) and RGB-composite (at left) of the pair STF 644 AB in Auriga (10 inch Newtonian, *f*/12). Numbers of superposed frames, exposure times, and calculated brightness differences are indicated.

The position angle and separation was also measured in the filtered images separately, and resulted in averages of 221.5 +/- 1.5 degrees and 1.58 +/- 0.04 arc seconds, respectively. This agrees with the recent entry in the WDS with date 2006 of 222 degrees and 1.6 arc seconds, taking into account that the system only changes very slowly.

Conclusion

The few examples discussed in the foregoing sections clearly demonstrate that with lucky imaging the resolution of double stars can be pushed to the limits of the respective telescopes. Diffraction limited images can be obtained almost routinely, even under non-optimum seeing conditions. However, considerable care has to be taken in selecting the best frames, as good looking frames can sometimes be misleading. It is also shown that the accuracy and reproducibility of position measurements can be much better than expected from the theoretical resolution of the telescope. Error margins can be reduced to below 0.05 arc seconds, which depends more on the resolution in the image, expressed in arc seconds /pixel, than of the telescope, given by the aperture. Similar values of standard deviations can also be obtained with other methods, in particular visual as is discussed in Observing and Measuring Visual Double Stars (Argyle 2004). An advantage of digital images is that they can be stored and later used for further processing, such as brightness measurements. Interferometric measurements are generally deemed to be more accurate, depending on the size of the telescope. Nevertheless, one can often find similar scatter of the data as above in the literature. On the other hand, as the sensitivity of modest amateur telescopes is limited, direct imaging is the method of choice when dealing with fainter stars to be recorded with short exposure times, in order to find "lucky" images.

Author's Note:

E-mail: rainer.anton@ki.comcity.de

References

Anton, R. 2002. Measuring double stars with video. *Sky and Telescope*, July, 117-120.

Anton, R. 2004. Double star measurements at the southern sky made on digital images recorded with a video camera and a 14-inch Schmidt-Cassegrain. *The Double Star Observer, 10* (1), 2-10.

Anton, R. 2006. Measurements of double and multiple stars in the southern sky in 2002 using an 8-inch Schmidt-Cassegrain and a CCD video camera. *Journal of Double Star Observations, 2* (3), 108-117.

Anton, R. 2008. Double star measurements at the International Amateur Sternwarte (IAS) in Namibia in 2007. *Journal of Double Star Observations, 4* (2) 40-51.

Argyle, B., ed. 2004. *Observing and Measuring Visual Double Stars.* London: Springer.

Hartkopf, W. I. et al. *Fourth Catalog of Interferometric Measurements of Binary Stars*. U.S. Naval Observatory. Internet<http://www.usno.navy.mil/USNO/astrometry/optical-IR-prod/wds/int4> accessed 10 December 2008.

Hartkopf, W. I. et al. 2006.5. *Sixth Catalog of Orbits of Visual Binary Stars*, 5. CD-ROM version, available from U.S. Naval Observatory: wds@usno.navy.mil.

Internationale Amateur-Sternwarte (IAS). www.ias-observatory.org.

Mason, B.D. et al. *The Washington Double Star Catalog (WDS)*. U.S. Naval Observatory. http://www.usno.navy.mil/USNO/astrometry/optical-IR-prod/wds/WDS, accessed 10 December 2008.

Chapter 28

High Speed Photometry

Detection and Analysis Techniques

John Menke, MenkeScientific
Gary A. Vander Haagen, Stonegate Observatory

Introduction

Many of us have spent hours imaging events that are steady or change slowly with periods of hours to years. Some of us have ventured into the area where events occur in minutes to seconds and require skill and agility to catch. Very few of us have explored the region where events occur in fractions of a second or faster, where these phenomena may be periodic, quasi periodic, or short pulses of electromagnetic energy forever lost. Researchers interested in the formation and evolution of stars have shown great interest in the signals emitted by white dwarfs that may provide information about their earlier lives. These short duration signals, occurring at microhertz to millihertz frequencies, are being studied by the Whole Earth Telescope (WET), a network of larger telescopes with wide bandwidth cameras located around the world (Kawaler 2000). Similar research is being carried out by the European Space Agency and its fast response CCD camera, the S-Cam (ESA Science and Technology 2009) and the European Southern Observatory's Very Large Telescope, ULTRACAM (PHYSorg.com 2009).

While these efforts are focused on short, fast pulses with longer periods, other researchers are pursuing what might be described as audio class signals emitted from X-ray binaries—signals in the 10 to 2000 Hz region or periods from 0.1 to 0.0005 seconds (Warner 1988, Uthas 2005). This chapter will explore techniques for detecting and analyzing these high speed signals, those less than 1.0 second in period or duration, well beyond what can be explored with conventional CCD photometry, a region where studies use the ULTRACAM, video, or high speed single pixel detectors, and fast data collection systems. This area is awaiting further exploration. (See Fig. 1 next page.)

Areas of Potential Study

There are a number of interesting and scientifically important opportunities for photometric study of events that are fast occurrences or periodic at frequencies greater than 1 Hz. Some of these events can be studied with telescopes smaller than 14 inches with a suitable self-manufactured or purchased photo detector. Because many of these events require long setup times, have limited viewing opportunities, and/or need long or uncertain acquisition times, the collection of data by amateurs is sought by astronomical researchers and other collaborating observatories. There are at least four areas where high speed photometry is wide open for additional study: occultations, solar activity, stars with high speed variability, and atmospheric scintillation. Each area is discussed below.

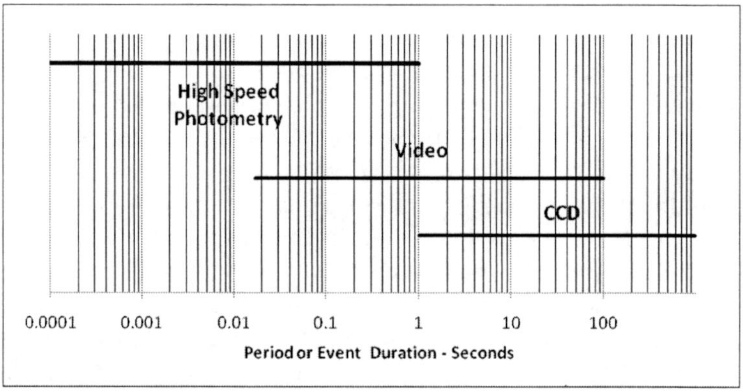

Figure 1: Spectrum for general and high speed photometric studies.

Occultations

Occultations are events where an object passes between the earth's observing point and another more distant object such as a star. The star's light is occulted, or blocked, from the earth's view. These are fun to watch and are highly organized events. Dunham (IOTA, 2009) and others have developed a network of observers with extensive website support on suitable targets, recommended methods, and a convenient place to send data to be analyzed and combined. Equipment may be portable telescopes in the 4 to 12 inch range and observation is usually visual or video.

By observing the star's light as it passes behind, we can learn about the movement of the nearby object, its atmosphere (or not), its surface, etc. Important discoveries such as lack of a Moon atmosphere, rings around Uranus, and accurate cross-sectional areas of asteroids are a few examples. A very interesting discussion on the many opportunities for general occultation studies are presented in the *Solar System Photometry Handbook* (Genet 1983).

Because asteroid occultations give direct information on the size and shape of the objects, this field is now being aggressively developed by Dunham and associates. The event is usually studied by a series of observers spread out over the relatively narrow (20-100 mile) predicted track. This is a very active area of work, both for fixed observatories and those willing to handle portable setups. Video is very popular in this field, but traditional photometers are also sometimes used. Ordinary CCD cameras can also be used, with accurate timing using drift scanning. Asteroid occultation work benefits from timing precision down to 0.001sec.

Lunar occultations are also very organized research events. Dunham (IOTA, 2009) and others have developed a network of observers with extensive website support including suitable targets, recommended methods, etc., very similar in support to the asteroid occultation studies. Equipment may be small portable telescopes in the 4 to 12 inch range and observation is usually visual or video.

There is increasing interest in observing mutual satellite occultations of Saturn and other planets. These can be relatively slow events, but are very difficult to observe, given the background lighting conditions. However, predictions are now being made for these events, so there should be an increase in activity in this area.

Detection of planetary occultations of Saturn ring gaps and ring structure, Uranus rings, inferring atmospheric structure, and similar studies can be done as a planet passes in front of a star. The issue is that the light of the planet tends to overwhelm the fainter star. It would be exciting to record the central flash from atmospheric focusing of light a billion miles away!

Solar Activity

Solar activity offers another exciting opportunity for study. While slow oscillations in the Sun's intensity have been reported, there are also periodic and quasi-periodic (short bursts of a periodic nature) signals that have been reported (Vander Haagen 2008). While further study is required for verification, periodic signals have been recorded in the frequency range of 30 to 1000 Hz. Such explorations are fruitful during the hours the telescope is unusable or skies are marginally clear. Here, separating the low level high speed signals from the over-powering Sun's intensity is an issue.

Stars with High Speed Variability

The General Catalogue of Variable Stars (GCVS) lists numerous variable star types of great interest in research. Observations of the longer period intensity variations are the principle area of study. Rotational periods, mass ratios, and other important physical characteristics can be determined using photometric study techniques. Many interesting study opportunities are available through participation in research programs coordinated by the American Association of Variable Star Observers, (AAVSO) and the Center for Backyard Astrophysics (CBA). Refer to their websites for current research campaigns.

Of the eight main types of variable stars, the "Intense Variable X-ray Sources" and "New Variability" classes have candidates that exhibit short or even extended periods of rapid variability. These star systems have reported low level starlight modulation at frequencies from micro Hertz to 1300 Hz or greater. It is postulated that in the case of x-ray binaries, these oscillations are caused by interaction between the spinning neutron star's magnetic field and the circulating accretion disc. Detection of these signals is challenging due to their low intensity and unpredictable occurrence. This necessitates good light collection and detector sensitivity, wide system bandwidth, and large data storage if analysis is not real time. Variables with activity have been reported that include Cygnus X-1, HD226868; X Per, HD24534; and AW UMa, HD99946, all with magnitudes V6-9 (Uthas 2005).

Atmospheric Scintillation

Scintillation is a limiting factor in fast photometry. Scintillation has been studied by many investigators for years in an attempt to better understand the mechanism and methods of reducing its undesirable effects (Young 1969). We know that the atmosphere is not uniform, and that there are small regions called cells in the 6 inch size range. These cells have non-uniform temperatures and densities, are located some 1-10 miles up in the atmosphere, and can cause small rapid variations in photometric data. This area offers many opportunities for study of both the phenomena and its minimization for wide bandwidth data collection.

Difficulties in High speed Photometry

It is helpful to consider the possible sources of fast or high speed signals. A 10Hz signal (0.1 second period) would likely be emitted in a region of a star that is 30,000 km across (the distance light travels in 0.1 sec). If a spot on the Sun were generating such a signal, the area of the spot would be about 0.1% of the area of the Sun. If the signal in such an area were a fluctuation of 10% in brightness compared to the Sun, the signal would be about 0.01% of the total sun's emitted energy. Observing from a long distance, this could not be detected with most equipment. If the localized brightness were much larger, it might be within the detection range. Or, if the signal were a periodic oscillation lasting for perhaps 10,000 cycles (1000 sec), a fast Fourier transform (FFT) analysis of a continuous data set might detect the signal. If the frequency fluctuated or was quasi-periodic during the 1000 seconds, the Fourier signal would be smeared and harder to detect. As the signal to noise (S/N) improves, detection probability increases and other techniques may prove useful. This process will be discussed later in the chapter.

Equipment

Telescope

Most phenomena of interest require a telescope to gather enough light to achieve a usable S/N. However, in fast photometry, the chief problem may not be that the objects are too faint but rather, even for bright objects, that atmospheric scintillation sets a limit on the precision of a brightness measurement at a particular speed. As usual, the bigger the scope, the more light there is to use and the more scintillation will be reduced. Telescopes of about eight inches or larger provide reasonable light gathering and some scintillation averaging. A high quality mount is as important as the telescope, since the viewing aperture is small for most high speed detectors.

Photometer/Optical Detector

With potential target events occurring with periods ranging from 1.0 to 0.0001 seconds, these periods are essentially your exposure times. It is obvious that very few photons will be collected by the sensor using modest size telescope apertures. Therefore, the need for the maximum sensor sensitivity and wide bandwidth is essential.

The photometer or optical detector uses a photosensitive device to convert light to an electric current that can be measured and analyzed. A photometer sensor is similar to a single large pixel in a CCD, but whereas the CCD accumulates charge for a delayed readout, the photometer detector reads out continuously. Unlike a photometer, a CCD can provide not only reference stars but also easy tracking, because the target can wander on the sensor without compromising the measurement. However, the photometer sensor must be kept aligned with the target. This can be difficult, especially if a small diaphragm, less than 1 arc minute, or small sensing area is in use.

The photometer sensor is most commonly a silicon photodiode with a typical area of 1 square mm. These devices are available in photovoltaic, PIN, and avalanche-type devices. The photovoltaic devices have the lowest bandwidth and are least costly. The PIN diodes are much faster, sensitive, and generally lower noise. Avalanche diodes are more costly but have the highest sensitivity and lowest noise. Cost can range as low as a few dollars for photovoltaic devices to $30 and up for PIN and avalanche diodes of commercial/research grade.

Avalanche diodes operate reverse biased at 100-300 volts and will provide a current gain of about 100 times. They are expensive and the multiplication process can introduce noise. The gain is also extremely temperature sensitive. Thus, unlike the photovoltaic and PIN diodes, the avalanche diode requires either close temperature stabilization or some other means of controlling the gain. Commercial controllers for this purpose are expensive; however, as with other equipment mentioned, inexpensive home-built options are also feasible. Photometers may also use vacuum photomultipliers or more recent solid state channel multipliers. As with avalanche diodes, these devices have their own electronic and physical requirements.

There is always a tradeoff of speed vs. sensitivity. Widely used silicon diode photometers in part achieve high sensitivity by foregoing operation faster than about 1 second response. However, it is entirely feasible to build a homemade fast photometer. The one shown in Figure 2 costs about $50 in parts, and uses a 1 mm square silicon photodiode. It was designed more for speed than sensitivity, achieving about 2 ms rise time with an internal noise roughly equal to a 9.5 magnitude star with a C11. The output of the photometer is digitized and sent over a serial link to a PC for recording and analysis.

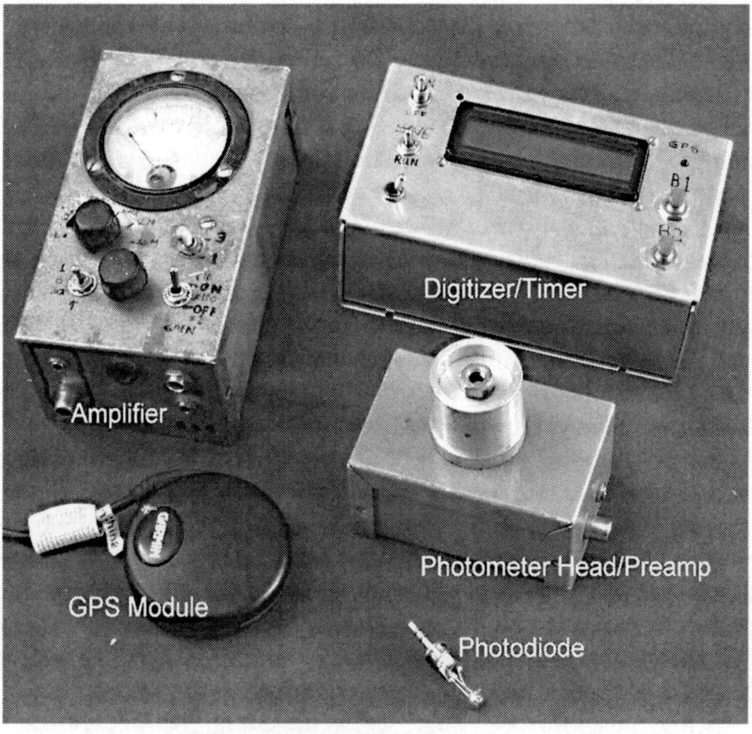

Figure 2: Home-built fast photometer equipment.

Commercial PIN Photodiode Photometer-Sensor

Let's take a closer look at the PIN diode just discussed, a good compromise with respect to speed, sensitivity, noise, and cost. These devices are P-Intrinsic-N semiconductors that are operated in a back-biased mode. They exhibit good quantum efficiency, wide bandwidths, and varied price-performance tradeoffs. This class of device dominates the high speed optical communications equipment industry. Figure 3 shows a typical circuit with PIN diode detector and amplifier. The PIN diode is back-biased. When a photon enters the intrinsic region, a carrier is created and the reverse bias sweeps out the carrier, resulting in a small current. This current is amplified by a wideband amplifier. Response of the detection system has the potential of GHz bandwidth. For good sensitivity and reasonable cost, bandwidths of 0.5 to 10 MHz are readily achievable.

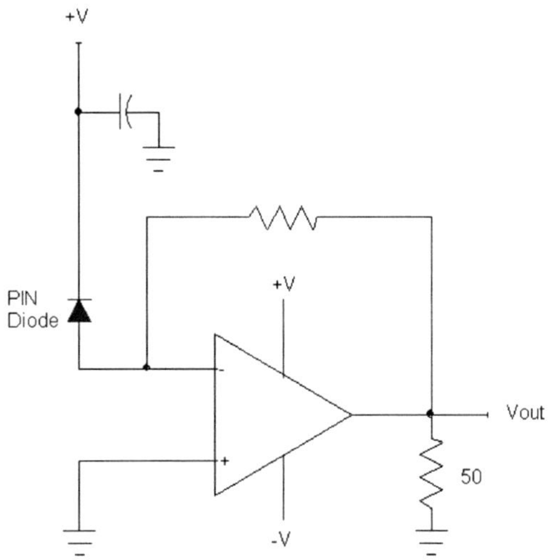

Figure 3: Typical PIN diode detector and amplifier; an operational amplifier with appropriate gain resistors.

A search for off-the-shelf detectors with amplifier found several good options. A Thorlabs PDA36A detector-amplifier was tested and proved acceptable for trial. Figure 4 show the detector amplifier. It has selectable gain of 10 to 70db and a bandwidth of 12 KHz to 12 MHz depending on the system gain: higher gain, lower bandwidth. In selection of the detector, the size of the active detection area is important. Since the capacitance of the detector increases with area, a compromise between detector size and bandwidth must be made. Table 1 shows this relationship. The 13 mm^2 size was chosen as a good compromise between bandwidth and ability to align the target during data collection.

Figure 4: Commercially available PIN photo detector with adjustable gain amplifier.

Active Area	Bandwidth
0.8 mm^2	150 MHz
13 mm^2	17 MHz
75.4 mm^2	1.5 MHz

Table 1: General relationship between PIN photo detector sensing area and bandwidth.

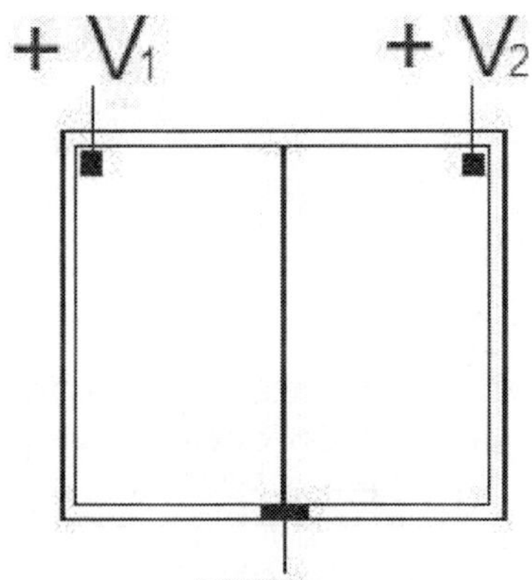

Figure 5: Bi-cell sensor showing two cells on one substrate with output terminals. Cells are available in either PIN or photovoltaic types. The bi-cell can be constructed from two separate small cells placed in close proximity.

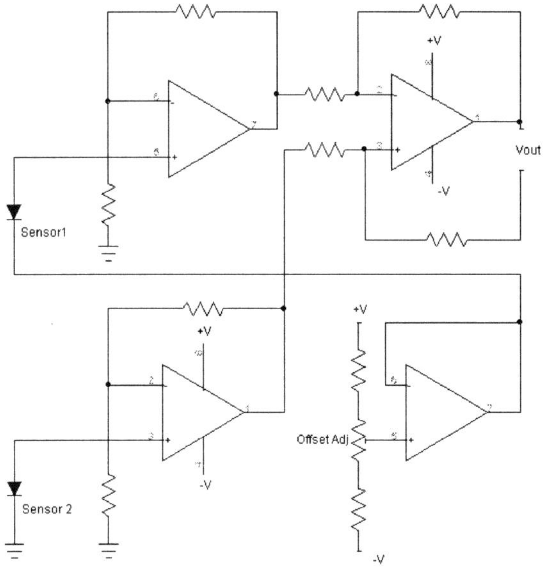

Figure 6: Photovoltaic bi-cell with individual amplifiers, differential amplifier for subtracting two signals, and adjustable offset control.

Bi-Cell Detector for Position and Amplitude Sensing

An extension of the single pixel PIN diode is the bi-cell detector shown in Figure 5. Here the single silicon substrate is sectioned so it forms two electrically isolated cells with a narrow gap between. The bi-cell is connected to a differential amplifier circuit as shown in Figure 6. For simplicity the circuit is shown using a lower performance photovoltaic sensor. The two non-inverting amplifiers take the low level sensor signals and feed a differential amplifier, thereby subtracting the two signals. An adjustable offset voltage compensates for DC cell sensitivity differences. When the entire target light is on sensor 2, the V out is positive. When the target light impinges on the sensor 1, the voltage generated subtracts from sensor 2 and makes the output voltage less positive. When the target is centered, the voltage is zero. This configuration offers help in positioning and focusing the target on the sensor. A further extension of this scheme is to have a quad-sensor: an X-Y pair of sensors and amplifiers, where position sensing in both axes is possible.

The split cell detection configuration, whether PIN diode or photovoltaic, present at least two intriguing research opportunities:

- A small telescope study of atmospheric scintillation with spectrum analysis over varying aperture, wind, and weather conditions.

- The use of separated detectors as shown in Figure 6 to study possible correlation conditions for atmospheric noise cancellation: one cell on a close proximity reference star, the other on the target star. Using back-scattered laser light as a reference, located in close proximity to the target, may also prove useful.

Analysis Alternatives

With high speed photometry, many of the events, particularly ones where the exact event occurrence is unknown, take hours of data collection before one may be recorded. Example: a rapid variability star could take 10 to 100 hours of data collection to catch an event. Assuming 0.1 millisecond exposures with 8 bit resolution, one generates about 0.6 Mbytes of data per minute.

Let's examine several different types of analysis alternatives useful for the events discussed earlier:

- Time based recording and analysis, useful for occultations.
- Frequency based recording and analysis using Fast Fourier Transform (FFT) for search of periodic events such as high speed star or Sun emitted signals.

Time Based Recording and Analysis

Time based recording and analysis, in its simplest form, amplifies the signal and feeds it to a strip chart recorder. Extending the concept, the recorder can be a PC based system with digitizer and software to scale the data, analyze, save, and print. The PC system is preferred since it is capable of processing and storing much faster events.

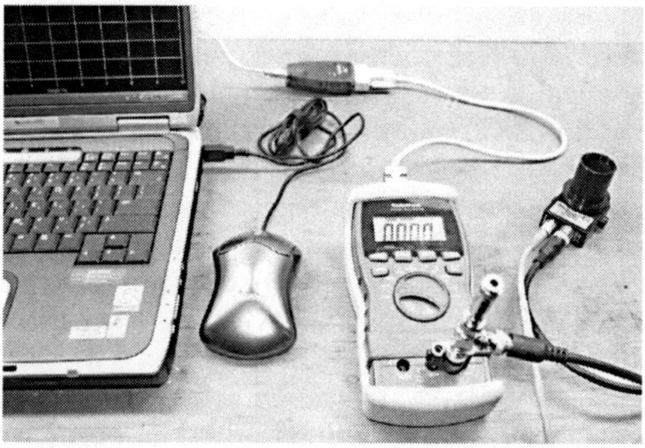

Figure 7: Typical digital multimeter (DMM) circuit showing sensor, 50 ohm termination resistor, DMM, serial to USB converter, and PC.

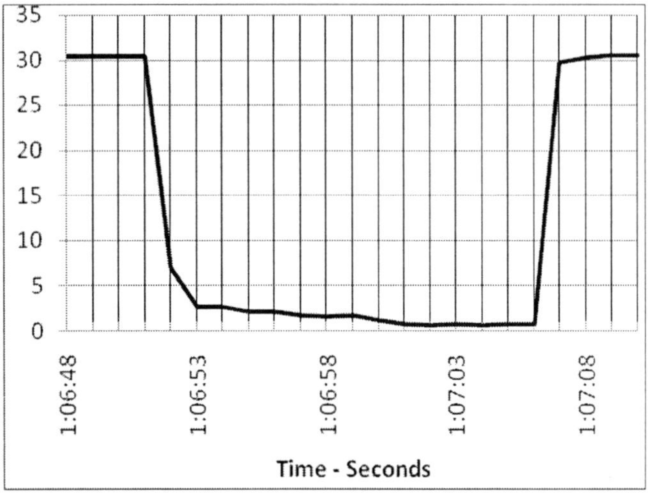

Figure 8: Simulated asteroid occultation graph recorded on PC connected DMM in 1 second intervals.

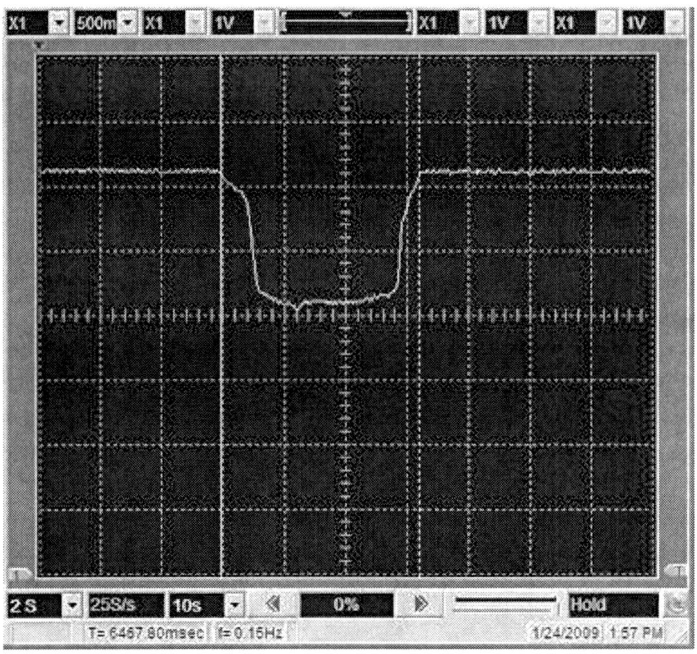

Figure 9: Simulated asteroid occultation graph using a PC based digital storage scope. Note the accurate time measurement of the period shown on the measurement window, 6467.8 milliseconds.

A good starting point system uses a standard digital multimeter (DMM) with serial output and PC software. A DMM has the capability of detecting signals into the microsecond range depending upon its sensitivity and bandwidth. A setup with a low end DMM is shown in Figure 7. The sensor is connected to the DMM with a serial connection to the PC. This low cost DMM, a RadioShack cat no. 22-812, logs data up to once every second with timing controlled by the PC's clock. The data can be plotted with the supplied software or exported for analysis in a spreadsheet. A simulated asteroid occultation graph is shown in Figure 8. In the example, sensor voltage is recorded every 1.0 second.

With a higher performance DMM or PC based Digital Storage Oscilloscope (DSO), the data recording interval could be millisecond or shorter intervals. Figure 9 shows a simulated asteroid occultation using a GAO Instruments Model- 2250 DSO. This unit offers a wide range of data collection and analysis capability at a cost only slightly higher than a high end DMM. In Figure 9, the data were collected in 0.01 millisecond intervals. The simulated occultation interval was measured as 6467.80 milliseconds. Note this is not the normal measurement point. Rise and fall times can also be accurately measured, as can other waveform features. Use of the second input for a time sync signal will allow very accurate correlation of the event to the UT clock. Either the PC connected DMM or PC based DSO are excellent devices for data archiving and printing of data in text or graphics form.

Video Camera Sensing and Analysis

For some work, a video camcorder can be used, either aimed at the target or through a telescope. The real time video image can be fed into the camcorder for recording, or directly into a computer for recording and analysis, frame by frame. In any case, the best time resolution is 1/60 sec (17ms) with 8 bit digitization implying about +/-10ms resolution of any particular event.

Recording video can consume memory very fast. For an uncompressed NTSC standard, VHS quality video signal, full motion video requires approximately 869 MB per minute. Modern compression schemes for a computer or a DVD recorder can reduce storage needs dramatically but also may reduce the S/N ratios for some applications. Furthermore, many inexpensive video cameras provide no gain control, so may be difficult to use for photometry. It is possible to perform photometry in real time on the video signal, thus greatly reducing data storage problems.

Other Time Stamping Techniques

There are pros and cons of using WWV, telephone, and computer based synchronizing using the Internet. Internet/PC-based timing sources frequently are off by as much as one second. While the display on a GPS is frequently wrong by many seconds, GPS modules are available that output a UTC pulse every second with an accuracy of a microsecond or better. One entrepreneur has developed a device for imprinting this time on raw video for use in asteroid occultation work (KIWI OSD). A related device was built for time-stamping digital data from the photometer at a 1 KHz sampling rate or for use in visual occultation work that proved very successful (MenkeScientific.com).

Frequency Domain Analysis

For frequency domain studies where the search is for low level continuous or short burst periodic signals, some serious issues arise. These events, particularly ones where the exact event occurrence is unknown, take hours of data collection before one may be recorded. For example, it could take 10 to 100 hours or longer to catch a cataclysmic variable event. This represents a real storage and analysis issue for a relatively narrow bandwidth. These data must then be analyzed by a PC based FFT program to search for possible periodic signals. A PC program suitable for this application has been identified—ScopeDSP Software (Menke 2007). This process requires sensing and digitization of the data, storage of the multiple 10 minute runs, and subsequent computer analysis producing Fourier spectra of the recorded event. While no periodic signals were detected on the eclipsing binary AW UMa, it was proof in principle of the process. This process is commonly used by the large research programs for data analysis.

The advantage of the storage data process is the ability to collect data over a long period and later analyze using the FFT software. The ScopeDSP software is a powerful tool and can process data very quickly.

Real Time Spectral Analysis

An alternative to the recorded data approach has been studied (Vander Haagen 2008). This study focused on real time analysis from 10 Hz to 20 KHz with a stretch goal of DC to 1000 KHz. The 10 Hz to 20 KHz range is common among audio and low frequency radio signal studies. The audio region is generally 10 Hz to 20 KHz and radio frequency region (RF) above 10 KHz.

A PC based system was investigated, since most stand alone audio spectrum analyzers are cumbersome and costly. The PC systems use a high quality sound card for amplification and digitization with FFT analysis done by PC software. The FFT software used was TrueRTA (True Audio). Several PC audio cards were tried and all proved satisfactory. The TrueRTA software employs FFT processing and covers 10 Hz to 50 KHz. However, the audio card performance limits spectral analysis to 20 KHz. The TrueRTA software requires substantial computing capacity. The desktop PC used had 1 GB memory, 3.0 GHz clock, and a dual-core processor. The RTA software normally requires 5-8% of the computer's capacity and would occasionally abort if other substantial processor loads were added.

The previously discussed DSO is also an outstanding tool for spectrum analysis. Most DSO's come with built in FFT systems. The model 2250 has a full capability FFT spectrum analyzer from DC to 100MHz. It is outstanding for real time signal studies. Figure 10 shows an example of a 990Hz signal superimposed on a sinusoidal 120Hz signal. In the first DSO time domain screen shot, the low level 990Hz signal is visible as "grass" riding on the larger 120Hz signal. If several additional periodic or random noise components were also present, the ability to discriminate the 990Hz would be nearly impossible in the time domain. The second screen shot shows the same signals in the frequency domain using the DSO's FFT system. The signals are easily discriminated. This technique is very powerful for many types of periodic signal analysis.

For frequencies above the audio region, a communications receiver was also investigated. A standalone receiver was available with coverage from 100 KHz to 2 GHz. This receiver has good sensitivity and is capable of scanning at 20 frequency stops/second in programmable increments over a selected frequency range. However, it does not provide spectral analysis other than by manually recording signal amplitudes versus frequency. In addition, short duration signals could only be detected if the receiver is tuned to the event frequency during the measurement time. Use of this receiver will be described later.

Recent developments in communications receivers have led to hybrid systems using a standalone communication system controlled by a PC. This receiver design greatly improves the analysis capability, including adjustable band pass, spectral analysis, data storage, statistical analysis, etc. One such receiver, WinRadio Model WR-G313e, covers the 9 KHz to 30 MHz RF region and offers substantial software-based analysis capability. The system is capable of scanning 400 stops/second, about 20 times the speed of typical receivers. This unit was not tried but is on the list for future study. The receiver is less costly than a modest CCD camera and appears very capable for analysis above 9 KHz where the signal duration is greater than one or two seconds.

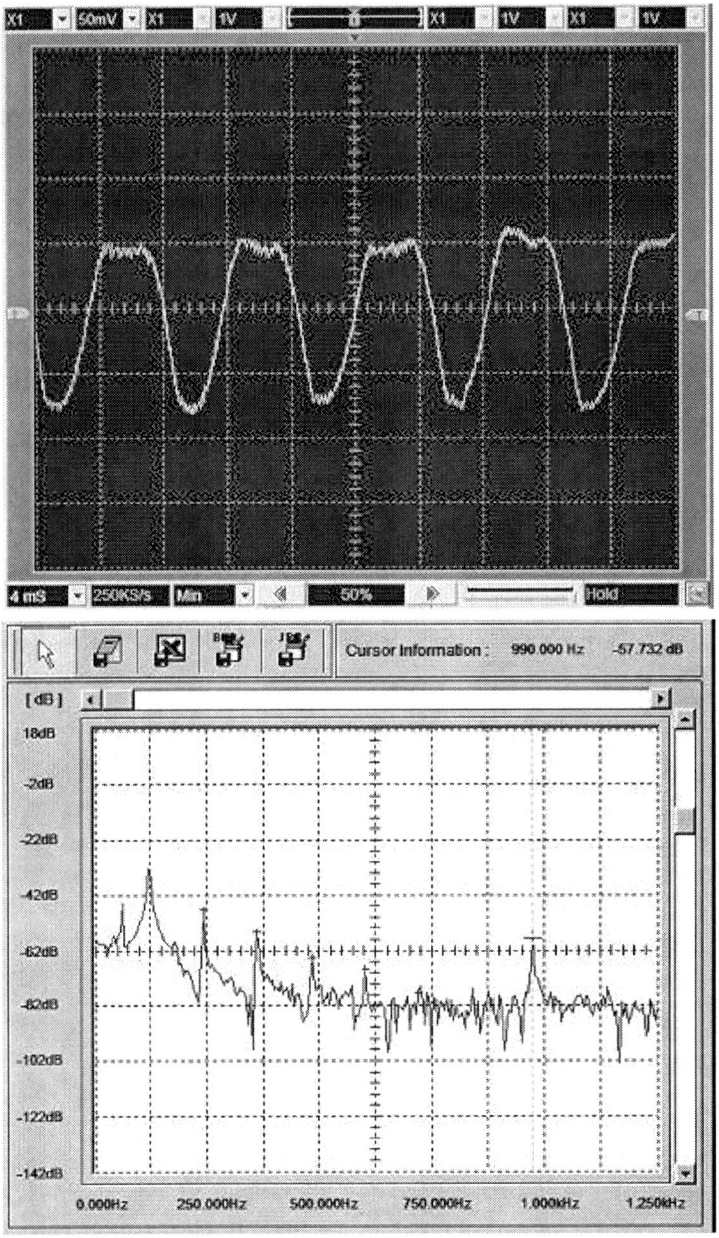

Figure 10: Using a digital storage oscilloscope (DSO) for data analysis. The top screen shot shows time based signal at 120Hz with low level 990Hz signal superimposed. The bottom screen shot of frequency spectrum shows all signals including 60Hz harmonics.

Table 2 shows the approximate sensitivity for the three real-time concepts. At the overlapping frequency of 10 KHz, the approaches are approximately equal. At higher frequencies the communications receiver exhibits a substantial sensitivity advantage. The two spectrum analyzer sensitivities are adequate for studies below 10 KHz.

Freq (Hz)	Audio Spectrum Analyzer	DSO Spectrum Analyzer	Comm. Receiver
	μV for 10db S/N		
10	40	20	*
100	30	20	*
1000	30	20	*
10 KHz	50	20	36
0.1 MHz	*	12	0.5
1.0 MHz	*	10	0.13

Table 2: Sensitivity for spectrum analyzer and communications receiver techniques. *out of normal detection range.

Comparing the stored data-FFT versus the real time approaches, they are approximately equal in detection sensitivity. The FFT processes for both are very similar and limited by the quality of the A/D conversion and noise within the system. However, at millihertz to microhertz frequencies or where data are quasi periodic, the stored data-FFT process permits correlation over a much longer period than is possible with most real-time analyzers. An interesting option at frequencies above one Hertz is to use real-time analysis and record the analyzer data output for further study.

Full System Sensitivity

The question arises: what is the dimmest star on which we can detect high speed modulation using the PIN diode sensor and FFT analysis? Since the reported high speed optical phenomena are below 2 KHz, the equipment of necessity will be the Thorlabs PIN Diode Detector using the audio or DSO spectrum analyzer. The audio spectrum analyzer was chosen initially because of availability.

A calibrated CCD camera and 650nm laser was used to determine the minimum magnitude star where 1% modulation could be detected (Vander Haagen 2008). Several different frequencies and trials indicated that a modulation level of 1-2% of the equivalent 7th magnitude laser signal was detectable at a 6 db S/N. This system sensitivity may be conservative based upon data developed on a similar sensing system, yielding 9th magnitude sensitivity (Menke 2007). The modulation level is defined as the ratio (V_{acpeak}/V_{dc}) X 100%. Audible and spectrum analyzer detection below that level was possible. This places the detection system sensitivity within the range of a number of high speed variable stars for study with an 11 to 14 inch Schmidt-Cassegrain scope.

One advantage of the audio spectrum analyzer is its ability to detect short periodic signals using the peak hold function. This capability was investigated. A 1000 Hz signal at minimum signal level and approximately 3 millisecond duration or 3 cycles was tried. It was readily detected by the spectrum analyzer, as shown in Figure 11. With moderate noise in the system, there is no problem detecting this short duration signal.

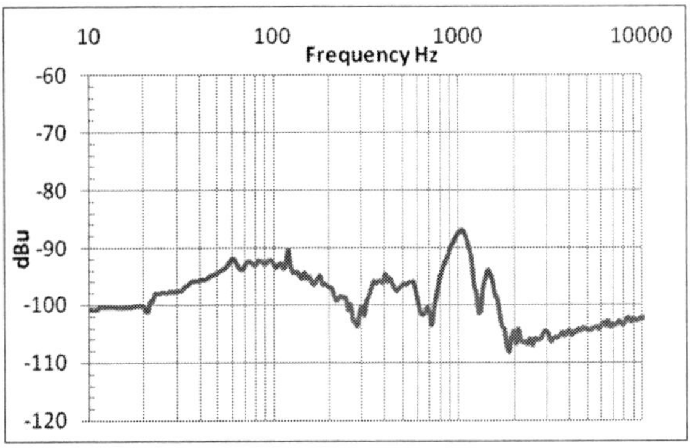

Figure 11: 1000 Hz signal of 3 milliseconds duration.

Atmospheric Scintillation

For high speed photometry, atmospheric scintillation is the limiting factor in the detection of low frequency signals. The atmosphere is not uniform and small regions or cells, in the six inch size range, exhibit refractive index changes due to temperatures and wind fluctuations generally occurring in the atmosphere 1-10 miles up. The basic mechanism is shown in Figure 12 (next page).

Scintillation affects stars of all brightness equally. Faint stars have additional noise sources including the shot noise from the sporadic arrival of small numbers of photons, and the risk of being submerged in other noise sources, whether in the detector element, electronics, or extraneous light sources. Using a small telescope aimed at a bright star, you can see the light and dark patches in the image if you run your CCD out of focus with short exposures or use a video camera.

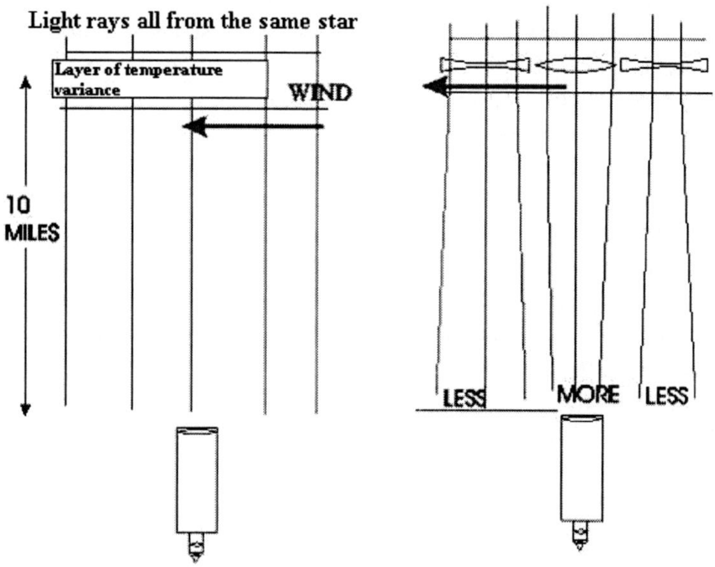

Figure 12: Cause of scintillation.

The distortions in the wave fronts by the over and under focusing caused by the cells produces bright and dim patches of similar sizes at the objective of the scope. This patch of light is moving at 100ft/sec and crosses the small aperture scope in about 10 milliseconds. As the cells cross the scope objective, more or less light is received and the apparent object brightness at the focus varies or scintillates.

A little geometry shows that scintillation is correlated over small regions of the sky, generally less than 30 arc seconds, often much less, and that these same variations produce spreading of the point image of the star, or seeing, on the order of several arc seconds.

Given these characteristics, the noise can be reduced by averaging measured brightness over time to determine the true value. Or, one can average over space: have a bigger scope receiving a larger number of bright and faint cells that all merge at the focal point for a more accurate average intensity for starlight in that direction. Or, take several telescopes, each with photometers or video cameras, and add the electrical signals or the numerical results together to produce the same improved source intensity. Adaptive optics is not the only answer.

How real is all this? Here are some data that show the effects. Figure 13 shows two data sets using two apertures on a C11 to simulate two different scope sizes. You can see that the bigger the scope, the more scintillation cells are averaged, the slower the time variation becomes, and the lower its amplitude. Using these data, one can see that detecting a 10% drop in a 5 millisecond signal is simply not feasible with a C11 and this level of scintillation, no matter how bright the star!

A Fourier transform can be taken on the intensity time data to determine the amplitudes of the different frequencies that make up the raw data. Excel will do this up to 32,000 points, but 20 minutes of data at 1000 samples per second is in excess of one million points. However, inexpensive software previously described can perform the FFT. Using FFT analysis, the data shows that most energy in the C11 trace is in the 10-100Hz range. This makes target signal extraction difficult over this frequency range.

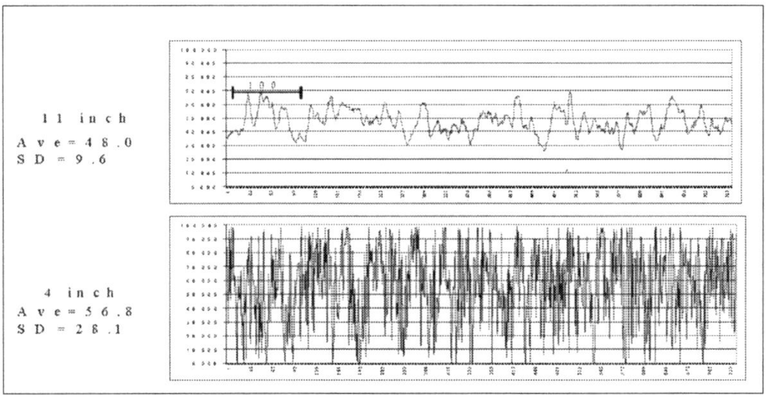

Figure 13: Scintillation effect with two different scope apertures.

This matches previous research results showing that scintillation power below a cutoff frequency (f ~½ π r, where r is the duration of the observation) contributes appreciably to photometric errors (Young 1969). For this example, a 1 millisecond intervals results in a cutoff frequency of approximately 160 Hz, below which scintillation becomes significant especially for apertures below 30 cm.

One interesting research project would be to determine the minimum modulation detection for frequencies above and below the scintillation cutoff frequency. This could be done using FFT analysis of the stored data or a real time spectrum analyzer.

Other Considerations

System Noise

It is important to note that there are many sources of electrical noise present that can steer one to misguided conclusions or muddy the data collection process. A troublesome interference is 60 Hz power line signals and its numerous harmonics. Multiple connections to the AC ground create ground loops. There must be only one connection to AC ground within the entire data collection system. This can be extremely difficult to achieve when the system includes both a remote observatory and a data collection and control center, as in this setup. All AC connections except one should use two pin isolation adaptors. Single ground connections between equipment are provided by shielded signal cables. Power line noise must be reduced to microvolt levels for studies of this type.

Other signal sources such as RF transmitters and receivers, video systems, monitors, and mechanical vibrations that modulate the detectors are sources for noise. Good cable shielding, removal of non-essential equipment, and attention to mechanical vibrations are all essential.

Trade-Offs

Research is all about trade-offs, so carefully consider what parameters are most important in a given investigation. For example, having relaxed guiding requirements and access to simultaneous comparison stars are some of the advantages of a CCD camera. In contrast, when using an aperture photometer, a guide scope, off axis guider, or semi-transparent pickoff mirror may be necessary to assure accurate centering and tracking of the target. Even with a photometer, accessing a comparison star may be possible. For example, use a second photometer-telescope setup either nearby or a distant reference with the data time stamped accurately. Or, move the telescope back and forth between the target and a comparison star, although this introduces timing uncertainties into the results which must be accounted for appropriately. Finally, simplicity is very desirable. Many different approaches theoretically may work, but amateurs and small observatories usually have both time and resource constraints that limit what can be done.

Signals from the Sun

With the PIN diode detector, amplifier, and real time detection, can we detect high speed signals on either the Sun or high speed variable stars? The Michigan weather puts an unusual damper on viewing, with several months of overcast skies. The Sun was the only opportunity for months. Both the audio spectrum analyzer and receiver were tried with the Sun as the optical source. Figure 14 shows the communications receiver attached to the Thorlabs detector. To keep the detector within its 5 volt maximum output voltage, a neutral density filter was placed in front of the sensor. This provided approximately 11X attenuation of the sun light. The receiver had a lower frequency limit of 100 KHz and a scanning speed of 20 stops/second. The stop increments were set at 1000 Hz, resulting in a scan rate of 20 KHz per second. It is unfortunate that the Sun was at the start of Solar Cycle 24, exiting a period of quiet. Little electromagnetic activity is expected or has occurred during the current period. No periodic signal activity was detected by the receiver other than some broadband noise at lower frequencies, 0.1 to 1 MHz.

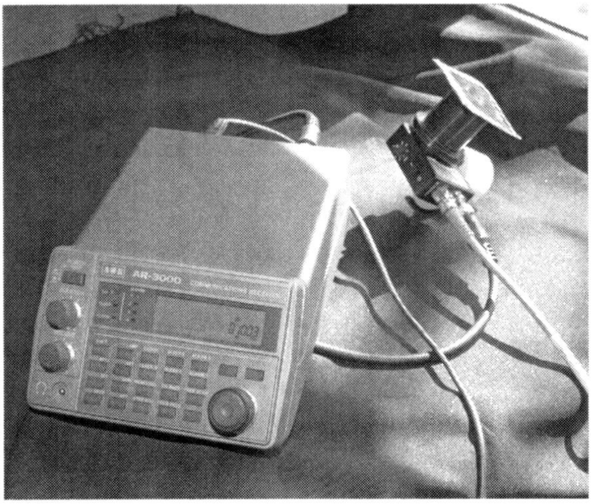

Figure 14: Receiver and photo detector setup for sun data collection.

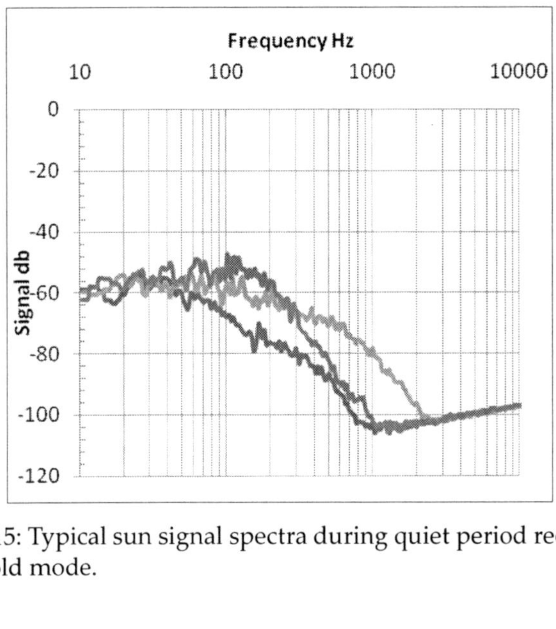

Figure 15: Typical sun signal spectra during quiet period recording in peak hold mode.

The detector was also connected to the audio spectrum analyzer. Figure 15 shows some examples of signals detected. Viewing during clear sky periods yielded widely varied signals. When viewing the real time spectrum there is evidence of broad band atmospheric noise at 10 to 100 Hz. The analyzer software was placed in peak mode to capture short duration high speed signals. Several such peaks are evidenced between 40 and 1000Hz. References indicate high speed solar activity that has been detected in the 0.5 to 2.5 Hz region, well below detection possible using the audio spectrum analyzer. (Williams 2001).

The atmospheric noise spectrum at low frequencies created the large low frequency humps. The effective collecting aperture was too small and should be increased to reduce scintillation effects. This is also a good opportunity for an external reference for instrumental and atmospheric noise cancellation. An extended study should also include use of a telescope with hydrogen alpha filtered light and/or heliostat mirror. Data should also be collected and analyzed using the stored data-Fourier technique.

As the Sun comes out of its low activity period, there will be opportunity to correlate new data with what was detected using this approach. Correlation of these data with known flares, electromagnetic storms, etc. will provide opportunities for future study.

A Look at X Persei (X Per)

Can high speed signals be detected on stars using a small telescope? X Per is cited as a potential candidate for emission of high speed signals (Uthas 2005, Warner 1988). X Per was also the brightest candidate in the highest sky position during the available viewing period. X Per is a 6.2 magnitude type GCAS+XP X ray binary system.

For data collection, the Thorlabs detector was mounted on a flip mirror with the second position occupied by the STV camera. The camera helped in the focusing and centering operation. Figure 16 (opposite) shows the setup on a C-14 telescope.

It was a clear night waiting game as X Per slowly set. Limited data was collected for 2 hour intervals in two cases just prior to reaching the lower altitude viewing limit of 30 degrees. With X Per, the most likely high speed phenomena are short periodic bursts of unknown duration, frequency, and amplitude. The audio spectrum analyzer was set to continuously monitor mode using the peak detection configuration. The result was a 10 Hz to 20 KHz spectrum saved to memory approximately every two minutes. Each spectrum resulted in 1/24th octave slices or 263 data points.

Figure 16: Thor sensor mounted with STV camera.

The detection system collected nine runs of X Per spectral data. These data arrays of 1/24th octave slices were analyzed by taking the peak intensity for each slice and subtracting the background data for that slice. Figure 17 shows X Per minus the average background and Figure 18, X Per minus the peak background.

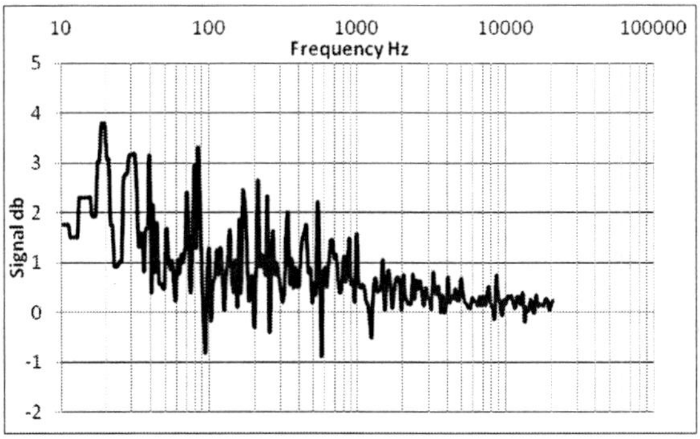

Figure 17: X Persei data with average background subtracted.

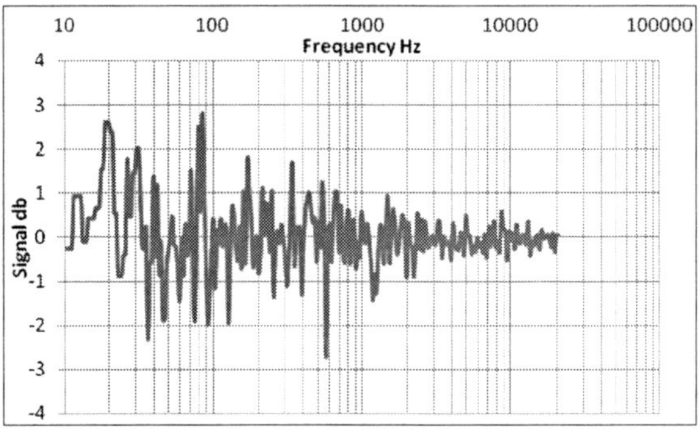

Figure 18: X Persei data with peak background subtracted.

Both plots show a number of periodic peaks below 600 Hz. Selecting only spectral data differences greater than 2 db above the average background yields the plot shown in Figure 19 and 2 db above the peak background plotted in Figure 20.

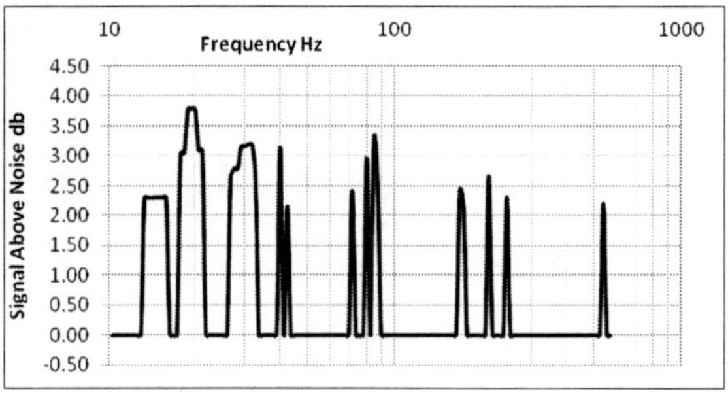

Figure 19: X Persei data 2 db or greater than average background noise.

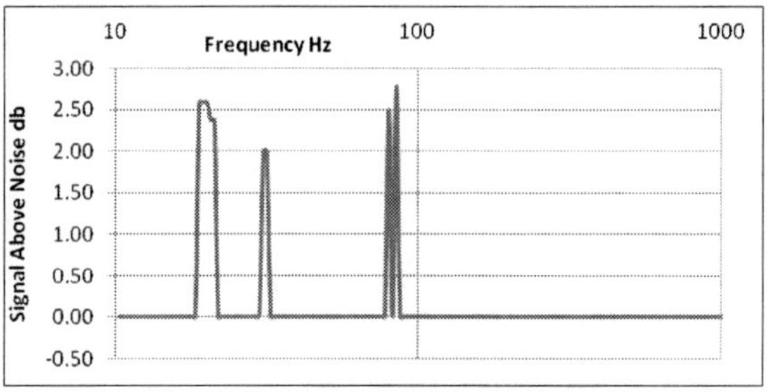

Figure 20: X Persei data 2 db or greater than peak background noise.

From Figure 19 and 20, the significant frequencies are shown in Table 3. The median frequency is computed for signal clusters. Table 3 shows common frequencies between the average and peak data in the 20 Hz, 30 Hz, 80 Hz, and 85 Hz bands. These frequencies do not correlate with more commonly reported observational data, the closest being 70 and 90 Hz (Uthas 2005). The detected signals are also within the range of maximum scintillation effect. Significantly more observations are required to draw any conclusions.

Conclusions

Amateurs and small observatories can use high speed photometric techniques and, in some areas, make contributions to astronomical science. The equipment can be reasonable in cost and at least some of the techniques are well within amateur capabilities. Relatively little has been done in this area, so there is an open opportunity to identify and develop new avenues of investigation.

Subtract Average Background	**Subtract Peak Background**
Hz Median	Hz Median
14.6	
19.4	20
29.5	31.3
40.0	
42.4	
71.3	
80.0	80
86.0	84.8
172.0	
213.6	
246.8	
536.2	

Table 3: Median frequencies greater than 2 db above average and peak background noise.

A PIN diode sensor with amplifier worked well with an experimentally derived 7th magnitude limit with approximately 1% minimum high speed modulation using a 14 inch Schmidt-Cassegrain telescope. A lower noise PIN diode amplifier and larger aperture telescope could extend signal detection to lower light levels.

The bi-cell offers opportunities for improved target alignment and numerous study options in scintillation and noise cancellation.

With scintillation being the limiting condition for a large range of interesting measurements, larger telescopes and good data extraction techniques are possible solutions. Telescopes in the 36 inch and larger range would be useful, even if the spatial resolution is only 20-40 arc seconds. Multiple telescopes on the same mount, nearby mount, or even far away mounts (if time stamped) are all feasible ways to reduce scintillation and improve fast photometry. A rough design of a 10x10 array of 5 inch simple lenses focused on inexpensive video cameras, and looks like a crude but possibly cost-effective approach. This subject would benefit from new thinking!

Collection of very low level, higher frequency signals is hampered by AC power line frequencies, wind and mechanically induced noise, and a myriad of spurious RF signals. This issue must be addressed systematically to reduce noise to acceptable levels prior to data collection.

Signal detection using a PC-based audio spectrum analyzer worked well at frequencies from 10 Hz to 20 KHz, providing real time data analysis for both Sun and star signals. The PIN diode sensor and communications receiver was used to look for Sun signal components above 100 KHz. No periodic signals were detected. However, the Sun is currently at a minimum for electromagnetic activity. PC-based receivers with integral spectrum analysis look favorable for work above 9 KHz but were not tried.

The use of a DSO proved useful for low level periodic signals, real time spectrum analysis, and general data collection. The approach is applicable for occultation studies and higher speed signals encountered with rapid variability star systems.

Observance of the X-ray binary system, X Persei, was hampered by months of cloudy skies. Two periods of short data collection were completed using the PIN diode sensor and the audio spectrum analyzer. Dominant frequencies at 20, 30, 80, and 85 Hz were detected. Considerably more data must be collected to draw any conclusions.

References

AAVSO: The American Association of Variable Star Observers. Internet <http://www.aavso.org/> accessed 17 January 2010.

CBA: Center for Backyard Astrophysics. Internet <http://cbastro.org/> accessed 17 January 2010.

ESA: European Space Agency. 2009. S-Cam 2 (search for papers on). Internet <http://sci.esa.int/science> accessed 1 March 2010.

GAO Instruments Inc. GAO USB PC based oscilloscope Model 2250. Internet <http://www.gaoinstruments.com/> accessed 1 March 2010.

Genet, R. M. 1983. Solar system photometry handbook. In *Occultations by Planets and Satellites*, R. L. Millis, 7.1-7.16; In *Asteroid Occultations A. W. Harris*, 8.1-8.11; In *Lunar Occultations* G. L. Blow, 9.1-9.25. Richmond, Virginia: Willmann-Bell, Inc.

GCVS: General Catalog of Variable Stars. Internet <http://www.sai.msu.su/groups/cluster/gcvs/> accessed 17 January 2010.

IOTA: International Occultation Timing Association. 2009. Mid-Atlantic occultation, eclipse, and lunar impactors. Internet <http://iota.jhuapl.edu/> accessed 1 March 2010.

Kawaler, S. D. and Dahlstrom, M. 2000. White dwarf stars. *American Scientist 88*, 498-500.

KIWI OSD. Imprints GPS time onto video. *PFD Systems*. Internet <http://www.pfdsystems.com/> accessed 17 January 2010.

Menke, J. 2007. Fast photometry. *Proceedings for the 26th Annual Conference of the Society for Astronomical Sciences*, 111-117.

MenkeScientific.com. 2010. Home-built fast photometer and time stamp system, information on. Internet <http://menkescientific.com/johnspage.html#EquipmentSoft> accessed 17 January 2010.

PHYSorg.com. 2009. Ultra-fast movies of the sky. Internet <http://www.physorg.com/news4466.html> accessed 17 January 2010.

ScopeDSP. Software for fast Fourier transformation. Iowegian International Corporation. Internet< http://www.iowegian.com/scopedsp> accessed 17 January 2010.

ThorLabs Inc. High speed optical detectors. Internet <http://www.thorlabs.com> accessed 1 March 2010.

True Audio. TrueRTA audio spectrum analyzer software. Internet <http://www.trueaudio.com/rta_down.htm> accessed 17 January 2010.

Uthas, H. 2005. *High-Speed Astrophysics: Rapid Events Near Compact Binaries.* Lund Observatory, Sweden.

Vander Haagen, G. A. 2008. Techniques for the study of high frequency optical phenomena. *Proceedings for the 27th Annual Conference of the Society for Astronomical Sciences*, 115-122.

Warner, B. 1988. *High Speed Astronomical Photometry*. Cambridge University Press.

Williams, D. R., Phillips, K. J. H., Rudawy, P., Mathioudakis, M., Gallagher, P. T., O'Shea, E., Keenan, F. P., Read, P., and Rompolt, B. 2001. High-frequency oscillations in a solar active region coronal loop. *Royal Astronomical Society, 326*, 428-436.

WinRadio Research Labs. PC based communication receivers. Internet <http://www.winradio.com/index.htm> accessed 1 March 2010.

Young, A. T. 1969. Photometric error analysis. VIII. The temporal power spectrum of scintillation. *Applied Optics, 8* (5), 869-885.

Chapter 29

Automated Time-Series Spectroscopy with a Small Telescope

Gary M. Cole

Starphysics Observatory, Reno, NV

Introduction

The primary tools of modern astronomy are the imaging camera and the spectrograph. Over the past year I have automated the operation of my C-14 based observing system to automatically select, locate, and take both spectra and photometric images. As of the time this paper was written, some 2500 spectra and accompanying photometric images have been obtained of some 300 target objects ranging to 13th magnitude. One of the benefits of this system is that it provides time series spectroscopy and coincident wide band photometry of a large collection of objects with very little supervision.

Hardware

A 0.35m Schmidt-Cassegrain telescope is supported by a Losmandy G-11 mount equipped with an Astrometric Instruments Smart Skywalker controller. The primary instrument package is shown in Figure 1. The telescope lives outside and is protected by a commercial all-weather cover. The control desk and computers are located inside about 20 feet from the instrument. This exposed arrangement is practical in our high desert, low humidity location and keeps the system at the ambient temperature. Dust, however, is a continuous problem.

The SBIG ST9XE imaging camera serves three purposes: target acquisition, photometric imaging, and as the controller for the remote guide head (RGH) that is used to autoguide the spectrograph. An Optec focal reducer provides a 14 arc minute field with 1.67 arc second pixels. Another reducer feeds the RGH whose sensor is normally binned 3x3 to obtain 2 arc second pixels.

The Meade 2″ flip mirror was automated by mounting a JMI Motofocus over the existing knob. An L-bracket was added to hold the Motofocus in position. The motor is connected to a JMI PCFC controller. This USB device is supported by CCDSoft's focuser programming interface. The duration of movement was determined experimentally. The two mirror positions are defined by mechanical stops in the mirror assembly. Switching takes about 5 seconds. The mirror shaft is intentionally overdriven to maintain tension on the mechanical stop. The imaging position is reproducible to better than one pixel.

The SBIG DSS-7 is a high efficiency classic spectrograph with a slit that subtends 2.1 arc seconds in my configuration. To put a star onto this slit, and keep it there for an hour or more, requires a precision target acquisition method as well as a long duration guiding solution.

For acquisition I use a combination of TPoint modeling and interactive plate solving. A pellicle beamsplitter with its own CCD camera forms an on-axis guider. This allows the system to guide on the reflection of the target object.

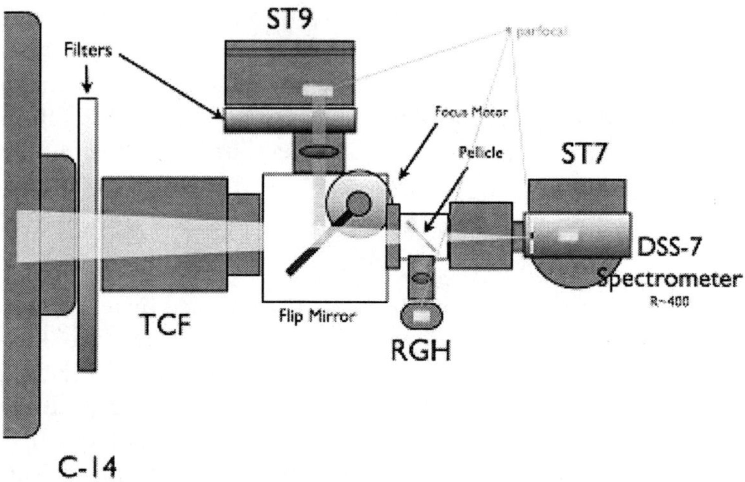

Figure 1: Instrument assembly.

The pellicle beamsplitter was fabricated from an off-axis guider assembly using an Edmund 1 inch uncoated pellicle mounted at 45 degrees. This assembly provides a clear aperture of 16 mm. In the f14 light cone, this is sufficient to fully illuminate the spectrograph slit located about 100 mm back. The pellicle reflects about 8% of the incident light onto the guider. In practice, the limiting magnitude of 10-second guide exposures is consistent with the limiting magnitude for 20-minute spectrographic exposures. An offline calibrating process locates the pixel corresponding to the slit center on the guide sensor. It is therefore essential that the mechanical coupling between the guider and the spectrograph does not allow movement greater than a fraction of a pixel. I have found it necessary to reinforce the threaded connectors with screw type hose clamps to maintain this level of rigidity.

Figure 2: Pellicle assembly.

It should be noted that there are several other ways to point and guide spectrographs. The solution used in the Tenagra telescope is to mount a fiber into the edge of the large primary image plane. This allows the primary camera to acquire the target by pointing to position that is offset from the target star by a known amount. In the SBIG SGS spectrograph and many others, the slit is slanted at 45 degrees. The front reflection off the slit plate is then directed to a guide camera. Both of these arrangements eliminate the need for guide-slit recalibration. Note that in some cases the slit view field is not large enough for reliable target acquisition, so an additional camera may still be required.

The ST9 imager, the Remote Guide Head (which is a TC237 imager that shares its electronics with the ST9), and the spectrograph slit have been adjusted to be parfocal. The entire assembly is behind an Optec TCF-S3 focuser that compensates for the change in focal position caused by temperature changes. This works so well that I have been able to go weeks at a time without refocusing.

Software Environment

There has been enormous progress in software since the first sustained automated photometry telescopes of the early 1980's. The automation software that I have created for this system manages the operation of a suite of commercial software products consisting of TheSky™, TPoint™, Orchestrate™ and CCDSoft™ from Software Bisque, Maestro™ from Astrometric Instruments, and utility programs from X10,

Optec, SBIG, and JMI. These programs in turn are running in virtualized Windows™ XP environments that are hosted by Mac OS X™, running Parallels™ virtualization on a dual processor Apple Macbook laptop. Essentially this one box functions as three independent software environments.

I have found that IP (Internet Protocol) communications are much more reliable and error tolerant than direct USB connections into the virtualized environments. Two external computers are used to convert the USB and parallel port SBIG cameras into IP devices using SBIG's Ethsrv software. An Aaxeon Serial to IP converter is used to provide connectivity to the mount and focuser controls.

CCDSoft is intended to control only one camera or guider. For my application, I needed to be able to operate at least two cameras independently. The solution was to use two virtual machines, each running a copy of CCDSoft under Windows XP. The first one controls the ST9 and the telescope, the other the spectrograph camera and flip mirror. This arrangement is easily extended to handle additional instruments.

Automation

My automation system was designed so that the only "user input" is a list of those objects to be observed and a concise description of the observations to be made of each object. Each target is entered as a single line of text in a "project" file using the format: "Objectname; Delay between observations; Observation List /comments"

Example: Vega; 72; DSS:1 V:1*3 R:1*3

This tells the system that I would like it to observe Vega every three days taking a 1 second DSS exposure and three 1 second V and R band images.

Multiple targets can be combined into a single project file, and one or more project files can be placed into a preset directory. The ordering of targets within the project file is unimportant, but the projects themselves can be prioritized.

This arrangement makes it possible to enter a large list of interesting objects without knowing when they will be observable. The only requirement is that the object name be found in the databases used by TheSky software. An additional syntax is supported to allow specific coordinates to be specified in lieu of an object name.

Note that data entry is done with a text editor. This makes it easy to construct observing projects from published catalogs.

The sequence of operation is shown in Figure 3.

The core of the system is a "dispatch" scheduling program I call "SelectTask". SelectTask evaluates the entire collection of targets to choose the next one to observe. The object name is resolved using the database of TheSky. From this, the current hour, angle, and altitude are obtained. Targets that are too far east or west, hidden by the house, too near the horizon, or too near the moon are rejected. The moon avoidance angle is determined by the current phase.

Once a candidate has passed this preliminary screening, the system estimates the required time from the object's observation list. Targets are rejected whose observations cannot be completed before the mount flips or the western horizon is reached. Remaining candidates are scored based upon airmass, remaining observing opportunities, time since the previous observation, and newness.

The original design of the system ran the entire selection process each time it was called. As I extended the target lists to include hundreds of objects, the selection time became significant. Hence the design was modified to create a secondary list of the next 4 highest scoring candidates. This batch of targets will be observed before the full selection process is used again.

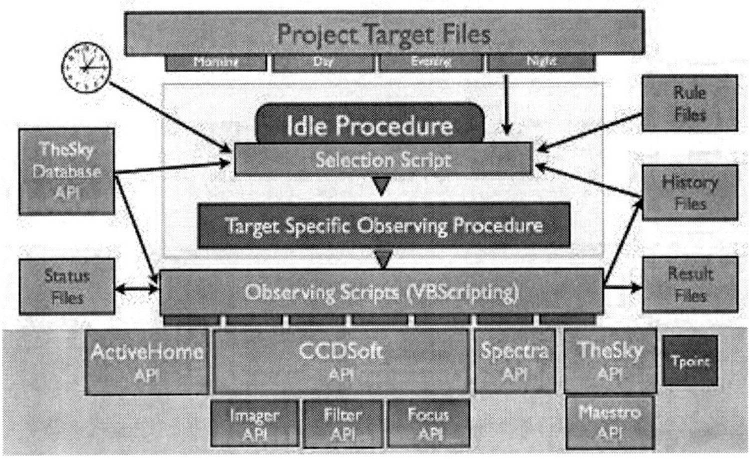

Figure 3: Observing workflow.

Once a target has been chosen, SelectTask creates an "observation procedure file". Each required measurement generates one or more observing commands in this file.

When SelectTask terminates, Orchestrate loads the newly created observation procedure file and begins executing the list of commands. Each observing command runs a VBScript program that I have written to control some action of the telescope or cameras. When completed, the selection process is restarted.

The selection process and target acquisition are done in the first virtual machine. The second is used to control the flip mirror and take spectra. Flags are maintained in common Mac OSX directories to coordinate the process.

The Observing Process typically proceeds as follows:

1. TheSky database is used to get the current target coordinates. Then the telescope is commanded to slew to these coordinates.
2. When the slew is complete, the ST9 camera is used to take an image. The CCDSoft function is used to match the image to the sky chart to determine the astrometric center of the image.
3. The CCDSoft Move function is then used to activate the guide relays of the camera to move the target to a preset x-y position that corresponds to the guider position of the spectrograph slit. This is an iterative process of imaging, plate solving, and moving.
4. A script is written into the Orchestrate inbox of the second Virtual Machine to command the flip mirror to be flipped to the inline position.
5. When the mirror position is confirmed, the ST9 Remote Guide Head camera is engaged as an autoguider. The guide point is preset by manual calibration.
6. Once per second, the guide error is measured. When the guide error is reduced to no more than 2 pixels, an observing script is inserted into the inbox of the spectrograph virtual machine.
7. The spectrograph camera is cooled if necessary and commanded to take an image. When that image is recorded, a previously recorded dark frame is applied. Next the image is processed to create a calibrated data set and graph as discussed below.
8. When done, the autoguider is stopped and the flip mirror returned to the right angle position.
9. One or more filtered photometry images are taken.
10. A final observing command issues an email reporting the observation

The entire process of acquiring a typical 10th magnitude target, taking a 5-minute spectrum, calibrating and graphing the data, and then taking a V band and an R-Band 15-second photometry image takes about 12 minutes to complete. Of this time, about 50% is actual data collection, 35% is target positioning, and 15% image processing and system overhead. Since the selection process is largely driven by sky position, the system observes each target several times over a period of months. As an example, 25 spectra and V-R photometric images of SS Cygni were obtained between August and November of 2008.

It is critical that spectroscopic data be accompanied by contemporaneous photometry because guiding variations make it very difficult to determine differences in brightness from the spectrum image.

An auxiliary function of SelectTask is to periodically initiate a procedure to collect median averaged dark images from the spectrograph camera. These are stored in a time and temperature stamped file so that darks can always be matched with the images they were applied to. These stored darks are used for exposures up to 5 minutes. Offset sky images are taken in conjunction with longer spectra for dark subtraction.

Data Pipeline

Unlike an image, you cannot judge the quality of a spectrogram without actually extracting the data values from the image, applying basic calibration, and graphing. In order to do this automatically, I ended up writing a fairly substantial amount of image processing code in Visual Basic. This code is invoked to process each spectrum immediately after it has been obtained. This allows the observer to review the results of a night's run as it progresses. It also makes it easy to review all of the spectra taken of a particular object as a slide show to look for possible changes.

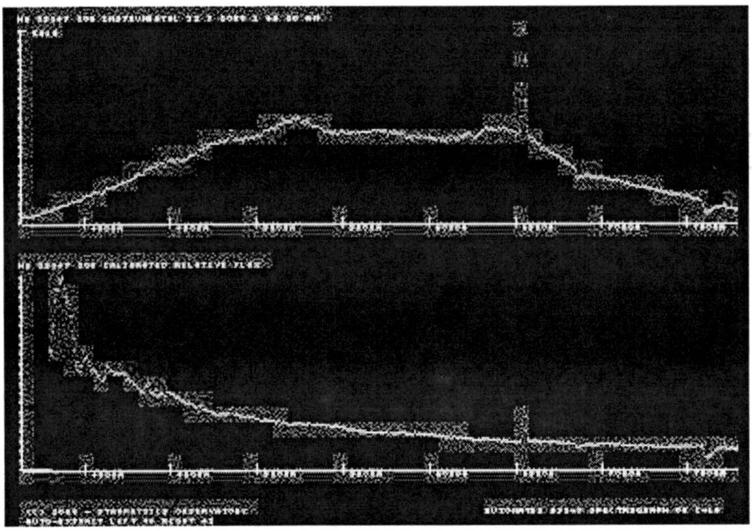

Figure 4: Spectrogram of HD35347 from 12/7/2007, instrumental above, flux calibrated below.

Useful as this is, it does not generate output that can be placed directly into scientific databases. Hence, my intention is to add an offline IRAF processing procedure to create calibrated data in publication formats.

Future Directions

Mounting multiple instruments on small telescopes is about to get a lot easier. Optec, Inc. is developing a four-way programmable optical manifold (rotating mirror). Using this, I anticipate having an imager, an infrared photometer, and both the low resolution (400) DSS spectrograph and the high resolution (10,000) SGS spectrograph integrated into an automated observing program by the end of 2009.

When doing spectroscopy of variable stars there is no point of spending an hour taking the spectrum of a star that has become too dim to observe. I intend to integrate real time photometry so that the system can skip spectrograph observations that would not be productive while maintaining time series photometry. This would make it much more worthwhile to include longer period and irregular variables into the observing program.

Conclusions

Converting a telescope to automated operation is possible with a few weeks of concentrated effort. What were once graduate school projects for telescope control and instrument operation are now fully scriptable functions of commercial products. What is still required is software to coordinate the functions so as to implement a complete observing program. My experience indicates that the results of creating such software easily justify the effort.

The system I have described in this chapter is generating more data every month than I have collected in past years of manual operation. One of the side benefits of programming the process is that I now have consistent file naming and directory structures to hold the thousands of new images.

It seems only a matter of time until complete systems like mine can be obtained off the shelf. This should lead to an explosion of productivity in small telescope science.

Meanwhile, anyone who would like to make any use of the software described herein is invited to email the author.

Author's Note

The author can be contacted at www.starphysics.com. This research has made use of the SIMBAD database, operated at CDS, Strasbourg, France.

References

Trueblood, M. and Genet, R.M. 1985. *Minicomputer Control of Telescopes*. Richmond, VA: Wilmann-Bell.

Kaye, T. 2006. SpectraShift extrasolar planet search project. *The Tenagra Spectrograph*. Internet <http://www.spectrashift.com/spectrographs_trenagra.shtml> accessed December, 2008.

Chapter 30

Spectroscopy for Eclipsing Binary Light Curve Analysis

Robert H. Nelson

Sylvester Robotic Observatory, Prince George, BC, Canada

Introduction

The last couple of decades have seen a revolution in hardware and software for data acquisition and analysis using small telescopes. Affordable telescopes with ever increasing aperture, often computer controlled, are now available to the serious amateur or semi-professional astronomer. Desktop computers of astonishing speed and capacity are now cheap and commonplace. The internet has revolutionized information exchange. But it is the camera using the charge coupled device (CCD), first constructed in 1969 by Boyle and Smith at AT&T Bell Labs and first used for astronomical imaging in 1983 that has truly revolutionized astronomy.

Various observing programs are now available to the observer possessing a CCD camera and access to telescopes ranging in aperture from 20 to 60 cm or more. Programs include supernova search, measurement of comet and asteroid positions, search for exoplanet transits, follow-up observations of gamma ray bursters, and variable star studies. For variable stars, the first two traditional categories (pulsating and eruptive stars) will not be dealt with here; it is the third category—that of eclipsing binaries—that will be the subject of this chapter.

While over one half the stars in the sky are thought to be part of binary or multiple systems, only a small fraction have orbits sufficiently inclined for eclipses to be visible for observers here on Earth (inclination $i = 0$ is in the plane of the sky, while $i = 90°$ is perfectly edge-on). Eclipsing binaries (EBs) are systems consisting of two or more stars in revolution about their common center of mass. In almost all cases, they are too far away for even telescopes of the highest resolution to resolve (but see Zhao, et al., 2008a and 2008b, for the first ever EB direct imaging of Beta Lyrae using the CHARA interferometer on Mt. Wilson). We therefore only know about their existence because during eclipses light intensity dips as one star passes in front of the other. By obtaining the light intensity over a complete cycle—the light curve—often using different pass bands, observers can determine fundamental parameters about the system.

The first is the period. Accurate times of minimum, gathered over long time periods and analyzed by Observed minus Calculated (O-C) diagrams yield precise periods (to 7 or 8 figures) and information about period changes over time. The latter have implications about possible mass interchanges and the evolution of EBs, the details of which are not fully understood at this time. There are numerous databases available on line giving easy access to information going back many decades (see Kreiner 2000, Nelson 2008a, and Kundera).

More fundamental quantities are orbital parameters (inclination, eccentricity, argument of the periastron, etc.) and physical parameters (stellar masses, stellar sizes, and surface temperatures). Light curve analysis can yield values for some or all of these quantities.

Light curve analysis routines have been around since as far back as 1952, when Russell & Merrill started with an unphysical rectified spherical star model (Kallrath & Milone 1998). The modern routine of choice is the Wilson-Devinney package (Wilson & Devinney 1971, Wilson 1990) which uses real physical models. This package consists of two parts: light curve synthesis (LC) and differential corrections (DC). The latest LC code uses actual physical stellar shapes (the Roche geometry), modern stellar atmospheres, and proper limb darkening. It allows for asynchronous rotation, stellar spots, a third star, and more, to calculate—based on the physical parameters that the analyst inputs—a theoretical light curve (or set of curves, for the different filters). By fiddling with the parameters (and this is a bit of an art), the analyst can make the theoretical light curve(s) approximate the observed one(s).

But how does one get an exact fit? That is the job of the differential corrections, DC—which uses the same resources as LC. The details will not be dealt with here, but essentially the routine makes incremental changes in chosen parameters, recalculates the light curve(s), determines the sum of squares of residuals for each point, $(O\text{-}C)^2$, and calculates the changes in each parameter to—theoretically—bring the residuals to zero (or at least minimize them). The routine can also use any observable parameter, including radial velocities, as discussed below. By this iterative process, one can narrow the parameters in n-space to obtain a solution. By convention, a solution is reached when the corrections are smaller than their estimated errors. Parenthetically, there are some serious questions of uniqueness that have been the subject of much debate.

The latest Wilson-Devinney code (in FORTRAN IV) may be obtained from Robert E. Wilson's ftp site at the University of Florida (Wilson 2005). It is suitable for a number of machines that run FOR-

TRAN, or it may be compiled to run on a PC (if you have a FORTRAN compiler for the PC). Unfortunately, to run this package, one needs to perform the tedious editing of input text files. The code uses formatted input, so an obscure (and difficult to find) mistake of even one column somewhere can cause execution to fail.

There are a number of packages that provide an easier path. For preliminary 'quick looks', see Bradstreet 1993 and Bruton 2008.

For full analysis, a Windows program, WDwint (written by the author—see Nelson 2008b) is popular because it completely circumvents the cumbersome and error-prone editing of text files—the program does that. It also displays the light output curves and parameters that need to be changed, greatly speeding up the overall process. Essentially WDwint is a front-end program that runs the compiled Wilson-Devinney DOS executables lcwin.exe and dcwin.exe. Extensive help is provided.

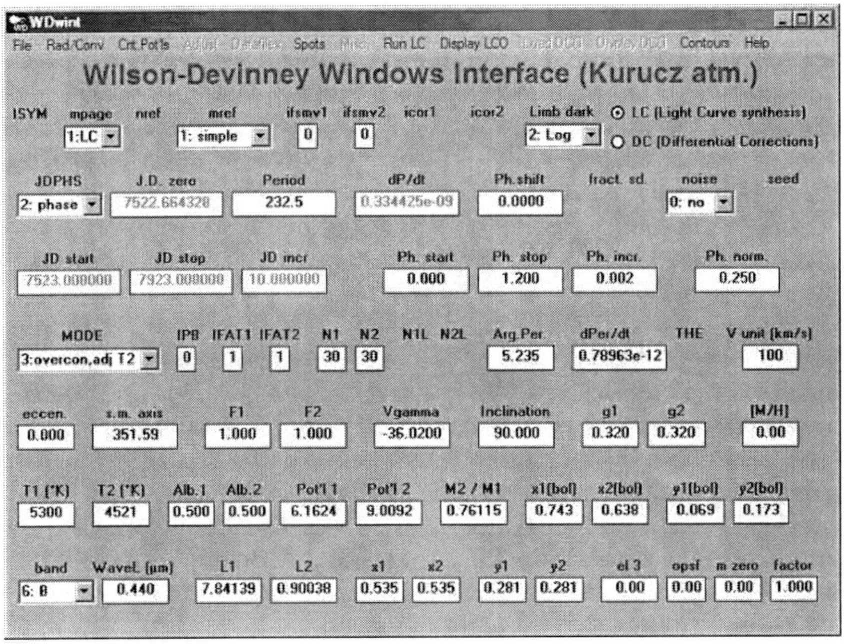

Figure 1: The main screen of the program WDwint.

The Need for Radial Velocities

Assuming all goes well, one gets a solution of stellar temperatures, orbital inclination, mass ratio, and stellar sizes (relative to the semi-major axis of the relative orbit). Using light curves alone, one cannot go any further. There is another problem: Terrell and Wilson (2005) have shown that it is impossible, in the case of contact binaries undergoing partial eclipses, to get a valid unique solution using light curves alone. What happens is that there is simply not enough information due to the inclination and mass ratio trade off effects. Increasing one and decreasing the other produces essentially the same light curve! The solutions are therefore indeterminate. A further problem is that for completely detached systems, the light curves do not constrain the stellar sizes, making solutions ambiguous.

The remedy for these problems is to obtain a different kind of information, the radial velocity (RV), defined as the component of the velocity vector in the line of sight from the observer. By convention, recession is taken as positive, and approach, negative. One then secures a pair of radial velocity curves, i.e. a representative set of RVs for each star over the complete cycle, usually concentrated near the quadratures—phases 0.25 and 0.75. Phase 0 is defined as the mid-point of the primary or deeper eclipse when the hotter star passed behind the cooler one, and phase 0.5 is defined as half a cycle further on; the latter coincides with the mid-point of the shallower eclipse if the orbit is circular. Quadratures are important because at these times the stars are moving at the maximum rate toward or away from the observer, and the best accuracy is attained there. RVs further away from quadrature, are useful, though, to define the shapes of the curves. In the case of elliptical orbits, the RV curves will not be double sine waves and extra RVs over the full range will be essential.

In practical terms, one rarely gets RVs exactly at quadrature—or any other specified phase for that matter—because to do so, one would need to keep the telescope idle, waiting for the exact time, and one does not want to waste precious time on a clear night. What happens is that one starts with a fairly large target list—say 15-20 stars—calculates phases for each throughout the night (there is software to do the job, Nelson 2008b), and picks out one or two useful targets for each hour, modifying the plan 'on the fly' as necessary. Reducing data during the night through dedicated software makes it possible to make optimum decisions to obtain effective data acquisition.

With the light curve(s) and the pair of RV curves, one can determine the mass ratio unambiguously and solve for absolute parameters, i.e. actual stellar masses and radii.

Obtaining Spectra

How does one get spectra good enough to extract radial velocities? First of all, you need a large telescope (that is, large compared to the telescopes required for securing light curves). For stars to magnitude 11 or so, an aperture of 1.5-2.0 metres is required.

The Dominion Astrophysical Observatory in Victoria, BC, Canada has a venerable 1.85 meter telescope ("The Plaskett") that is well set up for radial velocity studies with a state-of-the art Cassegrain spectrograph and detector, and dedicated staff to keep things running smoothly. There are undoubtedly other telescopes in this category around the world, but the author has not attempted to obtain observing time at them—yet.

The spectrograph needs to have a small enough reciprocal dispersion to yield high enough resolution. A grating with reciprocal dispersion of around 10 Angstroms/mm has been recommended (Rucinski, S. 2002) and seems to do the job.

The term *dispersion*, D, refers to how much the wavelengths in the spectrum are spread out. Therefore $D = dL/d\lambda$, where L = the distance along the wavelength axis at the detector, and λ is the wavelength of the light in the same units. It is more common to talk about the scale along the wavelength axis; this is *the reciprocal dispersion* $1/D = d\lambda/dL$. The DAO Cassegrain spectrograph on the Plaskett has available reciprocal dispersions of 10, 15, 30, 45, 60, and 100 Angstroms/mm, where the lower resolutions—higher values of 1/D—are used for classifying spectra. Therefore the value 10 Å/mm represents the highest resolution at this facility.

A related term is the *resolving power*. This is defined as $R = \lambda/\Delta\lambda$ where $\Delta\lambda$ = the error in the wavelength, which is a product of many factors. If the CCD pixel size is well matched to the spectrograph's resolution, we may take $\Delta\lambda$ as matching the pixel size. Then $\Delta\lambda$ = (pixel size) x $d\lambda/dL$. For the SITe-5 CCD chip at the Plaskett, $\Delta\lambda$ = 0.024 mm x 10 Å/mm = 0.24 Å, and therefore resolution R = 5000 Å / 0.24 Å = 21,000.

Typical Echelle spectrographs have much higher resolutions—between 25,000 and 85,000—but these values are not suitable for telescopes in the 1.5-2.0 metre range working on eclipsing binaries, as the light with these devices is spread out too much, resulting in count rates too low to be useable. Other considerations are that the star needs an autoguider to optimize the efficiency of data collection, a good reflectivity of the optics, and a well aligned optical train.

Before a run, the observer needs to have specified to the observatory staff the grating (mentioned above), the detector, and the wavelength setting. Because of the available grating blaze (the wavelength yielding the greatest intensity of the diffracted image) at the DAO (5000 Angstroms), and the peak in the CCD quantum efficiency (at about the same wavelength), the author has found it convenient to operate in the spectral range 5000-5250 Angstroms. The presence of two strong iron absorption lines there (Fe I of 5167.487 and 5171.595 Angstroms) is very fortunate indeed.

It is not clear that this is the only suitable wavelength range for eclipsing binary RV studies. If the CCD chip has a peak sensitivity at a wavelength other than 5000 Angstroms, and if the available gratings were blazed at close to that wavelength, then it would pay the observer to look for strong lines in that wavelength range. One is, however, always constrained by the equipment available at that observatory. On the other hand, if one is designing a new observatory, one can look around for the best combinations of the factors described above.

At the telescope then, one needs to do a number of things. Strictly speaking, one should take the usual bias, dark, and flat frames. The author's experience has been to omit these steps. The reasons follow.

For the bias frames, the pixel values consist of three components: the constant offset (necessary to prevent negative pixel values), the (random) noise component, and the structural component. The first is simply subtracted in the processing, and the second is different with each exposure and therefore cannot be suppressed. Only the third component can be dealt with by the master bias frame; however, the rms variation over the frame, in the author's experience at the DAO, is in the neighbourhood of only 25 counts and may be neglected. There is no pattern discernable anyway.

For the dark frames, cooling to -100°C reduces the dark count over hour-long exposures to only a few counts, so there is no appreciable dark component to the images (< 0.3 counts/minute).

For the flat-fielding, one is not concerned with brightness values as a function of wavelength, as one would if doing spectrophotometry. Later in the image reduction, it is necessary to normalize the continuum to a constant flat value, so any modifications made earlier by flat-fielding are pointless. It is only the wavelength shifts in which one is interested, not the intensity.

During the observing run, one needs to take three types of frames: standard star spectra, program star spectra, and arc spectra.

A list of RV standard stars is available in the *Astronomical Almanac* (Roberts & Boksenberg 1986, page H42-3, but any year will do). One generally selects stars that are near in spectral type and luminosity class to the target stars (typically A-F, luminosity class V) and as bright as possible. Typical exposures of standards, running from magnitude 2 to 8 on a 1.5-2.0 metre class telescope, run from a few seconds to perhaps 10 or 20 minutes.

Arc spectra (emission lines produced by a lamp containing low pressure gas) are taken by a source that is placed by observatory staff in the optical train so as to provide an accurate wavelength calibration with no spurious shifts due to poor alignment. Images are taken immediately before and after each program or standard spectrum in order to account for any flexure in the spectrograph and detector. These are, of course, emission spectra with sharp lines. At the DAO, a source containing Fe and Ar lines is used. Exposures are 5-10 seconds.

The exposures for the program stars must be long enough to allow for sufficient count rates and high enough signal-to-noise (S/N). For detached and semi-detached systems, S/N ~ 50 or higher are needed, while contact systems will often need S/N = 100 or more. S/N may be roughly calculated by taking the square root of the continuum level in the 1-dimensional spectral image. Therefore a continuum level of at least 10,000 counts is the goal.

Slit size is a trade-off. During times of good seeing and low wind speed, a 1 arc second slit will yield good images with low wastage of precious starlight. This presupposes the availability of an autoguider (the DAO has one) or conscientious manual guiding. In times of poor seeing and/or troublesome winds, however, a 2 arc second slit may be needed to secure useful data. Note that the velocity resolution scales linearly with slit width, whereas S/N scales with the square root of the slit width.

At locations plagued with clouds (such as Victoria), one is faced with a dilemma when clouds arrive and the star image disappears or grows faint. Should one keep going in the hopes that the clouds will go away, or should one terminate the exposure early? There is no easy answer here, but one should be guided by the fact that if the sky has been clear during the first part of the exposure and cloudy later on, there will be an error in the mid time of the exposure. Therefore, in the author's experience, it is better to quit early unless a visit outside reveals that the cloudiness is temporary. Don't clutch at straws!

Radial Velocity Reductions

Now how does one process the images? Specific comments will refer to the program RaVeRe (Radial Velocity Reductions, Nelson, 2008b). See Figure 2 (next page).

The first step is to deal with any bad columns that may be a documented problem with that particular CCD chip. RaVeRe uses a simple algorithm to deal with bad columns.

The next step is to cut out unused parts of the CCD frame, that is parts of the frame that are 'off the end' at either side of the wavelength (X) axis. RaVeRe does this easily. To find the limits, look at the FITS header or plot a row. Rows (Y) that are off the chip need not be dealt with because one selects the range for the background when the apertures are selected, and the rows that are off the edge are not selected.

Now for the cosmic ray hits: high energy particles hit the CCD chip, creating many electrons in a pixel and often several adjacent pixels.

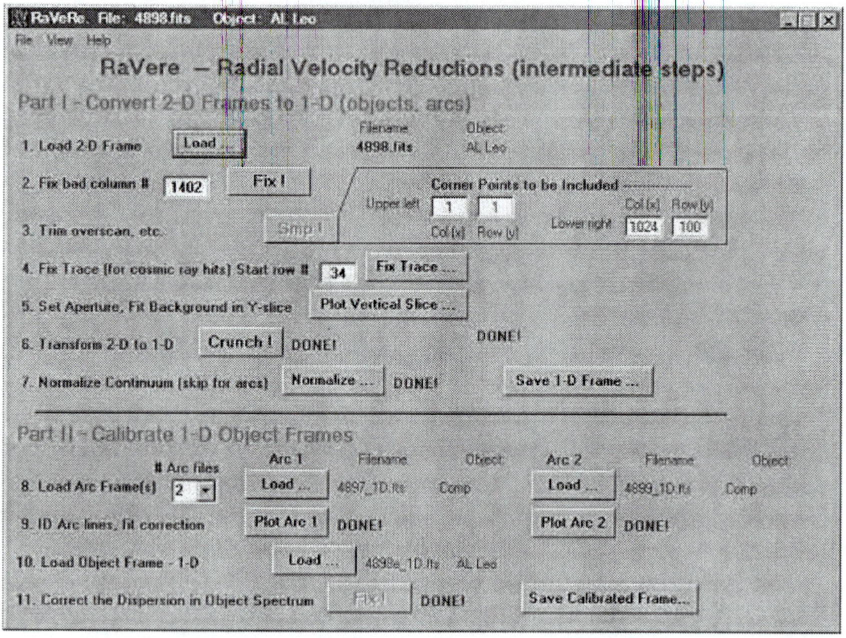

Figure 2: The main screen of the program *RaVeRe.*

These are simply a fact of life in long exposures. Observatories at sea level have half the level of those at elevation 1500 metres. Nevertheless, hour-long exposures will give typically 300 cosmic ray hits at sea level. Hits in the background (i.e., off the trace) are not a problem if the software uses a median fit for the background (RaVeRe allows for this). Extreme values get rejected. However, in the trace, automatic cosmic ray rejections are not appropriate since cosmic ray pixel values vary from highs of up to 10,000 counts down to zero (Monin 2007). Any automatic routine runs the risk, in the author's opinion, of degrading the spectrum. Manual methods are therefore preferred because one can use judgement about what to reject and what to accept. RaVeRe allows for manual rejection of cosmic ray hits in the trace, row by row. 'Bad' pixel values are replaced by a simple average of adjacent values from the same row.

Another process is the rectification for possible non-orthogonality between the slit and grating. If present, this would result in a trace that is not quite parallel to the CCD rows and problems during summation along the columns inside the aperture. In practice, this has not been a problem at the DAO with its well-maint

spectrograph. This feature will be added to RaVeRe if there is a demonstrated need for it.

The next process is to collapse the two-dimensional frame to a one-dimensional frame. One needs to do two things: a) set the aperture (rows to be included) for summation in each column, and b) set the apertures for the calculating of the background level and do the calculations to match the data (i.e., to do a background "fit"). RaVeRe gives the user a column plot that enables one to select the apertures and examine how well the calculated level matches the background data. The program then sums over the central aperture and subtracts the scaled median value for the background, arriving at a single value for each column (or wavelength bin). One now has the 1-D spectrum.

Once the one-dimensional spectrum has been obtained, one needs to normalize the continuum, i.e., make it flat (the following program *demands* this). The reason the program requires a flat continuum is that, once again, one is not interested in the intensities, just the wavelengths. RaVeRe does this by plotting the spectrum; the user then picks off likely points for the continuum fit either manually or by an automatic feature, fitting a polynomial of order up to degree 4, then normalizing the spectrum so that the continuum is flat. This completes Part I.

Representative samples of standard spectra (Figure 3) and program spectra (Figures 4-6) appear below.

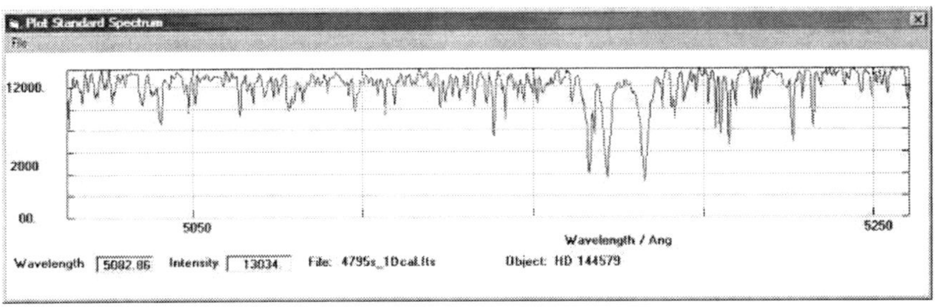

Figure 3: Spectrum of the RV standard HD 144579.

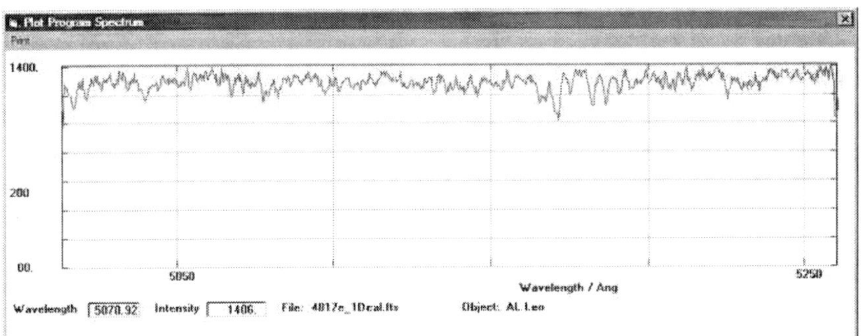

Figure 4: Spectrum of AL Leo (EA/D, P = 1.61 d, Sp: A8, S/N = 37).

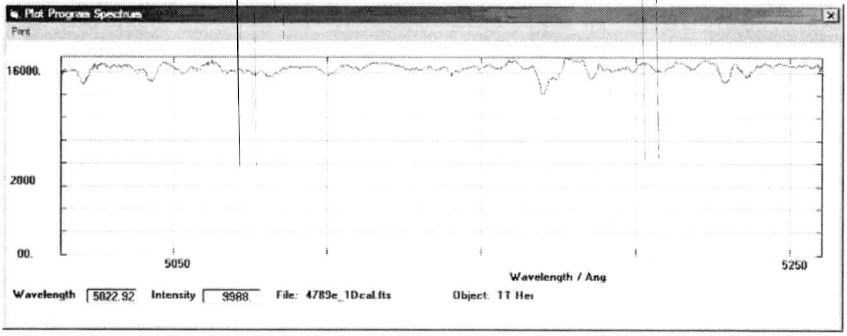

Figure 5: Spectrum of TT Her (EB/KE, P = 0.912 d, Sp: A7 V, S/N = 130).

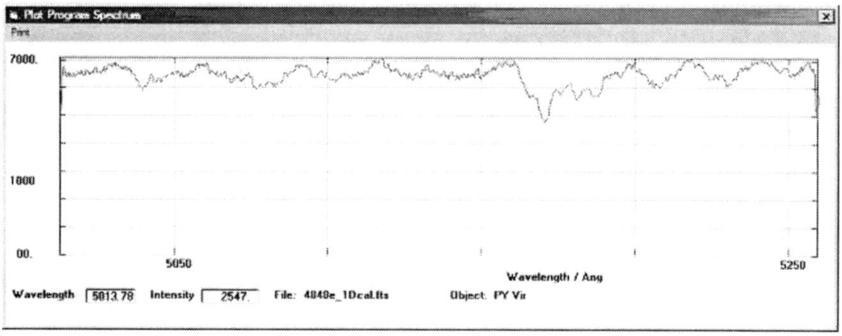

Figure 6: Spectrum of PY Vir (EW, P = 0.311 d, Sp: K0, S/N = 83).

Accurate wavelength calibration and rectification (remapping to a linear wavelength scale) are essential steps (this is the job of Part II in RaVeRe, see Figure 2). The first step is to identify several emission lines in the (FeAr) arc. Other sources such as cadmium are sometimes used—the key requirement is to have a bright source with many sharp lines. The standard source is the 100,000 line *MIT Wavelength Tables* (Harrison 1969), but the author has an Excel file of unknown source containing some 6800 lines that can be sent upon request. The identification of spectral lines can be prone to errors, but the plotting of tabular Fe and Ar wavelengths and strengths graphically on an Excel worksheet can be very helpful. See Figure 7 (next page).

RaVeRe allows for expedited identification of arc standard lines. For the first two lines only, one clicks on a line in the graphical display of the arc spectrum and a separate window (Figure 8) opens. One then types in the 7-figure wavelength in Angstroms or accesses tabular Fe/Ar tabular values by means of a scroll bar.

Once the first two lines have been identified, the wavelength scale is known roughly. For all subsequent lines, the program calculates approximate wavelengths and proposes an identification, together with deviations from a linear fit. The user can accept or scroll to a new value or cancel. In this way, identifications are built up throughout the range and a polynomial fit of the deviations calculated, up to degree 4. Of course, the user has full ability to delete bad identifications.

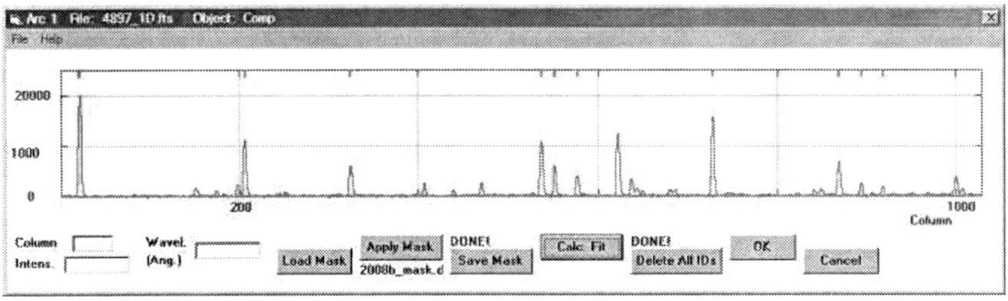

Figure 7: RaVeRe's plot of an arc spectrum. Identified lines are marked by the tic marks at the upper border.

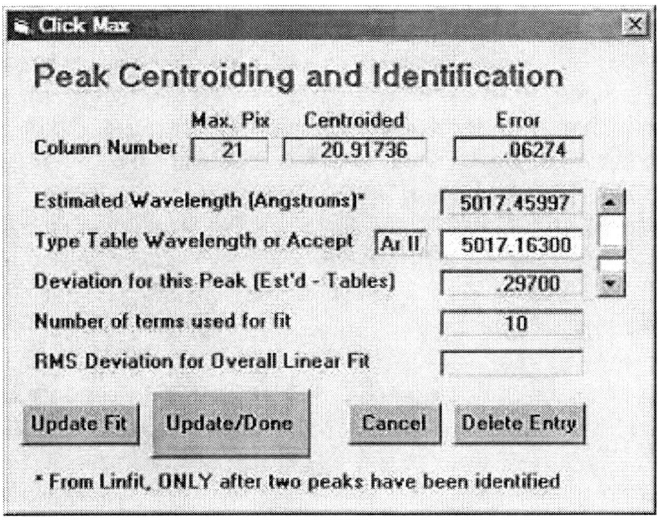

Figure 8: Centroiding and identifying lines.

Once a consistent set of identifications has been made, the user can save these in a *mask* file that circumvents the need for line identification each time. The mask is a simple text file containing the precise tabular wavelength values and the pixel number for each. The software uses it to identify the arc lines whose pixel values might differ by an integer or two from frame to frame. Only a few clicks are then necessary to identify everything and make the fit. See Figure 9 for a plot of the deviations and the 4th power fit made.

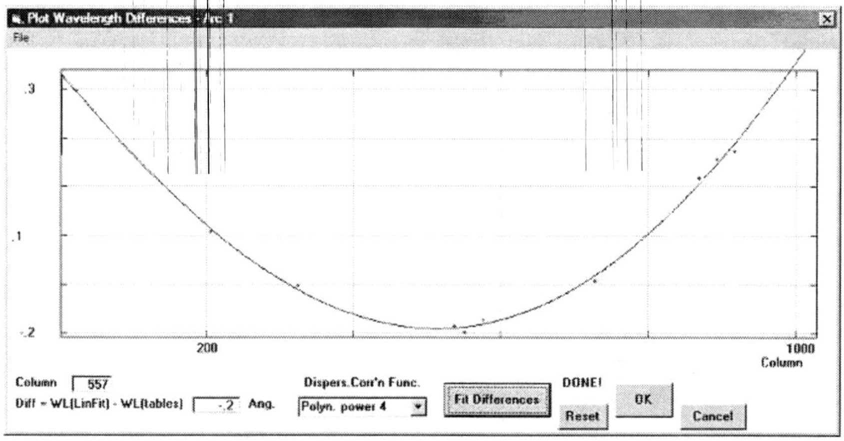

Figure 9: A plot of the deviations from a linear relation and the 4th power fit.

When the lines have been identified in the arc files taken immediately before and after the object exposure, and a fit of the deviations made, RaVeRe will rectify the wavelength scale in the object spectrum (standard or program star) so that it is strictly linear. The constants (the wavelength of the first pixel at the blue end and the scale in Angstroms per pixel) are then written to the fits header for use by subsequent software.

Now, given that one has a well-calibrated set of spectra, for both the target and standard stars, how does one extract the radial velocities?

Radial Velocity Extraction

Historically, the standard tool for the extraction of RVs has been that of the cross-correlation function (CCF), first used in astronomy by Simkin (1974); in her case it was for galactic spectra (Kallrath & Milone 1998, page 37): "Typically, the Fourier transform of the linearized and rectified spectrum of the program star along with the conjugate transform of the reference spectrum is calculated, and the product of the transforms is evaluated. The CCF can be described as the Fourier transform of this product."

For *detached* stellar spectra, CCFs can yield errors in the order of 1 km/s (Kallrath & Milone 1998). However, contact binaries are problematic because the single dumbbell-like structure has regions moving at RVs continuously spaced over a wide range of values at useful times in its orbit. Therefore the spectral lines are heavily broadened and do not respond well to cross-correlation with narrow-lined comparison spectra. The peaks in the CCFs are often weak and ambiguous.

An alternate method, that of broadening functions (BFs), was introduced by Rucinski (1992, 2002, 2005). Simply speaking, the procedure takes the observed spectra of program stars that are broadened and shifted, together with that of the radial velocity standard stars that are unbroadened but shifted by a known amount, and solves for the process that produced the broadening and shifting. Linear transformations are used throughout. For a more complete explanation, see the section near the end of this chapter titled "Broadening Functions."

The original implementation of Rucinski's broadening functions was written in the computer language IDL. As the latter is unavailable to many researchers, the author rewrote the complete package in Microsoft Visual Basic 6.0. In order to weed out the many errors involved in writing the extensive code, he tested the process line by line in side-by-side computers, one running VB 6, and the other IDL. See Figure 10 for the main screen of the program.

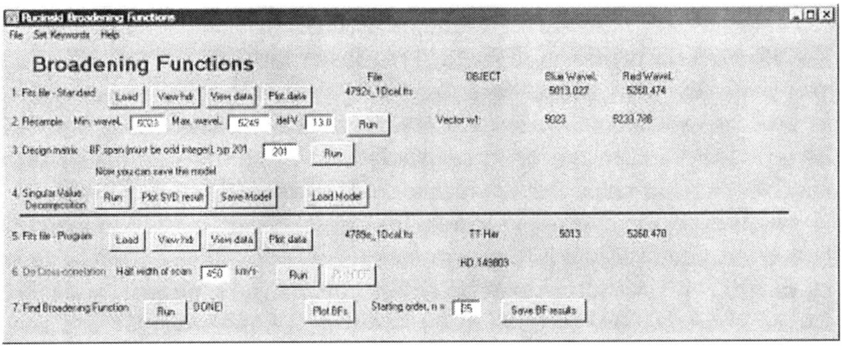

Figure 10: The main screen of program Broad.

The software package is called Broad. In it, the steps are laid out in a logical order. Most of the steps are just a simple mouse click, but there are some choices along the way.

Step 1: Load the standard file.
Step 2: Resample the intensity values and assign them to a scale that is linear in velocity (Rucinski, 1992).
Step 3: Assemble the *design matrix* **S**.
Step 4: Invoke the method of SVD to solve for matrices. This can in principle be time-consuming, depending on the computer and the values of m and n. However, for the author's Pentium computers, and for n = 1024 and m = 201, computation times in the order of 20 seconds were experienced. But for speed in processing many

> spectra, the author implemented a feature whereby the user can save the model to disk, later loading a number of models (up to 13) for rapid switching.

Once the model has been determined, one merely needs to multiply it by vector **b** (a quick process) to obtain the BF, which then can be plotted, peak velocities found, etc.

Step 5: Load the program file.
Step 6 (not implemented yet): Perform the cross correlation, a desirable but optional step.
Step 7: Plot the broadening functions.

Some typical BFs are shown below. The independent variable in the BF is radial velocity (km/s), while the dependent variable is intensity per resolution element (Rucinski 1992).

If the binary pair is detached, strong detached peaks will be seen in the BF, even for signal-to-noise (S/N) ratios as low as 50 or less. The peak velocities are then easily determined to excellent precision, ± several km/s. One can also determine the areas under the peaks whose ratio will be equal to the luminosity ratio, needed to find a unique WD solution for detached systems; this is a feature in the later versions of *Broad*.

However, for contact binaries, the peaks will not be detached and one often needs S/N = 100 or more especially for extreme mass ratios q, where the secondary will be small and its light output weak (See Rucinski 1992 with AW UMa where q = 0.08; he needed S/N = 345!).

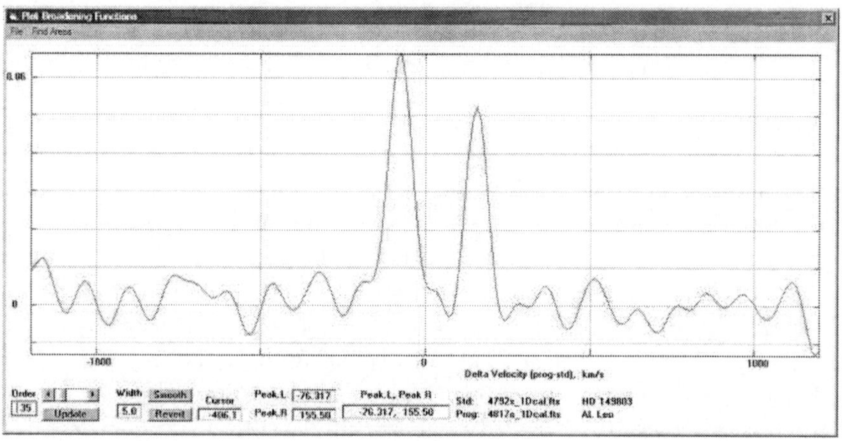

Figure 11: Broadening Function for AL Leo (EA; P = 1.61 d; q = 0.97; S/N = 37).

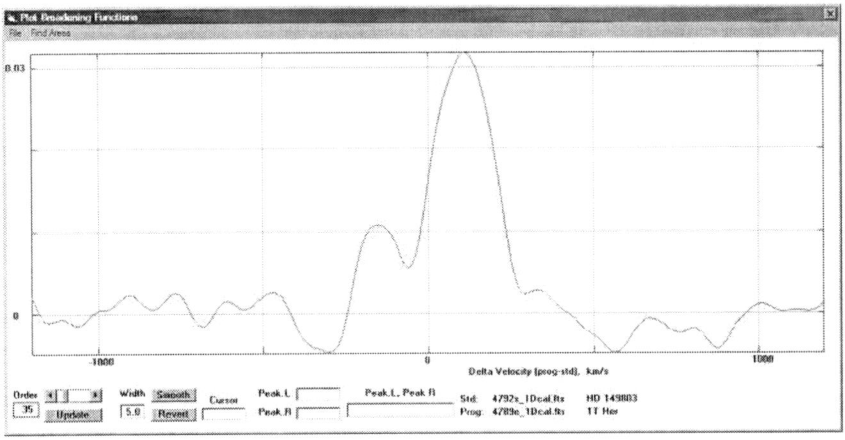

Figure 12: Broadening Function for TT Her (EB; P = 0.91 d; q = 0.37; S/N = 130).

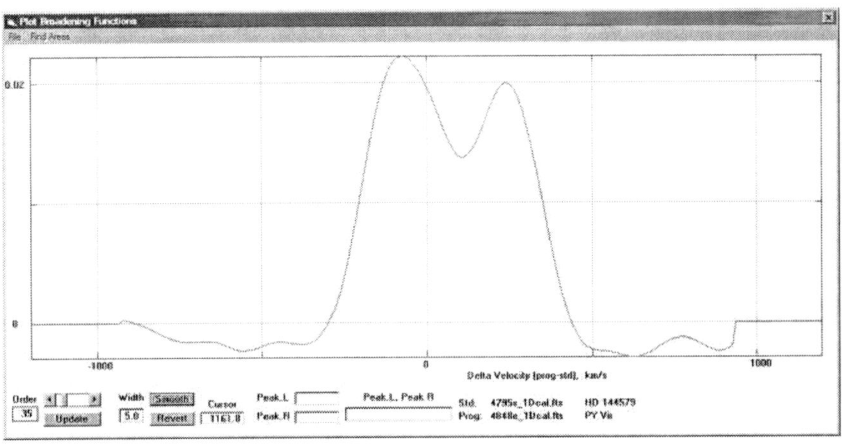

Figure 13: Broadening Function for PY Vir (EW; P = 0.31 d; q = 0.67; S/N = 83).

The differences between these three BFs should be immediately apparent. For AL Leo (Figure 11), a detached binary, the peaks are separate and high, as noted above, even for the fairly low S/N. For TT Her (Figure 12), a close but still detached binary, the peaks are closer and no longer separate. There is a large difference in the heights of the peaks, appropriate for the low mass ratio, and therefore greatly differing stellar radii, resulting in greatly differing luminosities.

In the case of PY Vir (Figure 13), a contact binary, the peaks are even closer, as noted above. This closeness can be problematic for low S/N and/

or low mass ratios where the secondary star is small and faint. In that case, it is difficult to sort out the secondary peak from the "bumps" due to noise.

A parameter that must be sorted out is the order number to be used in plotting the BF. This is essentially the number of terms in the summation. Clearly the velocities extracted should *not* be a strong function of the order number! Therefore it is incumbent on the user to determine a curve of growth of the velocities and pick an order number in the middle of a flat region. A useful feature is the Gaussian filter provided in Broad. Low widths, like 3 or 5 pixels, can filter out high frequency noise and make the peak velocities easier to determine, whereas values that are too large can smooth out the secondary peak entirely. Clearly, the determination of peaks can be a bit of an art; however, consistent methods should be used throughout the series of program frames. The reader should consult Rucinski (1992, 1995, and 2005) for further information.

Final Steps

The Earth, of course, is moving in its orbit at about 30 km/s; this movement must be accounted for with both the standard and program spectra. For the former, we need to find the relative velocity of the standard star relative to the Earth. We do not care about any transverse components of the velocity vectors, just the radial components. Therefore for the relative velocity of the standard star relative to the Earth we need only take the tabulated RV (V_{SH}) and subtract off the component of the Earth's velocity (V_{EH}) in the direction of the standard star at the instant that the spectrum was taken. That is:

$$V_{SE} = V_{SH} - V_{EH} \quad \text{where S = standard star, E = Earth, H = Sun (Helios)}$$

Next, one realizes that the radial velocities derived from the BF, ΔV, are in fact the relative, radial velocities of the program star relative to the same standard star placed hypothetically in the same part of the sky. Therefore the radial velocity of the program star relative to the Earth is

$$V_{PE}' = V_{PS}' + V_{SE} = \Delta V + V_{SE}$$

where P = program star and the prime emphasizes a different direction and a different coordinate system.

Then, finally, one converts the above velocity to a heliocentric one:

$$V_{PH} = V_{PE}' + V_{EH}'$$

Quantities V_{EH} and V_{EH}' must be determined somehow. Traditionally, astronomers used *The Astronomical Ephemeris* (Roberts & Boksenberg 1986), looking up values and interpolating as necessary. This was tedious and time-consuming, and required one to have an up-to-date copy of *The Astronomical Ephemeris*.

However, there are many sources of 'cookbook' formulas that can be used to calculate the Earth's velocity at any instant to a sufficient precision, that is, with errors less than other errors in RV determination (See Duffet-Smith 1979, for example). These formulas have been incorporated by the author into Excel template worksheets as Visual Basic macros so that values such as V_{SH} (standard star RVs), ΔV (values from the BFs), star coordinates, dates and times (UTC) can be entered into the worksheet and the results of the above equations found automatically. This saves much time and avoids mistakes. Copies of these template files will be sent upon request or found in the author's website (Nelson 2008c).

Examples of RV curves are given below (Figs 14-15 next page). The full analyses of these solved systems can be found in Nelson (2004, 2006).

The Broadening Function

To explain how the method of broadening functions works, one starts with a column vector **p** that is the already broadened and shifted spectrum of the program system of two stars. It will contain **n** pixels (typically 1000-2000), and be rebinned so that the wavelength scale is linear in velocity. One then needs to solve for the *linear* transformation which maps the sharp-lined spectrum of a standard star into spectrum **p**. This transformation is represented by another column matrix, **b**, having m elements (where m < n is typically 201, etc.—m must be an odd integer). It is a set of elements along the velocity axis that, for each velocity, represents the fraction of the spectrum that comes from that velocity. Then one sets up the (n-m+1) x m matrix **S** consisting of the sharp-lined standard spectrum in each row. However, each row is shifted a specified velocity interval from the one above (because one doesn't know what the program star shifts are—one is solving for them). Note that the center row is unshifted relative to the program spectrum. Also, matrix **p** is truncated by m-1 elements to clip off the non-overlapping values at each end. Then one has the system of linear equations

$$\mathbf{S} \bullet \mathbf{b} = \mathbf{p} \quad (1)$$

where the raised dot represents matrix multiplication.

In order to visualize what is happening, let us consider a highly simplified example. Imagine that the standard star spectrum in the region of interest contains just a single narrow absorption line, in this case, also located in the centre of the wavelength range of interest. It will in general be moved in wavelength from its laboratory value, but that amount will be known (it is, after all, a standard RV spectrum), and will be dealt with later in the analysis. This spectrum is now represented by any row in the matrix **S** on the left. (See Figure 16 next page.) The dots represent matrix elements; they have been omitted here and there for clarity. In this case, matrix **S** has 11 rows and 9 columns. Matrix **b** has 9 rows, as it must for it to be multiplied by **S**. The product of the two, matrix **p**, has 11 rows, corresponding to the number of rows of **S**, and one column.)

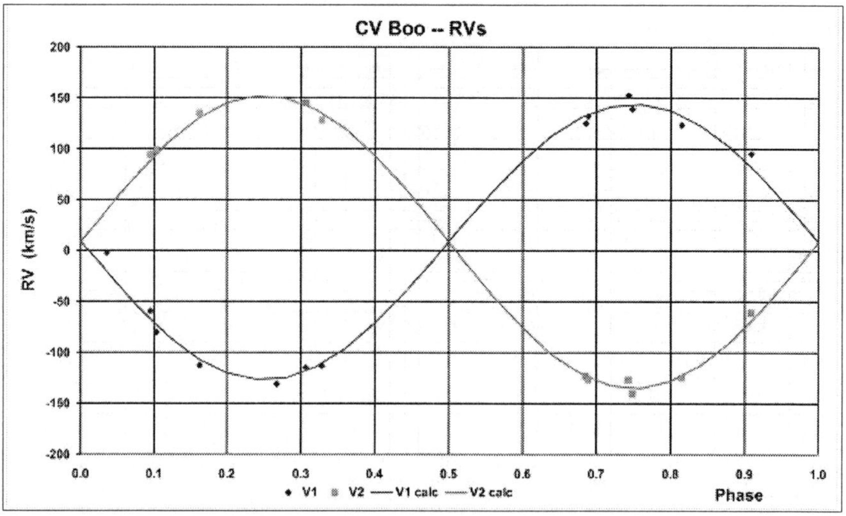

Figure 14: RV curve for CV Boo (EA, P = 0.847 d).

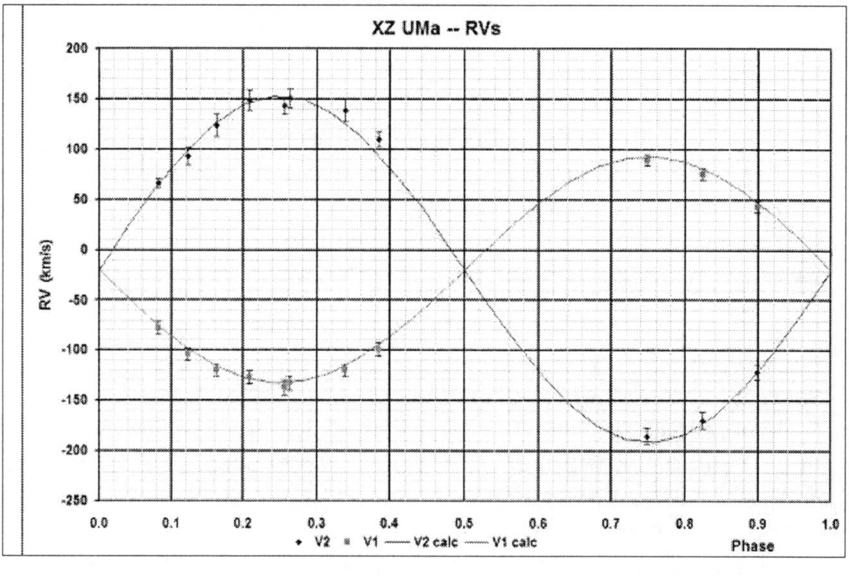

Figure 15: RV curves for XZ UMa (EA, P = 1.222 d).

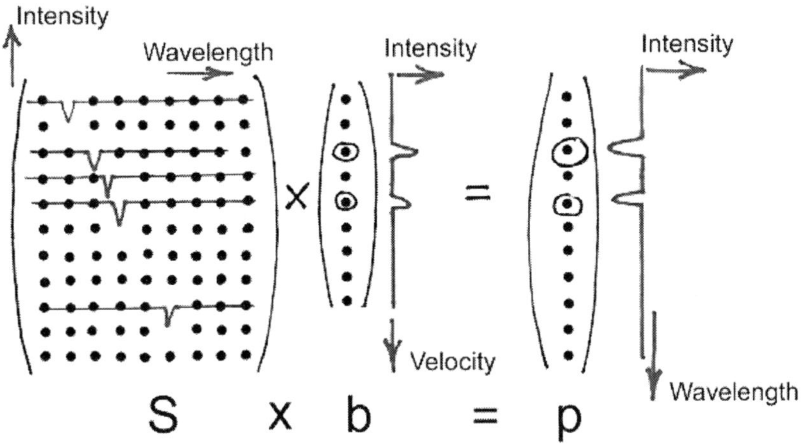

Figure 16: A schematic diagram of the defining matrix multiplication.

Remember that each row in **S** is shifted a specified wavelength (or velocity) amount from the row above. Little sketches are given to help the reader visualize the situation. Matrix **b** is the broadening function, which will be discussed in a moment.

Let us image also that the system of interest consists of two stars, well detached (that is, in orbits whose radii are large compared with the stellar radii), and that the spectrum was taken at quadrature. Imagine also that each star has the same single narrow absorption line as the standard star. At the instant that the spectrum is taken, one star will be moving toward the observer, and the other away, relative to the velocity of the centre of mass of the system. One of the stars will be moving at a velocity that corresponds to row 3 in matrix **S**, the other, 5. Multiplying matrix **S** by **b** then gives **p**, the observed spectrum, with dips corresponding to elements 3 and 5. Therefore matrix **b** represents the transformation applied by the Doppler shifts of the two stars. The velocities of the peaks will correspond to the velocities of each star.

In a system of close binaries (especially contact binaries), the situation is not so simple. Different parts of a given star at any given moment will be moving at quite different velocities. The algorithm allows for this, yielding a (broad!) broadening function that represents the amplitude or probability of getting a given velocity from somewhere in the system. Each peak corresponds to some sort of average velocity, but the exact nature of this average is unimportant, because the Wilson-Devinney routines will calculate the same average. In any case, it will usually, except in the case of highly distorted systems, correspond to the velocity of the center of mass of the star in question. Remember that because of the makeup of stellar interiors, each mass may be considered, to a high degree of accuracy, to be a point mass at its center.

For close binaries, then, the velocity distribution, as represented by the broadening function, will consist of two close broad peaks close together. In cases with a low mass ratio, that is where the secondary is much smaller (and as a rule dimmer) than the primary; this means that the secondary peak in the broadening function will be weak. A high S/N ratio in the program spectrum to separate the valid peaks from the background noise is therefore required.

Now, getting back to equation (1), inverting matrix **S** to solve for **b** is difficult and error-generating since n is in the order of 1000 - 2000 elements, and m ~ 100-500. In any case, the system of equations is over-determined in that there are more equations than unknowns.[1] It turns out that the method of *singular value decomposition* (SVD) is ideal for the purpose of finding a "best" solution (Press, *et al.* 1988). To use this method, one needs a square matrix on the left, and matrix S is, in general, not square.

The remedy is to multiply both sides by $\mathbf{S}^T$, the transpose of matrix **S**, getting

$$(\mathbf{S}^T \bullet \mathbf{S}) \bullet \mathbf{b} = \mathbf{S}^T \bullet \mathbf{p} \quad (2)$$ [2]

In the last reference, the authors quote a theorem which states that square matrix $(\mathbf{S}^T \bullet \mathbf{S})$ can be replaced by the product of matrices $\mathbf{U} \bullet \mathbf{W} \bullet \mathbf{V}^T$ where matrices **U** and $\mathbf{V}^T$ are orthogonal (that is, their inverses are equal to their transposes).

$$\mathbf{U} \bullet \mathbf{W} \bullet \mathbf{V}^T \bullet \mathbf{b} = \mathbf{S}^T \bullet \mathbf{p} \quad (2a)$$

1 That is, if $\mathbf{S}^{-1}$ is the inverse of matrix **S**, then, by definition $\mathbf{S}^{-1} \cdot \mathbf{S} = \mathbf{S} \cdot \mathbf{S}^{-1} = \mathbf{I}$ where **I** is the unitary matrix whose diagonal elements are all 1, and all else are 0. (Note that as a rule, one cannot alter the order of matrix multiplication and get the same result. One needs to pay strict attention to the order of multiplication which must be the same on each side of the equation.) Then, multiplying both sides of equation (1) on the left by $\mathbf{S}^{-1}$, one gets

$$(\mathbf{S}^{-1} \cdot \mathbf{S}) \cdot \mathbf{b} = \mathbf{I} \cdot \mathbf{b} = \mathbf{b} = \text{(right side of equation)} = \mathbf{S}^{-1} \cdot \mathbf{p} \quad (4)$$

and this is a solution for **b**.

2 The transpose of a matrix has the rows and columns interchanged. If matrix **S** has p rows and q columns, $\mathbf{S}^T$ will have q rows and p columns. Therefore, by the definition of matrix multiplication, the product matrix $\mathbf{S}^T \cdot \mathbf{S}$ will be square (with p rows and p columns). Note that multiplying both sides of the equation by $\mathbf{S}^T$ on the left is a valid algebraic manipulation of the equations that leaves the solutions intact. See the graphic below for a primer on matrix multiplication. Note too that the reverse product $\mathbf{S} \cdot \mathbf{S}^T$ would have q rows and q columns, and would be quite different.

Note that T denotes transpose and that here the (m x m) matrix **W** is a diagonal matrix having only positive elements. (Its inverse is easy to find—each diagonal element is the reciprocol of the corresponding original element.) Then it follows that, if a solution can be found, vector **b** can be found by multiplying both sides of the equation on the left by the appropriate matrices:

$$\mathbf{b} = \mathbf{V} \bullet [\mathrm{diag}\,(1/w_j)] \bullet \mathbf{U}^T \bullet \mathbf{S}^T \bullet \mathbf{p} = \mathbf{M} \bullet \mathbf{p} \quad (3)$$

and one may then examine the vector **b** (the BF) for 'peaks'. The meaning of the last equation is that the matrices **V, W,** $\mathbf{U}^T$ (and $\mathbf{S}^T$–easily found) are solved for and their product found by simple matrix multiplication. The result is then stored; this is the *model* $\mathbf{M} = \mathbf{V} \bullet [\mathrm{diag}\,(1/w_j)] \bullet \mathbf{U}^T \bullet \mathbf{S}^T$. Then, for each program spectrum, matrix **p** is determined, and one needs only to multiply by matrix **M** to get the broadening function **b.** Peaks in the latter will tell the analyst the desired radial velocities. For a more complete discussion of the above, see Rucinski 2005.

Conclusion

In this chapter, the author has laid out all the essential steps needed to derive RVs from standard and program spectra. It is to be hoped that an increasing number of observers will take up the cause of eclipsing binary studies using radial velocities. It is not difficult. The key lies in choosing systems that are bright enough for the telescope available, have strong enough lines, and are double lined. The requirements for the hardware are discussed above. In choosing targets, there are a number of lists available (see Nelson 2008c).

References

Bradstreet, D.H. 1993. Binary Maker 2.0 -- An interactive graphical tool for preliminary light curve analysis. In Milone 1993, below. A Windows version, 3.0, is available at binarymaker.com.

Bruton, D.. 2008. *Starlight Pro*. Free software available at midnightkite.com/binstar.html.

Duffet-Smith, P. 1979. *Practical Astronomy With Your Calculator.* Cambridge University Press.

Ghedini, S.. 1982. Software for *Photometric Astronomy.*

Harrison, G.R. 1969. *MIT Wavelength Tables, 2nd ed., Vol. 1.* Cambridge, MA: MIT Press. See Amazon.com.

Kallrath, J. & Milone, E.F. 1998. *Eclipsing Binary Stars - Modeling and Analysis.* New York: Springer.

Kreiner, J., *et al.* 2000. *An Atlas of O-C Diagrams of Eclipsing Binary Stars*. Internet <http://www.as.ap.krakow.pl/o-c/cont.html> accessed 18 January 2010.
Kundera, T., [date unknown], *Eclipsing Binaries Minima Database*. Internet <http://www.oa.uj.edu.pl/ktt/krttk_dn.html> accessed 18 January 2010.
Milone, E.F., ed. 1993. *Light Curve Modeling of Eclipsing Binary Stars*. New York: Springer-Verlag.
Monin, Dmitry. 2007. Private communication with author.
Nelson, R.H. 2004. The detached solar-type binary CV Boo. *IBVS*, 5535.
Nelson, R.H., Terrell, D., and Gross, J. 2006. The classical Algol XZ UMa -- observations and analysis. *IBVS*, 5715.
Nelson, R.H. 2008a. *Eclipsing Binary O-C Files*. Internet <http://www.aavso.org/observing/programs/eclipser/omc/> accessed 18 January 2010.
Nelson, R.H. 2008b. *Software by Bob Nelson*. Internet <http://members.shaw.ca/bob.nelson/software1.htm> accessed 18 January 2010.
Nelson, R.H., 2008c, *Spreadsheets for Astronomy*. Internet <http://members.shaw.ca/bob.nelson/spreadsheets1.htm> accessed 18 January 2010.
Press, W.H., Flannery, B.P., Teukolsky, S.A., and Vetterling, W.T. 1988. *Numerical Recipes in* C. New York: Cambridge University Press.
Roberts, C.K. and Boksenberg, A. 1986. *Astronomical Almanac*. London: H.M. Stationery Office.
Rucinski, S. 1992. *AJ 104* (5), 1968 ff.
Rucinski, S. 1999. Determination of broadening functions using the singular value decomposition (SVD) technique. IAU Coll. 170, *ASP Conf. Ser. 185*.
Rucinski, S.. 2002. private communication with author.
Rucinski, S. 2005. Internet <http://www.astro.utoronto.ca/~rucinski/SVDcookbook.html> accessed 18 January 2010.
Russell H.N. and Merrill, J.E. 1952. *Princeton U Obs. Contr., 26*,1-96.
Simkin, S.M.. 1974. *A&A, 31*, 129-136.
Terrell, D. and Wilson, R.E. 2005. *Ap&SS, 296*, 221.
Wilson, R.E. and Devinney, E.J. 1971. *ApJ, 166*, 605.
Wilson, R.E. 1990. *ApJ, 356*, 613.
Wilson, R.E. 2005. Internet <ftp://ftp.astro.ufl.edu/pub/wilson/lcdc2003/readme.1april2008> accessed 18 January 2010.
Zhao, M., *et al.* 2008. *astro-ph*, arXiv:0808.0932v1, online paper, or at Zhao, M., *et al.*, 2008. *ApJ 684*, L95-L98.

Chapter 31

RTS2

Open Source Autonomous Observatory Control

Petr Kubánek

Remote Telescope System, Spain

Introduction

Developing an autonomous observatory seems easy. One locates an appropriate site, selects and acquires all required hardware and software, and starts to produce thousands of images of the night sky. And by a simple analysis of the acquired images with existing software, one will easily publish results in prestigious journals.

But life is hard and setting up an autonomous observatory is not an easy job. To ease the job a bit, we started development of software for a fully autonomous observatory. The driving force was to produce software which can easily integrate various subsystems. Building an observatory control environment will then be similar to creating something from LEGO blocks. This software is RTS2, which stands for Remote Telescope System, 2nd version (Kubánek 2006).

This chapter describes the software and the results acquired with it. As you can freely download RTS2 from the Internet, you can easily test the software and see if it suits your needs. RTS2 is written for Linux, but can run on the Microsoft Windows operating system as well.

Reasons for a Fully Autonomous Observatory

One might argue that a fully autonomous observatory, which does not require any human presence, is not necessary. Yet the nature of and gradually developing understanding of Gamma Ray Bursts (GRBs) provided an excellent reason for such a setup (Ackerlof 2007). This discovery was a complete surprise for the scientific community. Until recently their observation was amazingly difficult.

Two main reasons for a fully autonomous observatory devoted to GRBs are GRBs' occurrence rate and fast variations of the associated optical emission. On a typical observatory location with a fair clear night and good equipment, one can expect to detect one event per month (Kubánek 2007). GRBs quickly change magnitude within seconds from the trigger. If the observation is controlled by an observer, he/she must be very patient and react properly once the trigger is received. It is clear that a properly programmed autonomous observatory is the solution one would like to have for proper GRB observations. RTS2 provides software which can be easily extended to fit the needs of any observing program, but it was and still is designed to provide services for the quick change of targets needed for GRB observations.

RTS2 Design

RTS2 design separates software for *devices* providing interfaces to the instruments and *services*, representing a higher level logic operating on *devices* (Kubánek 2006). All RTS2 components communicate with each other through a text protocol (Kubánek 2008).

Each RTS2 program is written in C++ using common API provided in the RTS2 library. It is a single thread code, based around select calls responding to external events. On RTS2 Wiki – rts2.org/wiki – readers can find evolving pages describing how to write a device driver and how to use RTS2 API. As the API uses quite extensive object oriented programming advantages, interested readers should be familiar with objects and their features.

Device drivers are available for almost all possible kinds of instrumentation useful in observatory automation. And it is quite easy to write a driver for a new device.

Installation Instructions

Installation instructions are included in the INSTALL file, shipped with distribution. It is recommended to start installation from the source code obtained from Subversion Repository. The INSTALL file provides a list of libraries and utilities necessary for RTS2 compilation. In short, compiling from Subversion source code is performed by calling:

```
user@host:svn co https://rts-2.svn.sf.net/svnroot/rts-2/trunk/rts-2 rts2
user@host:cd rts2
user@host:~/rts2$ ./autogen.sh
user@host:~/rts2$ ./configure
user@host:~/rts2$ make
user@host:~/rts2$ sudo make install
```

After building from the source code, one can play a bit with dummy devices to see how the system behaves. Afterward, codes for a real device can be much more easily used, and possibly written and tested.

Code Overview

Understanding the RTS2 code requires some knowledge. This and the following sections provide hints on how to read the code.

An RTS2 program usually contains one application class and multiple connection classes. If a system *select* call finds some incoming data, the connection reads them and if all data arrived (in the RTS2 protocol, commands are separated by new line characters), it passes the command for processing. After the *select* call, the *idle* routine is called, which can do anything that is not triggered by incoming events. The *idle* call is usually used to check various timeouts and cancels the operation if timeout expires.

RTS2 poses two basic event distribution mechanisms: one is based around the system *select* call. As described above, this reacts to a command coming from connections. The other is based on passing the *Rts2Event* class through all *Rts2Object* instances available to the application. This call is handy when it is necessary to inform multiple objects – for example when the telescope finishes movement to position and exposure on all present cameras and detectors can start.

Code structure

RTS2 uses its own class hierarchy for building the code, following graphic depict basic RTS2 classes. They are described in the following sections:

Rts2Object This is a base class. It provides one virtual method – *postEvent*. This method is passed as an *Rts2Event* instance. Based on a result of the event call, it does what the event originator thought.

Rts2Event Base class of an event. It provides a pointer for optional data passed with the event, and storage for event type.

Rts2Conn This class represents a connection. Usually its subclasses are used. Class provides various methods for adding file descriptors to set-check for read, write, and exception events, and responding when some of these events occur.

Rts2App Represents an application. This abstract class defines a pure virtual *run* method, which holds the main loop of every application. It also provides various functions for management of command line arguments.

Rts2CliApp Base class of command line application. It implements functions for running separate applications, which do not connect to RTS2.

Rts2Block Class acting as the common ancestor for programs connecting to RTS2. It can hold connections and provides functions for connections management.

Rts2Daemon Base class for both device drivers and services, for everything that can and usually runs as background (daemon) processes.

Rts2Client Client base class. This is for clients that connect to RTS and use its protocols to communicate with each other.

Rts2Config Holds various .ini style configuration file parsing routines.

Sites Using RTS2

RTS2 is currently used on sites on four continents – Europe. Africa, America, and Australia. The sites are described on the RTS2 website.

Important Results

GRB Observations

More than 100 GCN circulars describing GRB observation from RTS2 controlled observatories have been published. Among them, three events are worth noting: GRB 050904, GRB 060117, and GRB 070610.

GRB 050904 This event was the first high redshift GRB. RTS2 controlled the BOOTES 1 telescope, which observed this event location as world second and constrained its R-band emission. BOOTES 1 data did not show detection but significant limits for an early R-band emission, which could rule out a high-z origin.

GRB 060117 This event is one of the brightest GRBs, with recorded 9th magnitude brightness some 120 seconds after trigger. Its light curve shows a very steep decay, possibly with an early jet break. Its optical emission was observed by the *FRAM* telescope. The *Swift* satellite cannot observe this GRB position due to solar avoidance constraints. Observations from other ground instruments did not start until a few hours after the event.

GRB 070610 This event started as a strange GRB. Soon after receiving the trigger from the *Swift* satellite it became clear it was not a common GRB. The most plausible explanation of this event is a dim isolated neutron star (Castro-Tirado 2008). RTS2 controlled the *Watcher* telescope and provided valuable data confirming strange light curve behavior with multiple brightening periods, lasting less than 120 seconds. Combined with the other observations from ground and space based instruments, the conclusion is that most probably we witnessed behavior of a previously unknown object – an isolated neutron star.

RTS2 as Laboratory Master Controller

RTS2 also serves as the master control software at the LSST CCD characterization and calibration laboratory at Brookhaven National Laboratory. It helped with the acquisition of terabytes of data. Those data are used to properly characterize LSST CCDs.

Future development

In 2009 RTS2 was installed on Yock-Allen Bootes 3 in New Zealand and on the 1.23m telescope at *Centro Astronómico Hispano Alemán*. Apart from installing RTS2 on another telescope, we would like to further develop the software.

Significant improvements are expected in four areas: development documentation, graphical user interface, data processing, and network management functions.

Development Documentation Improvements

In late 2008, an effort was started to ease new developers entering the project. The RTS2.org site was setup. Also, Wiki pages (rts2.org/wiki) for documentation storage were created. So far there are a few guidance texts on how to extend existing RTS2 code. It is expected those instructions will be extended over the years as new developers join the effort.

Graphical User Interface Improvements

RTS2 was written primarily as a commend line set of programs. It now provides users with an *ncurses* based interface to monitor devices. This is sufficient for experienced users. But it is highly discouraging for new users, who are unable to quickly see results of their work. Currently there is a *PyGTK* based GUI under development. It is expected this GUI will reach maturity through 2009, with 1.23m CAHA telescope operations being a driving force for better GUI support.

Data Processing Improvements

RTS2 provides various hooks for data processing integration. The problem is that these hooks usually call in-house developed, sparsely documented scripts, which are very hard to understand. It is clear that the project will converge towards integrated, well documented, and easy to use programs for data processing. This field requires knowledge and support from experts dealing with other domains than GRB observation. We will be quite happy if those experts will join our effort.

Network Management Improvements

For network operation, there is a growing need for management software. This needs to provide easy access to the observatory's performance data and warning and errors coming from observatories and their instruments. GUI is very desirable. It is expected this task will be carried together with the GUI improvements mentioned above.

Conclusion

This chapter presented an open source autonomous observatory software. It highlights features of its design, problems encountered during development, and obtained results. It provides the reader with an overview of expected future development.

Author's Note

Petr Kubánek would like to acknowledge support from the Spanish "Programa de Ayudas FPI del Ministerio de Ciencia e Innovación (Subprograma FPI-MICINN)" and "el Fondo Social Europeo".

References

Akerlof, C.W. and Swan H.F. 2007. An estimation of the Gamma-Ray Bursts afterglow apparent optical brightness distribution function. *The Astrophysical Journal, 617* (2), 1868-1876.

Castro-Tirado, A. et al. 2008. Flares from a candidate galactic magnetar suggest a missing link to dim isolated neutron stars. *Nature, 455, 7212,* 506-509.

Kubánek, P. 2006. RTS2: a powerful robotic observatory manager. In *Advanced Software and Control for Astronomy*. Proceedings of the SPIE, 6274, 62741V.

Kubánek, P. et al. 2008. RTS2: open source package for autonomous observatory. Internet <http://www.rts2.org> accessed 1 February 2009.

Contributors

Billy Albertson worked on the design of the "Cal Poly 18" telescope while a mechanical engineering student at California Polytechnic State University in San Luis Obispo.

Rainer Anton is a retired physicist from the University of Hamburg, Germany, and has worked with video astronomy for about 15 years. His main focus is on high-precision measurement of double stars.

Jon Archambeau received his mathematics degree from the University of Hawaii at Hilo in 2007. Since 2006, Jon has been employed with Gemini Observatory and has held various positions with Gemini North. Among these are lead aircraft spotter, tour guide and virtual tour technician, safety projects assistant, and currently laser operator. Jon also is employed part time as a telescope operator for the University of Hawaii at the Hilo 88 inch telescope.

Andrew Aurigema is a full time rocket engineer down in the swamps of sunny Florida by day and inventor/astronomer by night. Having caught a bad case of aperture fever early on, ever larger telescopes have come from his private Hanger 51. Now the owner of OTF Designs LLC, makers of Starstone mirrors, this astronomy enthusiast has his sights firmly set on the 2 meter telescope target. Spaceflight and high energy field physics top out his free-time reading list.

Steve Badger is the director of technology for Pittsburgh Corning Corporation. He earned his doctoral degree in materials science and engineering from Pennsylvania State University. Steve has over 15 years experience in research/development and has authored numerous articles. He has spent the past few years researching glass products such as cellular glass, which has been used in the design of low cost telescope mirrors.

Howard Banich is a senior development engineer with Nike, and has been an ATM and visual observer since 1969. His 28 inch Newtonian portable telescope pioneered several features relevant to alt-az "aerospace" telescopes. Howard's computer-controlled telescope is being modified to include instrument de-rotation.
http://hbanich.googlepages.com/

Mel Bartels is the software development manager at FabTrol Systems. He ran the Amateur Telescope Makers listserv for six years, a worldwide group sharing the art of telescope making. In 1988 Mel began developing a freely distributed telescope control system that is in use worldwide. He has ground mirrors up to 30" in size, leads mirror-making classes, and conducted a workshop where attendees figured 16" mirrors. His interest in innovative mounting designs led to popularizing ultralights with single upper rings in 1994 and inventing the TriDob in 2003. A speaker at astronomical conferences each year, his telescope walkabouts at the Oregon Star Party are a highlight that attracts hundreds annually. The International Astronomical Union named asteroid 17823 Bartels for his contributions to amateur astronomy.

Richard Berry writes books and computer software from his home on the wet western side of the Cascade Mountains in Oregon. His books include *The Handbook of Astronomical Image Processing* with its AIP4Win software (with Jim Burnell), *Build Your Own Telescope*, *The Dobsonian Telescope: a Practical Manual for Building Large Aperture Telescopes* (with Dave Kriege), and *The CCD Camera Cookbook* (with Viekko Kanto and John Munger)—all published with Willmann-Bell. You can find more on his tests of the QSI 532ws plus pictures of Comet Holmes, etc., on his website at http://www.wvi.com/~rberry/index.html. And of course, he has another couple of books percolating….

Craig Breckenridge is the owner of Breckenridge Astronomical Ltd, building .8 m equitorial to 2.5 m alt-az telescopes and enclosures. He is also senior designer at Dynamic Structures Ltd, one of the leaders in telescope enclosures worldwide. An amateur observer with over 45 years of experience, he has also served as Vancouver chapter president of the Royal Astronomical Society of Canada.

Lisa Brodhacker received her doctoral degree in 2006 while working on *The Polymer Chemistry and Optical Testing of Epoxy Telescope Mirrors*. Currently, she is an assistant professor of chemistry at Lander University in Greenwood, South Carolina. Although her specialties are organic and polymer chemistry, she has an active interest in astronomy. Her main field of research is spin-casting polymer telescope mirrors, where she believes the key to success lies in manipulating the chemistry.

Rodolfo Canestrari received a Laurea degree in astronomy in 2006 at the Universita' di Bologna, Italy, on *Hard X-ray Astronomical Telescopes: Technological Development of Multilayer-Coated Mirrors*, while working at the INAF-Osservatorio Astronomico di Brera, Italy. From 2006, he has been a doctoral student at the Universita' dell'Insubria, Italy, and the INAF-Osservatorio Astronomico di Brera. The core activity of his research is new technologies for mirrors that will be manufactured for the next generation of ground-based telescopes, mainly glass slumping techniques. He is dedicated to prototypal mirror manufacturing and their optical characterization. He is working on an ion beam figuring technique for the very precise errors correction of optical surfaces.

Chris Carter is a telescope controls engineer at the Thirty Meter Telescope project in Pasadena, California, where he is working on the mount control and drive systems. Prior to joining TMT he was employed by the Gemini Observatory, where he worked on the commissioning and the subsequent operation of both Gemini North and South and gained considerable experience with secondary mirror systems. His areas of expertise are electronics and control systems. He holds a degree in electrical & electronic engineering from the University of Plymouth, U.K., and a masters degree in control systems from Imperial College, London.

Cary Chelborad is president of Optical Structures in Rancho Cordova, California. Optical Structures makes lightweight cellular optical substrates, LIDAR components, and telescopes.

Peter Chen received his B.S. from the University of Toronto, Canada, and Ph.D. in astronomy from Case Western Reserve University. He was an astrophysicist and team member for the ASTRO-1 and ASTRO-2 space missions at the NASA Goddard Space Flight Center. His current research interests include development of nanocomposite adaptive mirrors for space and ground applications, employing superconductor bearings for telescopes on the Moon, and using lunar dust to fabricate mirrors, structures, habitats, and shielding for a manned mission to Mars.

Wilson Chiu worked on the design of the "Cal Poly 18" telescope while a mechanical engineering student at California Polytechnic State University in San Luis Obispo.

Gary Cole (Starphysics Observatory, Reno) is a lifelong amateur astronomer and one of the early practitioners of amateur spectroscopy. Gary is a software designer and entrepreneur with a background in physics and a deep interest in astronomical instrumentation. He is developing an automated astrophysics observing system. Gary also teaches an annual seminar on spectroscopy at Western Nevada College.

Michael Connelley is a staff astronomer for the NASA Infrared Telescope Facility at the University of Hawaii. His research interests include near-infrared observations of very young stars to determine their binary statistics and monitoring their near-infrared spectroscopic variability. His observatory duties focus on maintaining and optimizing the IRTF's optical alignment and image quality. As an avid amateur astronomer and telescope builder, he enjoys building unusual telescopes and observing with them from Mauna Kea. Other interests include classic cars and photography.

David Davis is a member of the Oregon Scope Werks, an eclectic group of amateur telescope makers and astronomers dedicated to bringing astronomy to kids. He is also the founder of the Toledo Scope Werks, a commitment to discovery, testing the limits of design and materials in the pursuit of simple inexpensive telescopes.

Susanna Deustua is a professional astronomer at the Space Telescope Science Institute. She was, for a number of years, the education officer at the American Astronomical Society and has a continuing interest in education and smaller telescopes.

Michele Doro is member of the MAGIC telescope collaboration since 2003. For MAGIC, he followed the commissioning of the calibration system in 2004 for his diploma thesis and was responsible for the reflective surface of the 2nd MAGIC telescope recently completed during his Ph.D. thesis, obtained in 2009 at the University of Padua, Italy. Since 2008 he is also participating at the CTA project for a new generation of Cherenkov telescope to be operative after 2015. For CTA, he co-ordinates the mirror working group. For MAGIC, he is also convener of the Astroparticle and Fundamental Physics working group.

Chris Estrada has been a member of the Central Coast Astronomical Society for a number of years and is currently vice president of operations. He has been involved in amateur astronomy for the past 19 years. Along with his father, he constructed a large 22 inch Dobsonian telescope and is also involved in assisting Russ Genet with building the one meter photometric telescope. Chris is finalizing the results of a visual double star research project for publication. Chris is working on a degree in biochemistry.

Reed Estrada has been a member of the Central Coast Astronomical Society for several years and has been involved with amateur astronomy since 1991. He is a 23 year Air Force veteran pilot currently working as a test pilot at Edwards AFB in Mojave, California. Reed and his son Christopher built their first large Dobsonian telescope two years ago and are currently helping Russ Genet build a portable one meter telescope.

Vladimir Galogaza is a retired physicist. His current interest is in testing and promoting PSI-interferometry methods suitable for ATM use.

Cheryl Genet earned her doctoral degree in interdisciplinary science and theology. As director of the Orion Institute and its Science and the Human Spirit program, she focuses her research on scientific paradigms, cosmological stories, interfaith relations, and understanding our emerging global community. She taught at Central Arizona College, at Cal Poly's Osher Institute for Life-Long Learning, and is currently an adjunct professor of philosophy at Cuesta College. She has also authored several astronomy papers and facilitated astronomy conferences. www.OrionInstitute.org.

Russell M. Genet is research scholar in residence, California Polytechnic State University, and adjunct professor of astronomy, Cuesta College. The founder and former Director of the Fairborn Observatory, Russ and Louis Boyd pioneered the development of robotic telescopes. Author of several books on astronomy and telescope control, Russ was the 51st president of the Astronomical Society of the Pacific. www.OrionObservatory.org.

Mauro Ghigo received his master's degree in astronomy in 1985 at the University of Bologna, Italy. From 1986 to 1988 he was a paid associate in the European Southern Observatory, Germany. Since 1991 he has been a research astronomer with INAF-Osservatorio Astronomico di Brera, Italy. At INAF, he is responsible for the development and use of an Ion Beam Figuring facility for the correction of residual surface errors. His main field of interest is the development of innovative processes for mirror manufacturing and the use of ceramic materials for astronomical mirrors. He is the owner of several patents.

Dan Gray is president and director of engineering of both Technical Marine Service (TMS) and Sidereal Technology. Dan founded TMS, a marine controls company, in 1987, and Sidereal Technology in 2003. He has innovatively developed many types of control systems for ocean-going vessels as well as telescopes. An active telescope maker for 30 years, Dan created and popularized the "string" telescope. www.SiderealTechnology.com, www.tms-usa.com

Joe Haberman is vice-president of PlaneWave Instruments. An avid amateur astronomer, Joe started making optics in the early 1990s as an amateur telescope maker. He eventually started his own optics company, Haberman Optics, and built a reputation for making high quality, large aperture paraboloid mirrors. Joe went on to become the master optician at Celestron before co-founding PlaneWave Instruments in 2006 with Rick Hedrick. www.planewave.com.

Rick Hedrick is president of PlaneWave Instruments, which was founded in 2006 by Rick and Joseph Haberman. Rick has over 21 years of experience in telescope design. He worked for 11 years as the chief technology officer and previous owner of Celestron, where he was responsible for developing Celestron's award winning Schmidt-Cassegrain computerized telescopes. Getting a product like the CDK into the hands of serious imagers and amateur astronomers has been a shared vision of Hedrick and Haberman for many years. www.planewaveinstruments.com.

Arne Henden is the director of the American Association of Variable Star Observers (AAVSO) in Cambridge, MA. He is a professional observational research astronomer, with interests in hardware (CCD cameras, spectrographs, near infrared instrumentation), software (development of image processing systems and instrument control interfaces), and pro-am collaborations, in addition to his long-time love of variable stars. He has one textbook in print (*Astronomical Photometry*, with Ron Kaitchuck) along with over a hundred refereed papers. http://www.aavso.org

Bruce D. Holenstein is the president and CEO of Gravic, Inc., a computer software company located in Malvern PA, which manufactures pattern recognition, database replication, and imaging software products. He holds over a dozen patents and is the author of numerous articles and three books on high-availability computing. Dr. Holenstein attended the University of Pennsylvania for astronomy and astrophysics and is currently building Gravic Labs, an R&D Center at his company.

Dylan R. Holenstein is an infrastructure technical specialist at Gravic, Inc. in Malvern PA. He concentrates on high-performance computing platforms, high-speed networks, and web programming. Mr. Holenstein also develops software for large-database image identification and fast retrieval.

Kevin Iott is a design engineer for PlaneWave Instruments and a co-founder of Ascension Instruments. Kevin is a journeyman tool maker and has been designing and building German equatorial telescope mounts for almost 10 years. After finishing his apprenticeship, Kevin was employed by an industrial automation shop designing high speed automated equipment and robotic machinery for the automotive and solar industries. Kevin founded Iott Precision Instruments in 2007.

Jolyon M. Johnson is a student at California State University at Chico and previously attended Cuesta College in San Luis Obispo CA. He was enrolled in the astronomy research seminar at Cuesta College and is now majoring in geology with a strong interest in planetary science. Jo is currently concentrating his astronomical research on double stars, visual asteroids, and exoplanets, with more than a dozen papers in the field to his credit. He has co-chaired several Alt-Az Intitiative Conferences in Hawaii and San Luis Obispo, as well as assisted with the Pine Mountain Observatory Astronomy Workshops in Oregon.

Greg Jones is a long time amateur telescope maker with a penchant for the unusual. Primary instruments include an 8" Lurie Houghton and 6" Trischiefspiegler with a 20" corrected classical Cassegrain in the works. Current interests have turned to astronomical instrumentation, including adaptive optics, multi-object spectrographs, IFUs, and work in the near-infrared. Greg is the owner of Eclipse Technologies, a sales firm specializing in analytical imaging, semiconductor tests, and ultra-high vacuum equipment. He has a broad background in computer controls and software engineering, and is a life-long tinkerer.

Allan Keller is a systems designer for PlaneWave Instruments and a co-founder of Ascension Instruments. Allan has over 25 years of systems design experience and over 10 years of experience in telescope systems design and manufacturing. He assisted Dr. James Wray of SciTech Astronomical Research with the 32" Tenagra II Telescope and the designs of two observatory-class 24" equatorial telescopes, one of which is on Mt. Palomar, as well as other instruments. Allan designed and built the alt-az fork system for the CDK700 and oversaw the development of the mechanical systems and optical tube assemblies for the project.

Robert H. Koch is a retired astronomer who has worked on models for close binary photospheres and wind envelopes.

Tom Krajci, Major, USAF (retired), is an amateur scientist specializing in photometry. He operates the Astrokolkhoz Observatory at an elevation of 9,440 feet near Cloudcroft, New Mexico. Tom is translating several books on telescope making and optics design from Russian into English, including the works of Dmitry Maksutov. http://overton2.tamu.edu/aset/krajci/

Michelle Kirkup worked on the design of the "Cal Poly 18" telescope while a mechanical engineering student at California Polytechnic State University, San Luis Obispo.

Petr Kubánek is from Prague, Czech Republic. He is a Ph.D. student currently finishing his thesis on robotics observatories in Spain, at the University of Valencia and IAA CSIC Granada. After receiving his master's degree in computer sciences from Prague's Charles University, he worked for two years as a programmer and consultant. Before starting his Ph.D. in Spain, he worked for ESA Integral Science Data Center facility at Versoix, Switzerland. He is the main author of RTS2 (http://rts2.org), an autonomous observatory manager used on multiple sites world-wide. Not exactly an astronomer nor a computer scientist, he enjoys working on a wide range of topics, from assembling hardware up to artificial intelligence scheduling and telescope network management.

Tong Liu is president of Optel Engineering Group (Hubble Optics) Inc. Tong has developed Hubble Optics' innovative lightweight sandwich mirrors. He has many years of experience in optics and telescope design and manufacturing.

Eric Mendez worked on the design of the electronic controls for an active primary mirror control system while an electrical engineering student at California Polytechnic State University, San Luis Obispo.

John Menke is a retired physicist and long time amateur astronomer. In the early 1990s, he and his wife started Technical Innovations which supplied observatory domes and automation to users around the world. Since selling this business, he has constructed two observatories, one of which houses a built 18 inch *f*/3.5 Newtonian. While he concentrates on asteroid light curves and spectroscopy, he continues to design and construct astronomical devices, especially high speed photometers and electronics.

Richard J. Mitchell is a product delivery analyst at Gravic, Inc. in Malvern PA. He was a long-term electronics and optical technician for the department of astronomy and astrophysics and the Flower and Cook Observatory of the University of Pennsylvania. Mr. Mitchell designed and built many electro-optical assemblies for the detection of faint telescopic objects. His software developments include data analysis codes, image processing algorithms, and observatory automation programs.

Drew Murphy worked on the design of the "Cal Poly 18" telescope while a mechanical engineering student at California Polytechnic State University, San Luis Obispo.

Robert H. Nelson earned his doctoral degree at the University of British Columbia in plasma spectroscopy. Moving to Prince George, B.C., Canada, he taught university-level physics for 32 years before retiring. He is a founder of the Prince George Centre of the Royal Astronomical Society of Canada, and the builder of three observatories. He conducts research in eclipsing binaries and has written software for use by astronomers.

Tom Osypowski is a long time telescope maker, viewer, and astro-imager. In 1986 he started a company called Equatorial Platforms that specializes in custom tracking mounts for Dobsonian-type telescopes. The last 8 years have seen the development of complete metal alt-az telescope systems, with either equatorial platform drives or with our new alt-az SlipStream drives. Tom resides in the Sierra foothills of northern California with his wife Nan, where he builds telescopes and platforms for his customers and gets a round of golf in when he can. Equatorial Platforms 15736 McQuiston Lane, Grass Valley CA 95945. Email tomosy@nccn.net URL: www.equatorialplatforms.com

Giovanni Pareschi received his master's degree in astronomy in 1992 at the University of Bologna, Italy, and his Ph.D. in physics in 1996 at the University of Ferrara, Italy. From 1997 to 1998 he was an ESA postdoctoral fellow at the Danish Space Research Center of Copenhagen, Denmark. Since 1999 he has been an associate astronomer with INAF-Osservatorio Astronomico di Brera, Italy. He was a project scientist of the SIMBOL-X mission. His main field of interest is the development, implementation, and calibration of x-ray optics for astronomical missions. Since 2008 he is the director of the INAF-Osservatorio Astronomico di Brera.

Bob Peasley is a mechanical engineering consultant and former mechanical engineer for Celestron. Bob is also a professional photographer who provides images of products for Celestron's brochures and website as well as other companies in Southern California. Bob recently relocated to Taylor, Utah.

Terry Richardson is senior instructor of physics and astronomy at the College of Charleston, South Carolina, where he has taught for 28 years and holds the Mebane Chair for Teaching Excellence. In addition to optics, his academic interests include small bodies of the solar system and pre-college and public science education.

John Ridgely is an associate professor of mechanical engineering at California State Polytechnic University in San Luis Obispo, CA. He has advised several student projects involving the design of telescope components and subsystems at the undergraduate and graduate levels.

George C. Roberts is president and founder of INCA Corporation, which manufactures robotic and semi-robotic machines for displaying flat screen TVs, projectors, and similar electronic appliances in and out of concealments, and has been doing so since 1970. INCA occupies a 68,000 sq. ft. manufacturing facility in Gardena, CA. George has been involved with professional and amateur astronomy for a long time. Recent positions included the advisory board of the Lowell Observatory, and George was active in the development of Mauna Kea as a primary astronomy site, completing the first light pollution abatement ordinances for the Big Island and arranging for the Keck Telescope base facilities to receive their land as a gift from the Smart Trusts.

Gregory Rohde has been an active visual observer and amateur telescope maker since 2003. He is currently the senior technician at Sidereal Technology and president of Superior Optical Supply, which he founded in 2006. Greg is currently building an ultra lightweight airline portable string telescope, and is designing a 20" Gregorian.

David Rowe is chief technology officer and co-founder of Sierra Monolithics. An avid amateur astronomer, optical designer, and ATM, Dave has designed and fabricated many telescopes, including a corrected Dall-Kirkham, a flat-field concentric Schmidt-Cassegrain, and several Schmidt cameras and corrected Newtonians. His current interests are interferometry, telescope design, pointing and tracking models, motors, and ultra-high resolution encoders. dave@starryridge.com

Josh Schmidt worked on the design of the "Cal Poly 18" telescope while a mechanical engineering student at California Polytechnic State University in San Luis Obispo. Josh also did Finite Element Analysis (FEA) of structures for multiple mirror telescopes.

Kiran Shah is a mechanical engineer by training and is a co-founder and technical director of Chroma Systems, a company which develops quality control equipment, image processing software for the heavy engineering industry, and has developed, patented, and now mass manufactures low cost CNC machines, making them affordable for the small scale sector. He has been an enthusiastic amateur astronomer right from his childhood and is interested in mirror making, telescopes, and optics. Kiran is now deeply involved with a new start-up, Chroma Energy, which is primarily working on solar power generation using concentrator photovoltaic (CPV) cells. He is working on his own optic design for the solar concentrator along with a solar tracker. He spends his (now increasingly little) spare time reading, trekking, and paragliding.

Michael Sheehan is in charge of mechanical engineering at the two Gemini 8 meter telescopes. He was involved in the design of the telescopes and the continuing modifications and operational mechanical considerations. He spends time in both Hawaii and Chile.

Thomas C. Smith, director of the Dark Ridge Observatory, is a retired software and nuclear engineer and advanced amateur astronomer. Tom established the Dark Ridge Observatory as a non-profit organization in Weed, New Mexico, and has been working with students and faculty from several colleges and universities as a mentor for CCD photometry and image data reduction. Tom also conducts research on eclipsing binaries. www.DarkRidgeObservatory.org.

Matt Swanson worked on the design of the "Cal Poly 18" telescope while a mechanical engineering student at California Polytechnic State University, San Luis Obispo.

Hideki Takami graduated from Kyoto University, Japan, and obtained a Ph.D. degree in far infrared astronomy in 1988. He worked at the Communications Research Laboratory for the Ministry of Post and Telecommunications of Japan until 1994 and now is a professor at the National Astronomical Observatory of Japan, working with the Subaru Telescope. His current research field is infrared astronomy and adaptive optics technology.

Daigo Tomono is an astronomer and engineer working at the Subaru Telescope, an 8 meter telescope atop Mauna Kea, Hawaii. With colleagues, he maintains the telescope mechanisms, electronics, and software. As a hobbyist, he thinks that the Subaru telescope is rather too big and complicated to enjoy. http://dtomono.freeshell.org

Gary A. Vander Haagen is a retired CEO of Visteon Electronic Systems with extensive experience in electronic design and manufacturing for the automotive industry. He also spent a significant portion of his career in the electro-optics industry designing and manufacturing laser and holographic instrumentation. He has a number of publications in electronics, electro-optics and holography, and astronomy. Current focus in astronomical research includes asteroid light curves and H-G parameters, high speed optical detection and analysis, and exoplanets. Gary and his wife have seven grandchildren and live in Ann Arbor, Michigan.

Efrain Villasenor worked on the design of the electronic controls for an active primary mirror control system while an electrical engineering student at California Polytechnic State University in San Luis Obispo.

Vera Wallen is a retired public school district superintendent and teacher with 40 years experience in education. Dr. Wallen focused her career on leading special students to achieve their full potential and maximizing faculty application of the latest learning research. A life-long dreamer of space travel, she took her first astronomy class from Russ Genet at Cuesta College in her 10th year of retirement on the central coast of California and has served as copy editor for the first two volumes of the Alt-Az Initiative books.

Xiao-Hua Yu received her Ph.D. degree in electrical and computer engineering from University of California, Irvine. She is currently an associate professor in the department of electrical engineering at California Polytechnic State University, San Luis Obispo. Her major research interests include computational intelligence, adaptive control, and digital signal processing.

Index

S

T

U

V

W

Y

Z

Collins Foundation Press Books
www.CollinsFoundationPress.org

Astronomy Series

Edited Volumes
Small Telescopes and Astronomical Research
2009

The Alt-Az Intitiative
Telescope, Mirror, and Insturment Developments
2010

Website for the Alt-Az Inititative
www.AltAzInitiative.org

Other Books of Interest

Humanity Series

Humanity
The Chimpanzees Who Would Be Ants
by Russell Genet

Edited Volumes
The Evolution of Religion
Studies, Theories, and Critiques

The Evolutionary Epic
Science's Story and Humanity's Response

Science, Wisdom, and the Future
Humanity's Quest for a Flourishing Earth
(Mid-2010)

LaVergne, TN USA
24 July 2010
190698LV00001B/3/P

9 780978 844141